Biosimilars for Cancer Treatment

Shvetank Bhatt • Harish Dureja
Samir Gunvantbhai Patel
Archita Samir Patel • Kamal Dua
Editors

Biosimilars for Cancer Treatment

A Promising Approach

Springer

Editors
Shvetank Bhatt
School of Health Sciences and Technology
Dr. Vishwanath Karad MIT World Peace
University
Pune, Maharashtra, India

Samir Gunvantbhai Patel
Ramanbhai Patel College of Pharmacy
Charotar University of Science
and Technology
Changa, Gujarat, India

Kamal Dua
Discipline of Pharmacy
University of Technology
Sydney, NSW, Australia

Harish Dureja
Department of Pharmaceutical Sciences
Maharshi Dayanand University
Rohtak, Haryana, India

Archita Samir Patel
Ramanbhai Patel College of Pharmacy
Charotar University of Science
and Technology
Changa, Gujarat, India

ISBN 978-981-97-3000-1 ISBN 978-981-97-3001-8 (eBook)
https://doi.org/10.1007/978-981-97-3001-8

This Springer imprint is published by the registered company Springer Nature Singapore Pte Ltd.
The registered company address is: 152 Beach Road, #21-01/04 Gateway East, Singapore 189721, Singapore

Preface

Cancer is a leading cause of death worldwide, accounting for nearly ten million deaths in 2020, or nearly one in six deaths. Head and Neck, breast, liver, pancreatic, gastro-esophageal, bladder, lung, colorectal, blood, and prostate cancers are the most common cancers. Cancer occurs when cells of the body starts multiplying in uncontrollable manner. This is due to modification in apoptotic pathways and related mechanisms. The diseases normally start with small area and localized and in the later stages cells become metastasized in different locations in body.

A biosimilar drug is a very similar copy of an already approved biological drug. Biosimilars work the same way as biologics in our body. Now a day the high cost of various cancer therapies kept a significant burden on the health care system of the country and that's why the oncology research has been focused now more on the use of biosimilars as they are safer and cost-effective. The use of biosimilars might provide competitive, lower-cost alternatives to biologics used in cancer care.

Manufacturing of biosimilars are difficult as compared to synthesizing new chemical entities. Maintaining batch-to-batch consistency is very critical in the production of biosimilars. The most frequently used biosimilar drugs are Bevacizumab, Epoetin alfa, Filgrastim, Pegfilgrastim, Rituximab and Trastuzumab. Recently various biosimilar drugs are approved for the treatment of various solid tumors and blood cancers such as lung cancer, bladder cancer, melanoma, head and neck cancer and diffuse large B-cell lymphoma etc. In todays 'scenario biosimilars have an important role in the effective management of the different type of cancers i.e. colon cancer, lung cancer, gastric cancer, breast cancer, cervical cancer, liver cancer, ovarian cancer and blood cancer etc. These biosimilars can be used as monotherapy or in combination with other therapies like chemotherapy and targeted therapy.

We hope the book shall be a useful compilation for undergraduate, postgraduate, doctoral students, and researchers working in cancer and drug delivery research,

research and development, and national research institutes. We hope to receive feedback, suggestions, and inputs from researchers and students that will help improve the next edition of the book.

Pune, Maharashtra, India — Shvetank Bhatt
Rohtak, Haryana, India — Harish Dureja
Changa, Gujarat, India — Samir Gunvantbhai Patel
Changa, Gujarat, India — Archita Samir Patel
Sydney, NSW, Australia — Kamal Dua

Acknowledgments

I would like to express our sincere gratitude to the Honorable Founder-President, Dr. Vishwanath Karad, and Executive President, Mr. Rahul V. Karad, Dr. Vishwanath Karad MIT World Peace University, Pune, India, for their motivation. My heartfelt thanks go to Dr. Neeraj Mahindroo, Professor and Dean of the School of Health Sciences and Technology, Dr. Vishwanath Karad MIT World Peace University, Pune, India, for his continuous support during the writing of this book. Finally, I would like to extend our gratitude to the faculty members of the School of Health Sciences and Technology at Dr. Vishwanath Karad MIT World Peace University, Pune, India.

Pune, Maharashtra, India　　　　Shvetank Bhatt

Contents

Editors and Contributors

About the Editors

Shvetank Bhatt is currently working as an Associate Professor in the School of Health Sciences and Technology, Dr. Vishwanath Karad MIT World Peace University, Pune, Maharashtra, India. He has done M.Pharm. in Pharmacology from Manipal College of Pharmaceutical Sciences, MAHE, Manipal, Karnataka, India and Ph.D. in Neuropharmacology from Birla Institute of Technology and Science (BITS) Pilani, Pilani Campus, Rajasthan, India. His areas of specialization are CNS disorders, pain, inflammation and immuno oncology. He has published numerous papers. He is a life member of the Association of Pharmaceutical Teachers of India (APTI) and the Indian Pharmacological Society (IPS).

Harish Dureja is Head of the Department of Pharmaceutical Sciences, Maharshi Dayanand University, Rohtak, Haryana, India, and Director of the Centre for IPR Studies and Professional Consultancy Cell at M.D. University, Rohtak, India. He earned his Master's in Pharmacy from Punjabi University, Patiala, India, and Ph.D. in Pharmaceutical Sciences from Maharshi Dayanand University, Rohtak, India. He has published numerous articles in journals of repute. Dr. Dureja is also actively engaged in committees of various central and state universities, namely the Committee for Control and Supervision of Experiments on Animals (CCSEA), Pharmacy Council of India (PCI), All India Council of Technical Education (AICTE), National Testing Agency (NTA) and National Assessment and Accreditation Council (NAAC), etc.

Samir Gunvantbhai Patel works as Dean, Faculty of Pharmacy, Professor, Pharmaceutical Chemistry and Analysis, at Ramanbhai Patel College of Pharmacy (RPCP), Gujarat, India. He has more than 18 years of academic experience. His major research areas are design, synthesis and biological evaluation of cyclic peptides and synthetic derivatives for treating breast cancer, colon cancer and cancer metastasis. He has published several research articles and a few book chapters.

Archita Samir Patel is an Assistant Professor, Pharmaceutical Chemistry and Analysis, Ramanbhai Patel College of Pharmacy (RPCP), Gujarat, India. She has more than 18 years of academic experience. Her major research areas are the analysis of small molecules, development and validation of stability indicating analytical methods for small molecules and herbal actives, impurity profiling of synthetic drug substances and developing analytical methods using a chemometrics approach. She has published several research articles and a few book chapters.

Kamal Dua is an Associate Professor in the Discipline of Pharmacy at the Graduate School of Health, University of Technology Sydney (UTS), Australia. He has research experience of over 13 years in the field of drug delivery systems targeting inflammatory diseases. Dr. Dua is also a node leader of drug delivery research in the centre for inflammation at Centenary Institute/UTS, where the targets identified from the research projects are pursued to develop novel formulations as the first step towards translation into clinics. Dr. Dua's research in two complementary areas; drug delivery and immunology are evidenced by his extensive publication record in reputed journals. Dr. Dua's research interests focus on harnessing the pharmaceutical potential of modulating critical regulators such as Interleukins and microRNAs and developing new and effective drug delivery formulations for the management of chronic airway diseases. He has published numerous research articles in journals of repute and authored and co-authored a few books. He is also an active member of many national and international professional societies.

Contributors

Senthil Visaga Ambi Biopharmaceutical Research Lab, Anusandhan Kendra-1, School of Chemical and Biotechnology, SASTRA Deemed-to-be-University, Thanjavur, Tamil Nadu, India

Department of Bioengineering, School of Chemical and Biotechnology, SASTRA Deemed-to-be-University, Thanjavur, Tamil Nadu, India

Kuttiappan Anitha Department of Pharmacology, School of Pharmacy and Technology Management (SPTM), SVKM's Narsee Monjee Institute of Management Studies (NMIMS) Deemed-to-University, Shirpur, India

Pavithra Vimala Arulrajan Biopharmaceutical Research Lab, Anusandhan Kendra-1, School of Chemical and Biotechnology, SASTRA Deemed-to-be-University, Thanjavur, Tamil Nadu, India

Dhara Bhatt Medisynth Chemicals Pvt. Ltd., Navi Mumbai, Maharashtra, India

Shvetank Bhatt Department of Pharmaceutical Sciences, School of Health Sciences and Technology, Dr. Vishwanath Karad MIT World Peace University, Pune, India

School of Health Sciences and Technology, Dr. Vishwanath Karad MIT World Peace University, Pune, Maharashtra, India

Prarit Chandel Chitkara University, School of Pharmacy, Baddi, Himachal Pradesh, India

Pallavi M. Chaudhari Department of Pharmaceutics, Dr. D. Y. Patil College of Pharmacy, Akurdi, Pune, India

Santenna Chenchula Department of Pharmacology, AIIMS, Bhopal, Madhya Pradesh, India

Pranali Chimaniya School of Pharmaceutical Education and Research, Peoples University, Bhopal, Madhya Pradesh, India

Harita Desai Department of Pharmaceutics, Bombay College of Pharmacy, Santa-cruz East, Mumbai, India

Arghya Kusum Dhar School of Pharmacy, The Neotia University, Sarisa, West Bengal, India

Subas Chandra Dinda School of Pharmacy, The Neotia University, Sarisha, West Bengal, India

Dipali Dongare Department of Biotechnology, National Institute of Pharmaceutical Education and Research, Lucknow, Uttar Pradesh, India

Harish Dureja Department of Pharmaceutical Sciences, Maharshi Dayanand University, Rohtak, India

Arvind Ganpule Department of Urology, Muljibhai Patel Urological Hospital, Nadiad, Gujarat, India

Gayathri Gopal Biopharmaceutical Research Lab, Anusandhan Kendra-1, School of Chemical and Biotechnology, SASTRA Deemed-to-be-University, Thanjavur, Tamil Nadu, India

Department of Bioengineering, School of Chemical and Biotechnology, SASTRA Deemed-to-be-University, Thanjavur, Tamil Nadu, India

Nisha Gulati Department of Pharmaceutical Sciences, Maharshi Dayanand University, Rohtak, India

Daksh Sanjay Gupta Vivekanand Education Society's College of Pharmacy, Mumbai, Maharashtra, India

Dhruv Sanjay Gupta Department of Pharmaceutics, Shobhaben Pratapbhai Patel School of Pharmacy and Technology Management, SVKM's Narsee Monjee Institute of Management Studies (NMIMS) Deemed-to-University, Mumbai, Maharashtra, India

Girdhari Lal Gupta Department of Pharmacology, School of Pharmacy and Technology Management (SPTM), SVKM's Narsee Monjee Institute of Management Studies (NMIMS) Deemed-to-University, Shirpur, India

Pawan K. Gupta Amity Institute of Pharmacy, Amity University, Gwalior, Madhya Pradesh, India

Ekta Gurnany Department of Pharmaceutics, B. Pharmacy College, Rampura, Godhra, Gujarat, India

Ananya Nithin Kanade Biopharmaceutical Research Lab, Anusandhan Kendra-1, School of Chemical and Biotechnology, SASTRA Deemed-to-be-University, Thanjavur, Tamil Nadu, India

Neha Kanojia Chitkara University, School of Pharmacy, Baddi, Himachal Pradesh, India

Sidhartha S. Kar Faculty of Pharmacy, C. V. Raman Global University, Bhubaneswar, Odisha, India

Payal Kesharwani Ram-Eesh Institute of Vocational and Technical Education, Greater Noida, Uttar Pradesh, India

Anoop Kumar Delhi Pharmaceutical Sciences and Research University (DPSRU), New Delhi, India

Pallavi Manish Lavhale Ram-Eesh Institute of Vocational and Technical Education, Greater Noida, Uttar Pradesh, India

Saurabh Morparia Department of Pharmaceutical Analysis, Dr. Bhanuben Nanavati College of Pharmacy, Mumbai, Maharashtra, India

Manikanta Murahari Department of Pharmacy, Koneru Lakshmaiah Education Foundation, Vaddeswaram, Andhra Pradesh, India

Gayathri Devi Muthukumarasamy Biopharmaceutical Research Lab, Anusandhan Kendra-1, School of Chemical and Biotechnology, SASTRA Deemed-to-be-University, Thanjavur, Tamil Nadu, India

Lokesh Nagar Department of Pharmaceutical Sciences, Maharshi Dayanand University, Rohtak, India

Manish Nandpal Ramanbhai Patel College of Pharmacy, Charusat, Changa, Gujarat, India

Aryaa Nigade School of Health Sciences and Technology, Dr. Vishwanath Karad MIT World Peace University, Pune, Maharashtra, India

Shireen Nishad Department of Biotechnology, National Institute of Pharmaceutical Education and Research, Lucknow, Uttar Pradesh, India

Narhari N. Palei Amity Institute of Pharmacy, Amity University, Lucknow, Uttar Pradesh, India

Sachchida Nand Pandey Department of Pathology, Muljibhai Patel Urological Hospital, Nadiad, Gujarat, India

Nikunj Parekh Ramanbhai Patel College of Pharmacy, Charusat, Changa, Gujarat, India

Alkeshkumar Patel Ramanbhai Patel College of Pharmacy, Charusat, Changa, Gujarat, India

Archita Samir Patel Ramanbhai Patel College of Pharmacy, Charusat, Changa, Gujarat, India

Meghana Patel Ramanbhai Patel College of Pharmacy, Charusat, Changa, Gujarat, India

Priyal Patel Ramanbhai Patel College of Pharmacy, Charusat, Changa, Gujarat, India

Samir Gunvantbhai Patel Ramanbhai Patel College of Pharmacy, Charusat, Changa, Gujarat, India

Yash Patel Ramanbhai Patel College of Pharmacy, Charusat, Changa, Gujarat, India

Gauri Pathak School of Health Sciences and Technology, Dr. Vishwanath Karad MIT World Peace University, Pune, Maharashtra, India

Bhagwat Patil Department of Pharmaceutics, Dr. D. Y. Patil College of Pharmacy, Akurdi, Pune, India

Ravindra Babu Pingili Department of Pharmacology, School of Pharmacy and Technology Management (SPTM), SVKM's Narsee Monjee Institute of Management Studies (NMIMS) Deemed-to-University, Shirpur, India

Shiv Kumar Prajapati Institute of Pharmaceutical Research, GLA University, Mathura, Uttar Pradesh, India

Dishank Purandare School of Health Sciences and Technology, Dr. Vishwanath Karad MIT World Peace University, Pune, Maharashtra, India

Neha Raghuvanshi Department of Pharmaceutics, Bombay College of Pharmacy, Santacruz East, Mumbai, India

Anika Rana Department of Biotechnology, National Institute of Pharmaceutical Education and Research, Lucknow, Uttar Pradesh, India

Karthika Rangasamy Biopharmaceutical Research Lab, Anusandhan Kendra-1, School of Chemical and Biotechnology, SASTRA Deemed-to-be-University, Thanjavur, Tamil Nadu, India

Annu Saini Department of Pharmaceutical Sciences, Maharshi Dayanand University, Rohtak, India

Khyati Saini School of Pharmaceutical & Populations Health Informatics, DIT University, Dehradun, India

Keerthana Saravanan Biopharmaceutical Research Lab, Anusandhan Kendra-1, School of Chemical and Biotechnology, SASTRA Deemed-to-be-University, Thanjavur, Tamil Nadu, India

Pranav Shah Maliba Pharmacy College, UkaTarsadia University, Surat, Gujarat, India

Suhashini Shamuganathan Biopharmaceutical Research Lab, Anusandhan Kendra-1, School of Chemical and Biotechnology, SASTRA Deemed-to-be-University, Thanjavur, Tamil Nadu, India

Kangan Sharma School of Pharmaceutical & Populations Health Informatics, DIT University, Dehradun, India

Saritha R. Shetty Department of Pharmaceutics, Shobhaben Pratapbhai Patel School of Pharmacy and Technology Management, SVKM's Narsee Monjee Institute of Management Studies (NMIMS) Deemed-to-University, Mumbai, Maharashtra, India

Satish Shilpi School of Pharmaceutical & Populations Health Informatics, DIT University, Dehradun, India

Abhishek Singh Amity Institute of Pharmacy, Amity University, Lucknow, Uttar Pradesh, India

Neeta Solanki Department of Pharmaceutical Sciences, Maharshi Dayanand University, Rohtak, India

Ashwin Subramanian Biopharmaceutical Research Lab, Anusandhan Kendra-1, School of Chemical and Biotechnology, SASTRA Deemed-to-be-University, Thanjavur, Tamil Nadu, India

Vasanti Suvarna Department of Pharmaceutical Analysis, Dr. Bhanuben Nanavati College of Pharmacy, Mumbai, Maharashtra, India

Komal Thapa Chitkara University, School of Pharmacy, Baddi, Himachal Pradesh, India

Vaishnavi Thorat School of Health Sciences and Technology, Dr. Vishwanath Karad MIT World Peace University, Pune, Maharashtra, India

Pratima Tripathi Department of Biotechnology, National Institute of Pharmaceutical Education and Research, Lucknow, Uttar Pradesh, India

Saraswathi Venkataraman Biopharmaceutical Research Lab, Anusandhan Kendra-1, School of Chemical and Biotechnology, SASTRA Deemed-to-be-University, Thanjavur, Tamil Nadu, India

Nitin Verma Chitkara University, School of Pharmacy, Baddi, Himachal Pradesh, India

Seema Yadav Amity Institute of Pharmacy, Amity University, Lucknow, Uttar Pradesh, India

Overview of Biosimilars

1

Dipali Dongare, Anika Rana, Shireen Nishad,
and Pratima Tripathi

Abstract

Biotechnological drugs are increasingly crucial in modern medicine, which anticipated to occupy half the pharmaceutical market soon. Patent expirations have spurred the emergence of biosimilars, akin to follow on biologics aiming to replicate original biotech medicines. While a couple of biosimilars gained approval in the EU, others are in the pipeline. Despite their potential to reduce treatment costs, our understanding and long-term safety data for biosimilars, including immunogenicity, remain limited. Comparing biosimilars to classic chemical drug generics is vital, demanding focused discussions among physicians. Regulatory clarity, safety concerns, pharmacovigilance, automatic substitution policies, nomenclature, and labeling rules constitute crucial areas requiring attention as biosimilars evolve in the healthcare landscape.

Keywords

European Union · Tevagrastim · Biopharmaceuticals · Molecular generics

1.1 What Are Biosimilars?

A biosimilar is a biologic product that has been created to closely resemble the reference product, a biologic that has already received FDA approval. A biosimilar cannot differ from the reference product in any way that is clinically significant. If a biosimilar meets all the criteria for safety, potency, and purity for the intended use under the prescribed circumstances, it is deemed to be highly comparable to the

D. Dongare · A. Rana · S. Nishad · P. Tripathi (✉)
Department of Biotechnology, National Institute of Pharmaceutical Educatioṇn and Research, Lucknow, Uttar Pradesh, India

S. Bhatt et al. (eds.), *Biosimilars for Cancer Treatment*,
https://doi.org/10.1007/978-981-97-3001-8_1

reference product. The safety, purity, and potency of a biosimilar are demonstrated through a battery of tests and studies conducted by the producer with the aim of contrasting the biosimilar with the reference product (Fig. 1.1).

A product must have the same mechanism of action (how it functions in the body), dosage form (such as liquid), mode of administration (such as injection or infusion), and strength as its reference product in order to be authorized as a biosimilar by the FDA. The clinically inactive parts of the medication may differ slightly between the reference product and the biosimilar, among other possible small variations. In contrast, the FDA assesses these variations to make sure they are reasonable and do not affect the biosimilar's efficacy in comparison to the reference product.

The inventor never discloses the proprietary manufacturing process of reference drug. So, while balancing innovation, competitiveness, and regulatory compliance, biosimilar manufacturers seek to produce biosimilars that are comparable to already available reference biologics using different methods and resources (Afzali et al. 2021). The FDA closely monitors the biosimilar's manufacturing process, just as it does with the reference product. Minor deviations are anticipated (Kurki et al. 2017). But according to FDA regulations, these variances must be closely regulated, tracked, and maintained within reasonable bounds (Food and Administration 2019).

The absence of clinically significant variations between the biosimilar and the reference product is another prerequisite for biosimilars. This implies that a patient will react to the biosimilar in the same way as they would to the original medication physically (Derbyshire 2017). In order to demonstrate this, the biosimilar's manufacturer, or sponsor, compares elements like immunogenicity—the body's reaction to the drug, pharmacokinetics—how the drug is metabolized, broken down, or eliminated by the body, and pharmacodynamics—the impact of the drug has on the body and the illness.

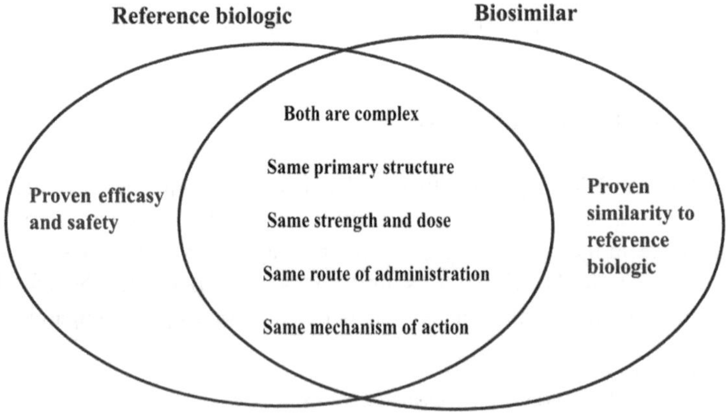

Fig. 1.1 Similarities and differences between reference biologic and biosimilar. Reference biologic undergoes extensive clinical trials to prove efficacy and safety, whereas biosimilar aims to prove similarities with reference biologic with no clinically meaningful differences

The EU established the biosimilars approval process in 2006, and these rules were updated in 2015. The recommendations have rapidly evolved for biosimilars, with the majority of nations adopting the overall framework of the US FDA, EMA, or WHO, while some nations developed their own guidelines based on these concepts (Aladul et al. 2018).

1.2 Biosimilar Development Process

Compared to the original reference biologic, biosimilar development process meets different development requirements. It gives stronger emphasis on analytical characterization and clinical equivalency with the reference biologic. The foundation for the developing of biosimilars is the thorough analytical characterization along with the development of processes necessary to create adequate biosimilarity to the original biologic, as the originator has previously proved the safety as well as effectiveness of the reference medication (de Mora et al. 2019). A biosimilar product may support significantly less extensive clinical testing with sufficient comparability testing, which means that the development expenses are considerably lower than those needed for an original medication (Glintborg et al. 2017) (Fig. 1.2). According to the US FDA guidelines, following are the steps for the approval of the biosimilars:

1. **Analytical characterization:** Analytical research shows the biological product's great similarity to the reference product despite very slight variations in its therapeutically inert constituents. Important factors for consideration in assessing analytical studies include stability, impurities, functional activities, physiochemical properties, and immunochemical properties (Rumore and Vogenberg 2016).
2. **Animal studies:** It includes an assessment of toxicity. Animal toxicity studies' scope and extent will be determined by the degree of known similarities or dif-

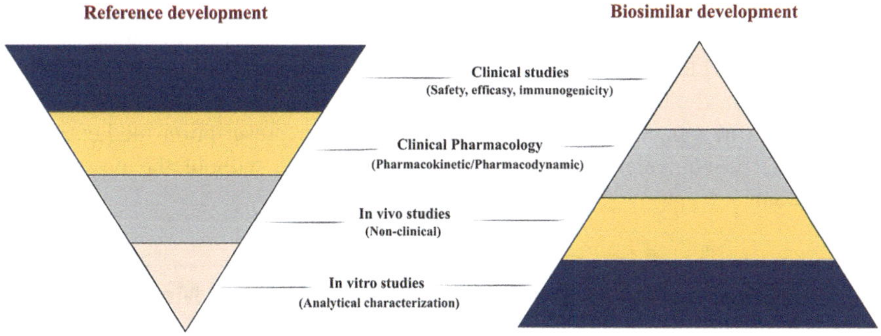

Fig. 1.2 Biosimilar development process. Biosimilars extensively focused on analytical characterization to closely resemble the reference drug without meaningful clinical difference than reference biologic

ferences between the reference product and the proposed biosimilar product, as well as by data submitted in the biosimilar application and made publicly available. In this, pharmacokinetic and pharmacodynamic studies are also done in animal models.

3. **Clinical studies:** To verify the safety, potency, and purity of the proposed biosimilar product in one or more of the indications for which the reference product is approved, a clinical study is conducted. In addition to evaluating the immunogenicity, pharmacokinetics as well as pharmacodynamics, this often involves a comparative clinical investigation. After clinical trials, FDA will approve the drug as biosimilar when all requirements have achieved.

1.3 Interchangeable Biosimilars

A biological drug that is extremely similar to another biological medication that has received prior approval is known as an interchangeable biosimilar. The reference product can be replaced with interchangeable biosimilars at the pharmacy without the help of prescribing physician. At the pharmacy level, the substitution known as an "auto-substitution" is used in clinical settings in the US, Canada, and Australia. Physician-directed substitution, or "switching," is a procedure that is carried out in certain EU member states as well as in Japan, India, and other countries.

"Interchangeability" is a term used in regulations, when further requirements for interchangeability are satisfied (Food U 2015). The US FDA grants a biosimilar the designation of interchangeable. Not every biosimilar can be used interchangeably. An interchangeable biosimilar must also fulfill additional requirements specified by the US Food and Drug Administration (FDA) based on an examination of the product known as the "switch trial design."

The FDA concludes that a reference biologic can be substituted with an interchangeable product when compared to reference drug, the biological drug "is biosimilar" and may be anticipated to result in the same clinical outcome as the reference product in any particular patient. According to the European Medicines Agency (EMA), decentralized agency of European Union (EU), the term "interchangeability" describes the ability to switch out a medication with another that is expected to provide the same therapeutic effect (McDonald 2015). According to the Health Canada, the drug regulatory agency of Canada, "interchangeability" commonly refers to a patient's ability to switch from one prescription medication to another that is equivalent, at the pharmacy's authority, without the prescriber's involvement.

1.4 How Are Biosimilars Different from Generic Medicines?

Pharmaceutical drugs that are meant to offer more affordable options to their respective reference products are known as biosimilars and generics. The two, however, differ greatly from one another, based on the complexity of their development, the

Table 1.1 Difference between reference biologic, biosimilar, and generic drug

Reference biologic	Biosimilar product	Generic drug
Generated in host cell lines with biological processes	Generated in host cell lines by a biological process	Produced by means of chemical synthesis
These are larger and complex molecules	These are larger and complex molecules	These are small and simple molecules
8–10 years required to market reference drug	7–8 years required to market the biosimilar product	2–3 years needed to market generic drug
Extensive clinical trials are required in case of reference drug	Biosimilar development involves at least one clinical trial	Generic drugs don't involve clinical trials
Analytical phase is less extensive in this case	Analytical phase is more extensive in biosimilar development	Analytical phase is less extensive and complex than in biosimilar
Periodic safety updates and pharmacovigilance updates are required	Periodic safety updates and pharmacovigilance updates are required	Requires brief period of time for generic drug approval
Reference biologic should provide evidence of "comparability"	Biosimilars need to show "similarity" with the reference biologic	Generics must demonstrate "bioequivalence" with reference drug

kind of medications they mimic, the time period they required to market the drug, etc. Following are some differentiative points which shows the distinctions between biosimilar and generic with the reference drug to which they mimic (Joshi et al. 2023a) (Table 1.1).

1.5 A Biosimilar Has a Biologic (Natural) Source

Biologics, derived from natural resources like humans, animals, or microorganisms, are manufactured using advanced biotechnology methods such as recombinant DNA technology and controlled gene expression. These medications have significantly benefited patients dealing with a range of conditions, including rheumatologic diseases, inflammatory bowel disease, cancers, dermatological issues, and other connective tissue disorders. They offer the potential to halt disease progression, alleviate symptoms, and improve patients' overall quality of life (Ghosh et al. 2019). However, a major drawback of biologics has been their high cost, rendering them unaffordable and inaccessible, particularly in developing nations where healthcare coverage remains limited. Nonetheless, when the innovator company's patent protection expires, it creates opportunities for other companies to produce similar versions of these drugs at lower costs, known as biosimilars. Biosimilars are highly similar to FDA-approved reference biologics, exhibiting no clinically meaningful differences in safety and effectiveness (Laekeman 2013). However, due to the complex structure of biologics, slight alterations in sequences and posttranslational modifications mean that biosimilars are not exact replicas of their reference counterparts. Despite these nuances, biosimilars offer a promising avenue for making crucial treatments more accessible while maintaining stringent standards for safety and efficacy.

There is widespread anticipation among healthcare professionals and experts that biosimilars will exert a positive influence on drug pricing, potentially reshaping the accessibility of crucial biologic medications. The optimistic outlook stems from the belief that the adoption of biosimilars could result in a reduction in the cost of biologics, thereby enhancing patients' access to these essential and life-saving treatments. A significant study conducted in the United States sheds light on this potential, estimating that over a decade, the utilization of biosimilars has the capacity to yield savings of approximately 54 billion US dollars (Niazi 2022). This compelling data underscores the substantial potential of biosimilars to contribute to more affordable healthcare by significantly reducing the overall treatment expenses associated with biologic therapies. As such, the collective impact of widespread biosimilar adoption could pave the way for enhanced accessibility to critical medications, potentially transforming the landscape of healthcare affordability and patient care.

1.6 A Biosimilar Needs Extra FDA Approval to Be Used Interchangeably

A biosimilar is not always interchangeable with its biologic brand name after receiving its initial FDA approval. Once biosimilars receive initial FDA approval, they can be used to treat diseases; but before they can be used interchangeably with brand-name biologics, they must first receive additional FDA approval (Joshi et al. 2023b). If a biosimilar is not authorized for usage in place of its brand-name biologic, a prescription must be issued from the prescriber for the biosimilar.

The FDA is the only regulatory body that can approve biosimilars that are equivalent to their original and has a statutory definition of interchangeability. The FDA released draft interchangeability guidelines in January 2017. According to these guidelines, a biosimilar must meet two requirements in order to be considered interchangeable: it must be "biosimilar to the reference product" and "can be expected to produce the same clinical result as the reference product in any given patient" (Tóthfalusi et al. 2014). Additionally, if a biological product is administered to a patient more than once, there is no greater risk of safety or diminished efficacy from using the biological product and the reference product alternated or switched between times (Declerck et al. 2017).

All biosimilar and interchangeable biosimilar products are authorized using a streamlined process that verifies biosimilarity by comparing the product to the reference product. Manufacturers must submit supplementary information detailing the potential uses of the interchangeable biosimilar with patients in the marketplace in order for it to be approved as an interchangeable biosimilar. Similar to generic drugs, individuals who get their prescriptions filled through their pharmacies have the option of switching between brand-name biologics and interchangeable biosimilars (Kim and Bindler 2016).

Clinical trial data has demonstrated that each biosimilar that has been licensed for usage is just as safe and effective as the biologic sold under its original brand name when treating a particular illness. The biosimilar manufacturer can decide to submit data to the FDA alone for this preliminary approval. However, the company has to provide the FDA with further data from clinical trials if it wants its biosimilar to be certified interchangeable, meaning that it can be automatically used in place of its brand-name medication (Wolff-Holz et al. 2019).

An interchangeable biosimilar is not necessarily safer or more effective than other biosimilars, even if the FDA can evaluate the safety of pharmacy-level substitution with the aid of this extra information.

For example: The first interchangeable biosimilar insulin that has been approved by the US FDA is Biocon-Viatris's biosimilar Insulin Glargine (Semglee®) (Joshi et al. 2023b). According to the INSTRIDE-3 Phase-3 Switch study, done in 2017, participants who switched between reference and biosimilar insulin glargine showed comparable safety, immunogenicity, and efficacy. This suggests that those on reference insulin glargine can safely switch to the biosimilar insulin glargine (Joshi et al. 2023b).

1.7 Are Biosimilars Safe?

A biosimilar must undergo clinical trial testing and FDA approval, just like other medications, before it may be used to treat a disease to ensure that it is safe to use in people. The biosimilar and its reference drug, first-developed biologic, are compared in clinical trials (Weise et al. 2012). The Food and Drug Administration (FDA) carefully examines the clinical trial data to ensure that the biosimilar medication is equally safe and effective as the biologic brand-name medication. A biosimilar medication will satisfy the stringent safety requirements if the FDA approves it. Testing is done during the biosimilar studies to ensure that it is identical to the reference under brand name in specific aspects (Kang et al. 2023). Testing must demonstrate that both medications:

- Come from the same source.
- Possess the same dose and potency.
- Can be given to patients via same route of administration.
- Offer comparable clinical advantages when treating a clinical condition.
- Possess the same adverse effects.

1.8 Why Are Biosimilars Being Developed?

The development of biosimilars serves a critical societal role by expanding patient access to transformative biological therapies that might otherwise be financially prohibitive. These drugs represent crucial treatments for various complex diseases like cancer, autoimmune disorders, and chronic illnesses, significantly enhancing the standard of care (Ahmad et al. 2016). Biosimilars act as a gateway, mitigating the barriers caused by the high costs associated with original biologics, ensuring that patients receive timely and affordable treatment options. Encouraging the streamlined regulation and fostering competition among multiple developers is paramount. This approach not only prevents unnecessary hindrances in drug development but also maximizes the societal benefits of biosimilars by stimulating a more competitive landscape, leading to increased availability and affordability of these life-changing therapies (Weise et al. 2012). By optimizing regulatory pathways, biosimilars can effectively offer hope and improved quality of life for a broader spectrum of patients, aligning with the fundamental principle that healthcare access should not be hindered by financial constraints.

Biosimilars hold their utmost value, and they are accessible to every patient who necessitates these therapies, ensuring that no individual is deprived of potentially life-changing treatments due to financial constraints. The fundamental purpose of biosimilars lies in providing a cost-effective alternative to expensive biologics without compromising therapeutic efficacy or safety. However, for this value to be fully realized, it is essential to foster an environment of robust competition in the marketplace based on fair and accurate comparisons. Unfounded suggestions or misconceptions about differences in efficacy or safety between biosimilars and their reference biologics can distort this competition. Therefore, ensuring a clear understanding and trust among healthcare providers, regulators, and patients regarding the high similarity and equivalent clinical performance of biosimilars to their reference products is crucial (McCamish and Woollett 2012). This approach not only supports fair pricing but also guarantees that patients have access to these vital treatments, thus optimizing the value and impact of biosimilars in improving healthcare accessibility and outcomes for all who require them.

The distinction between biosimilars and generics is crucial, primarily due to the inherent complexities in manufacturing biologics versus chemical drugs. Biologics are intricate molecules produced from living organisms, making their manufacturing process more complex, delicate, and costly compared to small molecule generics (Ridgway et al. 2013). Consequently, the development and approval of biosimilars entail more extensive studies and stringent regulatory assessments to demonstrate their similarity and equivalence to the reference biologic. This rigorous regulatory oversight, driven by the unique manufacturing challenges of biosimilars, instils confidence in their quality and safety. Despite these differences, both biosimilars and generics share a common economic objective: driving down costs and enhancing patient accessibility to essential treatments. While the manufacturing challenges and regulatory scrutiny for biosimilars are distinct, the overarching goal remains the same—to offer more affordable alternatives to expensive biologics, thereby improving access to vital therapies for patients in need. This alignment in the economic rationale underscores the importance of both biosimilars and generics in contributing to healthcare cost reduction and broader patient accessibility to essential medications.

1.9 Biosimilar Drugs: Current Status

Biosimilars a new class of drugs, intended to any off patients, offer comparative safety and efficacy. Their active protein structure enables them to respond in various immunological conditions as well as acute and chronic conditions. The main reason of biosimilar development is their early expiry of patent protection for many biopharmaceuticals (Nowicki 2007). The year 2006 marked a significant milestone in the pharmaceutical realm with the approval of Omnitrope, a biosimilar iteration of Pfizer's Genotropin. Omnitrope, developed and brought to market by Sandoz, introduced a recombinant human growth hormone called somatropin. Subsequently, further expanding options in the European Union (EU), another biosimilar somatropin named Valtropin, crafted by Bio partners, received regulatory approval.

In Europe, two biosimilar versions of recombinant human erythropoietin have gained approval, collectively marketed under five different brand names. The first, HX575, available as Binocrit (Sandoz), Epoetin alfa HEXAL (Hexal), and Abseamed (Medice Arzneimittel), is a biosimilar of the reference product Eprex (epoetin alfa) and shares the international non-proprietary name (INN) epoetin alfa. The second approved biosimilar, SB309, marketed as Retacrit (Hospira) and Silapo (STADA), is also a biosimilar of Eprex (Schellekens 2004). However, it differs by utilizing the INN epoetin zeta instead of epoetin alfa. These approvals offer multiple alternatives with different brand names for the same biosimilar erythropoietin products in the European market (Kumar and Singh 2014).

In European market, three distinct biosimilar filgrastim products obtained approval, collectively offered under six different brand names, primarily for treating neutropenia. Ratiograstim, filgrastim ratiopharm (both from Ratiopharm), and biograstim (CT Arzneimittel) stem from the same manufacturer and exhibit similarity to the reference product Neupogen.

Additionally, Tevagrastim (Teva Generics), the second among the three approved biosimilar filgrastim products in Europe, shares similarity with Neupogen. More recently, another biosimilar filgrastim variant, EP2006, received approval. EP2006 is available under two distinct brand names—Zarzio by Sandoz and Filgrastim Hexal by Hexal (Roger 2010). Like the others, EP2006 also references Neupogen as its originator product. These approvals have expanded the range of available options, offering multiple brands for biosimilar filgrastim products in Europe, each designed to closely resemble and function similarly to the original Neupogen for treating neutropenia.

In the United States, biopharmaceutical medications like insulin and growth hormone (e.g., Omnitrope) have gained approval through an expedited process. Yet, for the FDA to authorize more products, specific legislation aimed at facilitating the introduction of biosimilars will be necessary. A vast array of 100 biopharmaceuticals is presently in active clinical usage, with hundred more undergoing development (Crommelin et al. 2003). This extensive pipeline predominantly focuses on addressing criterial areas such as oncology, infectious disease, autoimmune disorders, and respiratory ailments. This robust development landscape indicated significant potential for the future emergence of biosimilar medicines (Table 1.2).

Table 1.2 Biosimilars approved in India

Product name	Active substance	Company	Launch year
Basalog	Insulin glargine	Biocon	2009
Biovac-B	Hepatitis-B vaccine	Wockhardt	2000
Cresp	Darbopoetin Alpha	Dr. Reddy's Laboratory	2010
FostiRel	Follitropin beta	Reliance life science	2010
Neukine	Filgrastim	Intas Biopharmaceuticals	2004
Wepox	Epoetin alpha	Wockhardt	2001
Wosulin	Human insulin	Wockhardt	2003

Of particular interest is the regulatory stance concerning biosimilars, especially those pivotal in saving lives, notably in oncology. There exists a keen interest in understanding whether the guidelines governing the development and approval processes for these life-saving biosimilars are notably more rigorous compared to the standards applied to current biosimilars utilized in supportive-care settings (Duerden 2007). The evaluation and potential differences in regulatory stringency can significantly impact the development, approval, and accessibility of biosimilars crucial for treating life-threatening conditions, warranting a careful examination of regulatory approaches in these specific contexts.

1.10 Pharmacovigilance and Biosimilars

Pharmacovigilance is defined by the World Health Organization (WHO) as the science and practices associated with the identification, assessment, understanding, and prevention of adverse drug reactions (ADRs) or any other drug-related issues (text IPCJBtc 2018). Any suspected reaction, suspected drug–drug or drug–food interaction, adverse drug reactions linked to drug withdrawal, medication errors or overdose, and lack of efficacy must be reported to regulatory bodies in accordance with acceptable pharmacovigilance practices. Pharmacovigilance also requires the aggregate reports including periodic safety update reports (PSURs) and risk management plans (RMPs). The PSUR is an essential instrument for identifying arising safety signals, evaluating changes in the benefit–risk profile, communicating risks to regulatory bodies, and identifying when risk management actions are required. It also serves as a tracking mechanism to evaluate how well these initiatives are working. The RMP records the risk management system deemed required to identify, characterize, and reduce a medicinal product's significant risks over the course of the product's life cycle, ensuring benefit–risk balance (Oza et al. 2019). Global regulatory bodies, such the US Food and Drug Administration and the European Medicines Agency, have established well-organized pharmacovigilance systems. In January 2018, Indian pharmacovigilance guidelines for marketing authorization holders (MAHs) was released by Indian Pharmacopoeia Commission (IPC) and Pharmacovigilance Programme of India (PvPI) (text IPCJBtc 2018; U.S. Department of Health and Human Services et al. 2005).

For any medicine that has been approved, systematic and continuous safety monitoring is required to identify and assess post-approval safety signals. These signals could be the result of inherent variations in immunogenicity among products or the identification of rare events that might be especially linked to a specific product (Seidl et al. 2012). In order to analyze and characterize a product's risk profile and to make well-informed decisions on risk minimization, the FDA guidelines on appropriate pharmacovigilance state that post-marketing safety data collecting and risk assessment based on observational data are essential (Casadevall et al. 2013).

Since biosimilar medications are manufactured by different companies than reference products and are not reference medicines in the conventional sense, pharmacovigilance is more crucial for them. A biosimilar medication may not show many side effects until it is taken more widely, for longer periods of time, and in a larger number of patients. Prescribers and manufacturers both need to be aware of the need for post-marketing surveillance and careful when managing patients who are using biosimilars (Kumar and Singh 2014).

1.11 Applications of Biosimilars

Biosimilars have gained prominence due to their potential to offer more affordable alternative to expensive drugs while providing efficacy and safety to the patient. Biosimilars also play an important role in various disease conditions such as in autoimmune disorders like Rheumatoid arthritis, Psoriasis, and Crohn's disease (Kaida-Yip et al. 2018). These drugs target the tumor necrosis factor-alpha (TNF-alpha) such as in adalimumab, infliximab provide cost-effective alternative to expensive biologics (Fig. 1.3).

Biosimilars in oncology also offers a potential therapeutic role in disease condition. Biosimilar version of drugs such as trastuzumab, bevacizumab, and rituximab have been developed that offer similar safety and efficacy at potentially reduced costs (Ecker et al. 2015). In disease like diabetics, bone disorders, and kidney

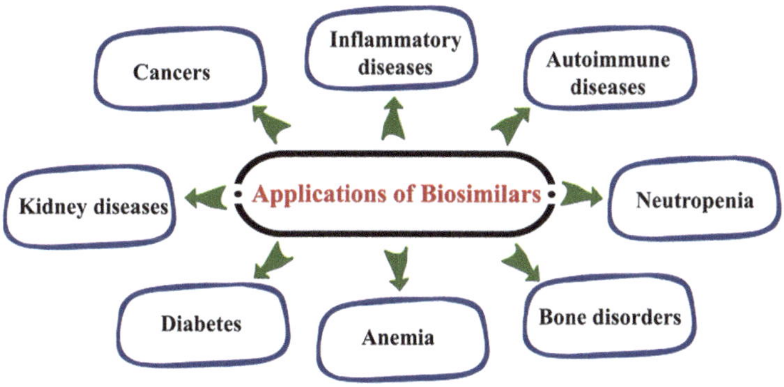

Fig. 1.3 Various Applications of Biosimilars. Various applications in different diseased conditions where biosimilars have been used efficiently

diseases, they offer similar version of therapeutics, aiming to provide affordable options for the patient requiring insulin therapy and erythropoietin stimulating agents are effective versions of biosimilars used as alternative therapy in chronic kidney diseases (Schellekens and Ryff 2002). Although still several emerging biosimilars for multiple sclerosis and certain rare disorders are being researched to expand the lifespan of the patients, they are also used in treating various hematological conditions, such as Filgrastim which is used to boost white blood cell count and epoetin, used to stimulate red blood cell production, these have been developed to potentially increase the accessibility of therapeutic approach to the patients with conditions like anemia and neutropenia (Rathore 2011).

Overall, in each of this diseased area, biosimilars hold a promise of cost-effective and safety profile of the patient. However, it is crucial to note that adaptation and acceptance of various biosimilars in different fields are important in understanding the effects of the bio-phytochemicals in diseased conditions.

References

Afzali A, Furtner D, Melsheimer R, Molloy PJ (2021) The automatic substitution of biosimilars: definitions of interchangeability are not interchangeable. Adv Ther 38:2077–2093

Ahmad A-S, Olech E, JE MC, Kirchhoff CF (eds) (2016) Development of biosimilars. Seminars in arthritis and rheumatism. Elsevier

Aladul MI, Fitzpatrick RW, Chapman SR (2018) Healthcare professionals' perceptions and perspectives on biosimilar medicines and the barriers and facilitators to their prescribing in UK: a qualitative study. BMJ Open 8(11):e023603

Casadevall N, Edwards IR, Felix T, Graze PR, Litten JB, Strober BE et al (2013) Pharmacovigilance and biosimilars: considerations, needs and challenges. Expert Opin Biol Ther 13(7):1039–1047

Crommelin DJ, Storm G, Verrijk R, de Leede L, Jiskoot W, Hennink WE (2003) Shifting paradigms: biopharmaceuticals versus low molecular weight drugs. Int J Pharm 266(1–2):3–16

Declerck P, Danesi R, Petersel D, Jacobs IJD (2017) The language of biosimilars: clarification, definitions, and regulatory aspects 77:671–677

Derbyshire M (2017) Interchangeability of biosimilars in the US and around the world. GaBI J 6:97–98

Duerden M (2007) Prescribing advice needed for new biosimilar biological drugs, vol 18. John Wiley & Sons, Ltd., Chichester, pp 7–2

Ecker DM, Jones SD, Levine HL (eds) (2015) The therapeutic monoclonal antibody market. MAbs. Taylor & Francis

Food U, Administration D (2019) Considerations in demonstrating interchangeability with a reference product: guidance for industry. US Food and Drug Administration, Silver Spring

Food U (2015) Drug Administration guidance for industry: scientific considerations in demonstrating biosimilarity to a reference product. FDA, Rockville

Ghosh S, Bose S, Gowda S, Mukhopadhyay P (2019) Biosimilar insulins–what a clinician needs to know? Indian J Endocrinol Metab 23(4):400

Glintborg B, Sørensen IJ, Loft AG, Lindegaard H, Linauskas A, Hendricks O et al (2017) A nationwide non-medical switch from originator infliximab to biosimilar CT-P13 in 802 patients with inflammatory arthritis: 1-year clinical outcomes from the DANBIO registry. Ann Rheum Dis 76(8):1426–1431

Joshi SR, Mittra S, Raj P, Suvarna VR, Athalye SN (2023a) Biosimilars and interchangeable biosimilars: facts every prescriber, payor, and patient should know. Insulins perspective. Expert Opin Biol Ther 23(8):693–704

Joshi SR, Mittra S, Raj P, Suvarna VR, Athalye SNJEOoBT. (2023b) Biosimilars and interchangeable biosimilars: facts every prescriber, payor, and patient should know. Insulins Persp 23(8):693–704

Kaida-Yip F, Deshpande K, Saran T, Vyas D (2018) Biosimilars: review of current applications, obstacles, and their future in medicine. World J Clin Cases 6(8):161

Kang HN, Wadhwa M, Knezevic I, Ondari C, Simao M (2023) WHO guidelines on biosimilars: toward improved access to safe and effective products. Ann N Y Acad Sci 1521(1):96–103

Kim AP, Bindler RJ (2016) The future of biosimilar insulins. Diabetes Spectr 29(3):161–166

Kumar R, Singh J (2014) Biosimilar drugs: current status. Int J Appl Basic Med Res 4(2):63

Kurki P, van Aerts L, Wolff-Holz E, Giezen T, Skibeli V, Weise M (2017) Interchangeability of biosimilars: a European perspective. BioDrugs 31(2):83–91

Laekeman G (2013) Guideline on non-clinical and clinical development of similar biological medicinal products containing recombinant human insulin and insulin analogues. ESCP Newsl (159):6

McCamish M, Woollett G (2012) The state of the art in the development of biosimilars. Clin Pharmacol Ther 91(3):405–417

McDonald C (2015) Biosimilars ramp-up: five themes for the future. Pharm Exec 40-2

de Mora F, Balsa A, Cornide-Santos M, Carrascosa JM, Marsal S, Gisbert JP et al (2019) Biosimilar and interchangeable: inseparable scientific concepts? Br J Clin Pharmacol 85(11):2460–2463

Niazi SK (2022) Biosimilars: harmonizing the approval guidelines. Biologics 2(3):171–195

Nowicki M (2007) Basic facts about biosimilars. Kidney Blood Press Res 30(5):267–272

Oza B, Radhakrishna S, Pipalava P, Jose V (2019) Pharmacovigilance of biosimilars–Why is it different from generics and innovator biologics? J Postgrad Med 65(4):227

Rathore A (2011) Development and commercialization of biosimilars in India. BioPharm Int 24(11)

Ridgway A, Ritter N, Schiestl M, Schreitmuller T (2013) Biosimilar products. BioProcess International. http://www.bioprocessintl.com/manufacturing/biosimilars/biosimilar-products-347979

Roger SD (2010) Biosimilars: current status and future directions. Expert Opin Biol Ther 10(7):1011–1018

Rumore MM, Vogenberg FR (2016) Biosimilars: still not quite ready for prime time. Pharm Ther 41(6):366

Schellekens H (2004) How similar do'biosimilars' need to be? Nat Biotechnol 22(11):1357–1359

Schellekens H, Ryff J-C (2002) 'Biogenerics': the off-patent biotech products. Trends Pharmacol Sci 23(3):119–121

Seidl A, Hainzl O, Richter M, Fischer R, Böhm S, Deutel B et al (2012) Tungsten-induced denaturation and aggregation of epoetin alfa during primary packaging as a cause of immunogenicity. Pharm Res 29:1454–1467

text IPCJBtc (2018) Pharmacovigilance guidance document for marketing authorisation holders of pharmaceutical products. 2022(1)

Tóthfalusi L, Endrényi L, Chow SC (2014) Statistical and regulatory considerations in assessments of interchangeability of biological drug products. Eur J Health Econ 15(Suppl 1):S5–S11

U.S. Department of Health and Human Services, Food and Drug Administration, Center for Drug Evaluation and Research (CDER), Center for Biologics Evaluation and Research (CBER) (2005) Guidance for industry: good pharmacovigilance practices and pharmacoepidemiologic assessment

Weise M, Bielsky M-C, De Smet K, Ehmann F, Ekman N, Giezen TJ et al (2012) Biosimilars: what clinicians should know. Blood 120(26):5111–5117

Wolff-Holz E, Tiitso K, Vleminckx C, Weise MJB (2019) Evolution of the EU biosimilar framework: past and future. BioDrugs 33(6):621–634

Immunobiology of Cancer

2

Nitin Verma, Komal Thapa, Neha Kanojia,
and Prarit Chandel

Abstract

One of the results of three decades of very fast research into the mechanisms of cancer pathogenesis has been the advent of mechanism-based targeted medicines for the treatment of human tumours. The complex and dynamic interaction between cancerous cells and the immune system is reflected in the immunobiology of cancer. Over the past 15 years, there has been a renewed focus on cancer immunosurveillance, which has expanded to include cancer immunoediting. The latter contends that the immune system both shapes tumour immunogenicity and shields the host from the emergence of primary nonviral malignancies. It is backed by compelling correlative data from research on human cancer as well as robust experimental data from mice tumour models. Three stages make up the process of cancer immunoediting: equilibrium, escape, and elimination (also known as cancer immunosurveillance). We provide a summary of the evidence for each of the three cancer immunoediting steps below. The main ideas of immunobiology in cancer are briefly reviewed in this chapter, along with the most recent developments in using the immune system to treat cancer. We also explore the function of inflammation and its dual effects on tumour growth: acute inflammation may aid in immune responses against cancer, but chronic inflammation may promote carcinogenesis.

Keywords

Cytokines · Cytotoxicity · Immunobiology · Inflammation · Vaccines

N. Verma (✉) · K. Thapa · N. Kanojia · P. Chandel
Chitkara University, School of Pharmacy, Barotiwala, Himachal Pradesh, India
e-mail: nitin.verma@chitkarauniversity.edu.in

S. Bhatt et al. (eds.), *Biosimilars for Cancer Treatment*,
https://doi.org/10.1007/978-981-97-3001-8_2

2.1 Introduction

Coley showed that using bacterial products might cause some tumours that were inoperable to recede. For many years, it has been understood that the immune system is crucial to the development of cancer, chronic inflammation, and inflammation (Oiseth and Aziz 2017). As a result, scientists kept trying to connect those processes in order to comprehend the significance of Virchow's discoveries. The initial immune-related treatments were nonspecific immunotherapies that did not target a particular cancer cell culture and instead attempted to stimulate the immune system. Understanding the nature of this connection could throw light on the mechanisms of carcinogenesis, tumour growth, and metastatic dissemination. It may also present chances to devise treatment plans that improve prognosis, diagnosis, and the overall cancer treatment experience (Decker and Safdar 2009; Decker et al. 2017). The goal of further research was to pinpoint the antigens on cancer cells and create monoclonal antibodies (MABs) that specifically target those antigens (Gross 1943). This has led to a new direction in research: identifying the pathways and signalling molecules involved in immune suppression processes, which may be targets for anticancer treatments.

2.2 The Hallmarks of Cancer

The acquired abilities to maintain proliferative signalling, elude growth suppressors, fend off cell death, permit replicative immortality, induce/access vasculature, trigger invasion and metastasis, reprogram cellular metabolism, and elude immune destruction are currently included in the eight hallmarks (Smith Jr. and Stehlin Jr. 1965). Eleven years later, it is clear that deregulating cellular metabolism and preventing immune destruction, like the original six, can be regarded as core hallmarks of cancer. These two concepts were separated as "emerging hallmarks" in the most recent iteration of this notion. While a series of oncogenic events characterize the progression of cancer, the characteristics listed below only apply to aggressive cancer that has reached its full potential and not to its precursors. The fact that several noncancerous cell types can be found within malignancies and may contribute to the cancerous phenotype suggests that not all malignant cells share the characteristics of the disease. Another concern is whether they apply to all cancer cells. In fact, the idea of a cancer stem cell (CSC) is in direct opposition to the idea of a "cancer cell" (Chow et al. 2012; Halliday et al. 1995). The latter idea, which postulates a qualitative, deterministic, and irreversible change from CSC-tumour propagating cell (TPC) to derived cells, has come under scrutiny and is now refuted in the context of human colon carcinoma and a mammary cancer cell line. Without getting into this discussion, it follows logically that only a small percentage of cancerous cells will exhibit all the characteristics required to cause the cancer, given that the majority of cancerous cells are derived cells (Fife and Bluestone 2008; Brunet et al. 1987). This would particularly apply to the immortality attribute. All of the hallmarks cannot be attributed to all cancer cells, even if the ill-defined and contentious concept of a

CSC is ignored. It is true that a significant percentage of cells in tumours die, enter quiescence or senescence, and so are unable to further the growth of the tumour (Dariavach et al. 1988). These cells most definitely would not display the telltale signs of immortality or avoidance of cell senescence. Furthermore, it is commonly known that cellular heterogeneity occurs in cancer and that even in populations of homogeneous cells, there is a significant degree of cell-to-cell variability in phenotype, which is influenced by a number of factors including history, location, and size. The fact that breast tumours are classified as HER2 positive or oestrogen receptor positive when the percentage of these cells in the tumour surpasses 30% or 11%, respectively, serves as an example of this heterogeneity (Harper et al. 1991). Therefore, not every tumour cell, or even most cancer cells, exhibits the characteristics of cancer. In contrast, certain normal cells already show certain characteristics; normal free-moving cells that naturally avoid tissue restrictions or contact inhibition include lymphocytes and blood cells generated from bone marrow and can enter blood without the need for an epithelial-mesenchymal transition (EMT). They can also survive anoxia without the aid of Warburg metabolism or angiogenesis. Various hallmarks of cancer are shown in Fig. 2.1.

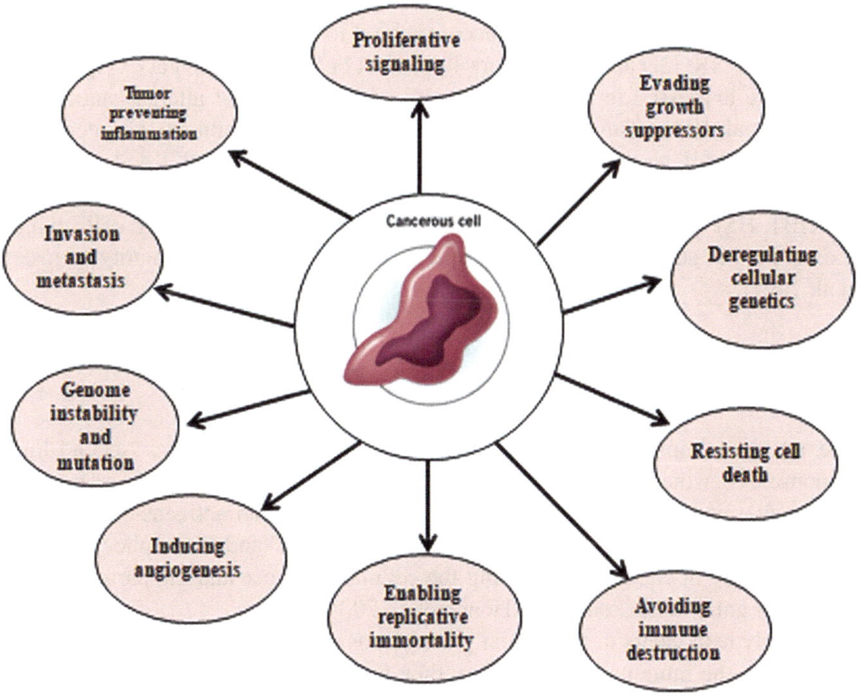

Fig. 2.1 Various hallmarks of cancer

2.3 The Principles of Immune Response in Solid Tumours

Innate and adaptive immune responses comprise the two main categories of the anticancer immune response. They function as an organism's first line of defence by identifying stressed or mutated cells and initiating effector mechanisms that attempt to eradicate them. The adaptive immune response and specific immunological activation of B cells is then set off.

2.3.1 Innate Antitumor Response

Cells inside the organism may undergo malignant genetic and epigenetic changes, resulting in the acquisition of new traits that allow for unchecked growth, survival, and spread (Walunas et al. 1994). Such genetic damage activates the innate immune system, which typically acts as an early warning system and responds quickly. Through a complicated process of expressing several inhibitory and stimulatory chemicals, the NK cells are able to discriminate between normal and tumour cells. The identification of particular MHC inhibitory receptors has provided insight into the molecular mechanisms behind the natural cytotoxicity of the innate antitumor response and the subsequent activation of natural killer cells (Lo et al. 2015). NKp46, NKp30, NKp44, and NKG2D are among the many receptors that are expressed on the surface of NK cells and are commonly referred to as natural cytotoxicity receptors (NCRs). These receptors bind to ligands of the MHC class I. It appears that NKG2D and NCRs work in concert or in complementary ways. These ligands are expressed in response to stress on the surface of altered cancer cells (Linsley et al. 1991). Tumour cytotoxicity and IFN-γ production are inhibited when MHC-I-related ligands bind to NKG2D and activate NK cells, NKT and $\gamma\delta$ T cells, and CD 8 T cells. The release of extracellular cytoplasmic stress molecules, such as HMGB1, HSP-70, and uric acid, activates macrophages and dendritic cells, which results in the generation of IL-12 and a shift to adaptive immunity (Linsley et al. 1994).

2.3.2 Adaptive Antitumor Responses

The adaptive immune response is sometimes referred to as the "second-line" response. After the initial fast innate response, this comparatively late response is extremely unique to a single pathogen. By providing tumour antigens to class I and II MHC, digesting them, and then presenting them to T and B lymphocytes, dendritic cells are in charge of triggering the cellular response and the particular production of antibodies (Pentcheva-Hoang et al. 2004).

As early responders to a potentially dangerous antigen identify foreign cells and take part in the immune response, they then break down the proteins into peptide, which are then presented back at the extracellular membrane surface of the APCs in relation to MHC class II. Additionally, they generate a variety of cytokines in great quantities, which strengthens the immune system. Dendritic cells in particular stop further autoimmune reaction when the immune system is not directed appropriately against self-antigens (Krummel and Allison 1995). Endogenous cross talks, also referred to as "immune checkpoints," are utilised by the immune system," which generally cease immune responses following T-cell activation by antigens. Other cells in the immune complex are exposed to the foreign antigen once the immune response has been initiated. When the foreign antigen is seen by the more specialised cells, the lymphocytes proliferate and differentiate into various subpopulations as a response (Krummel and Allison 1996). As mature, functioning cells, B-lymphocytes develop and mature in the bone marrow before entering the bloodstream. They have an antigen receptor expressed on their surface, and when they come into contact with that particular antigen, they begin to divide, differentiate into plasma cells, and create soluble immunoglobulin molecules or antibodies, which are then released into the bloodstream. The thymus, after which it gets its name, is where T cells mature after emerging from the bone marrow. After emerging from the thymus, immunocompetent T cells go into the bloodstream. T cells are categorised according to their phenotypic and function. Most T cells are classified as T helpers because they express the CD-4 glycoprotein (Waterhouse et al. 1995). Through the secretion of cytokines and direct cell-to-cell interaction, they augment the immunological process. We refer to a large number of additional distinct functional T-lymphocytes as cytotoxic (CTLs). They are able to directly destroy antigen-containing cells (cancerous or virally infected) because they express CD8 glycoprotein. When they come into contact with their target, they induce apoptosis in the malignant or diseased cell. A portion of the lymphocytes continue to function as long-lived, sensitised memory cells that can only identify one antigen and are primed to react if it is encountered again. A tiny subset of T cells known as regulatory T cells (T regs) express CD 25 glycoprotein, which aids in the detection of self-antigens and averts autoimmune responses (Tivol et al. 1997; Mandelbrot et al. 1999).

2.4 Targets of Immunotherapy in Solid Tumours

2.4.1 Active Immunotherapy

Active clinical research has lately been conducted on active immunotherapy. The immune system should react to tumours that express several tumour-associated antigens or neoantigens by activating. Active immunotherapy is defined as any process that triggers the immune system (Tai et al. 2007). The goal of developing active immunotherapy is to activate and trigger the patient's own immune response. This still-emerging science of immunotherapy is exemplified by Sipuleucel-T, the first active cellular immunotherapy that the American Food and Drug Agency (FDA) approved for clinical use in the treatment of prostate cancer. According to study data, patients with castrate-resistant prostate cancer (CRPC) who were either asymptomatic or had little symptoms fared better in terms of survival (Lee et al.

2016). PAP peptides work with MHC I and II, whereas GM-CSF promotes the development of APCs into mature DCs. An immunological reaction to PAP-containing cells is elicited upon reinfusion in the patient (Kuehn et al. 2014; Intlekofer and Thompson 2013).

Cellular adoptive immunotherapy, which uses the patient's own T-lymphocytes and is presently being studied in phase I/II trials, is another example of active immunotherapy. Additionally, there are numerous trials involving autologous or donor dendritic cells, autologous tumour cell lysate that are administered as a monotherapy or in conjunction with other passive immunotherapy options like PD-1 MAB, anti-PD-L1 MAB, and anti-CTLA 4 MAB (Darlington et al. 2002; Schneider et al. 2006).

2.4.2 Passive Immunotherapy

As of right now, the term "passive immunotherapy" is widely used to describe the administration of previously manufactured substances that the immune system may utilise. Early clinical trials showed that in certain solid malignancies, the use of immunomodulatory cytokines such interleukin 2 (IL-2) or interferon alpha (IFN α) may generate antitumor immune-mediated effects like tumour shrinkage (Qureshi et al. 2011; Greenwald et al. 2001). For many years, cytokines have been utilised as cancer immunotherapy. They either directly inhibit cancer growth or indirectly boost the immune system's ability to fight cancer. Numerous studies conducted in vitro have demonstrated that TNF-α and IL-6 have a direct antitumor effect, inhibiting the development and survival of cancer cells. However, because of their high toxicity and the contentious influence of individual molecules like TNF-α and IL-6, the therapeutic use of these cytokines in cancer patients has produced less satisfactory outcomes. IL-6 may also have immunosuppressive effects, and while they can inhibit the growth of certain tumours, they actually encourage the growth of other tumours (Chuang et al. 1999). As such, the application of cytokines' direct antitumor action is yet solely a research *endeavour*. Conversely, other cytokines are more frequently employed in clinical practise. IL-2 and IFN-α, on the other hand, promote the proliferation and activation of T-lymphocytes and NK cells. These cytokines are now well-established cancer immunotherapies; for instance, IL-2 is used to treat metastatic renal cell carcinoma and melanoma, and IFN-α is licenced to treat malignant melanoma (Marengere et al. 1996; Chuang et al. 2000). Recombinant IL-2 has also been utilised to treat various solid tumour malignancies, such as neuroendocrine tumours, according to findings in the literature. As a result, immunotherapy was first used to treat metastatic melanoma in 1998 and metastatic renal cell carcinoma in 1992 as an anticancer treatment. Following this, immunotherapy using interferon was also approved as a useful adjuvant treatment for patients with high-risk malignant melanoma (Fraser et al. 1999).

The application of MAB is another common instance of passive immunotherapy. Several MABs are used to treat solid tumours, including trastuzumab, an antibody that targets the Human Epidermal Receptor type 2 (HER 2), and cetuximab, an

MAB that targets the Epidermal Growth Factor (EGFR). These monoclonal antibodies (MABs) selectively attach to their receptor on the surface of cancer cells, blocking the intracellular signal cascade that would otherwise promote tumour growth or replication (Jain et al. 2010; Wing et al. 2008). Additionally, soluble circulatory factors that are significant for the tumour may be targeted by MABs. For example, bevacizumab, an MAB, targets VEGF.

Tumours use the PD-1/PD-L1 relationship as one of their main mechanisms to inhibit immune response. Under normal circumstances, PD-1's usual role is to downregulate excessive or undesired immune responses, including autoimmune reactions (Takahashi et al. 2000; Tai et al. 2012). Preclinical anticancer activity is mediated, and T cell responses are enhanced in vitro by inhibiting the interaction between PD-1 and PD-L1. PD-L1 causes tumour-specific T cells to undergo apoptosis, inhibits T-lymphocyte proliferation, survival, and effector functions (cytotoxicity, cytokine release), and encourages CD4+ T cells to differentiate into regulatory T cells. In patients with advanced cancers, blocking PD-1/PD-L1 causes a strong and long-lasting tumour reduction and extended stabilisation (Walker 2013). Thus, blocking PD-L1 binding to PD-1 seems like a good way to get tumour-specific T cell immunity back. Although PD-1 and CTLA-4 both regulate a comparable set of signalling proteins, their mechanisms for downregulating T cell responses are different. Following their activation, T cells, B cells, monocytes, and natural killer T cells have all been demonstrated to express PD-1 (Chikuma and Bluestone 2007). In lymphoid organs, PD-L2 regulates immunological T cell activation, while PD-L1 shields healthy tissues from unjustified T-cell immune-mediated harm. While few, if any, healthy organs express PD-L1, this T cell inhibitor is highly expressed in many malignancies. Poor prognosis and survival have been linked to high expression of PD-L1 on tumour cells (and to a lesser degree, PD-L2) in a number of malignancies, including ovarian carcinoma, pancreatic carcinoma, hepatocellular carcinoma, and RCC (Read et al. 2006, 2000; Friedline et al. 2009). Moreover, it has been proposed that PD-1 controls the growth of tumour-specific T cells in melanoma patients (Sojka et al. 2009). The association between PD-L expression and clinical outcome in many malignancies indicates that the PD-1/PD-L1 pathway is crucial for tumour immune evasion and, as such, should be a desirable target for therapeutic intervention.

2.5 Role of Inflammation in Cancer

Inflammation is a characteristic of innate immunity, which differs from adaptive immunity in that it activates quickly and is mediated by various receptors. Aside from being more evolutionary old than adaptive immunity, innate immunity is triggered by normal cellular components released upon injury and cell death, which are referred to as damage-associated molecular patterns (DAMPs), or by foreign microbial and viral structures, which are known as pathogen-associated molecular patterns (PAMPs) (Hou et al. 2015). When innate immunity is triggered, it leads to the overexpression of MHC class I and II, costimulatory molecules, and a variety of inflammatory chemokines and cytokines that draw in and stimulate T cells for activation via various antigen receptors (Grosso and Jure-Kunkel 2013). T and B cells will become activated during adaptive immunity, intensifying the original inflammatory response even more. Accordingly, Th1 helper T cells (Th1 cells) stimulate macrophages through cell-to-cell contact and the release of interferon (IFN)-γ, Th2 cells release cytokines to stimulate eosinophils, and B cells secrete antibodies that activate the complement cascade in addition to phagocytes, NK cells, and mast cells through Fc receptors (Leach et al. 1996). Nonetheless, some immune cells—Tregs in particular—may prevent inflammation.

It is well known that inflammation and the growth of tumours are related. Chronic inflammation in an organ aids in the promotion of cancer growth and emergence in precancerous stages. The inflammatory context can be caused by external bacterial infections, such as Helicobacter pylori in gastric cancer and non-Hodgkin gastric lymphoma, or viral infections, such as hepatitis B and hepatitis C viruses in hepatocellular carcinoma and cervical and head and neck cancer; or by increased permeability of the epithelial barrier to commensal microbiota promoting the development of colorectal cancer (CRC); or the presence of toxic chemicals, like smoke for lung cancer (Kwon et al. 1997; Fecci et al. 2007; Yang et al. 1997). Tumour necrosis factor (TNF)-a, transforming growth factor-b (TGF-b), adenosine, prostaglandin E2, and cytokines like interleukin (IL)-6, IL-1, or TNF-a can all be produced in an inflammatory environment and contribute to the formation of tumours and suppressor myeloid or T cells that draw chemokines such as CCL12, CXCL5, CXCL1, CXCL2, and c-x-c motif ligand (CXCL)1 (Mangsbo et al. 2010). Anti-inflammatory medications including aspirin and Cox-2 inhibitors have been demonstrated to be effective in preventing colorectal cancer (CRC). More remarkably, the immunisation and medications both treat Helicobacter pylori infections. Tumour-associated macrophages (TAMs) are the primary leukocyte population that infiltrates malignancies and aids in the progression of tumours (Hurwitz et al. 1998). Despite the fact that macrophages are capable of attacking and killing tumour cells, TAMs interfere with the growth and function of immune effectors and display a number of protumoral characteristics that are partially shared by macrophages participating in tissue repair. A high number of TAMs is linked to a poor prognosis in many human tumours (van Elsas et al. 1999). Non-autonomous regulation plays a significant role in the carcinogenesis of cancerous cells. Cancer cells require humoral factors including VEGF, FGFs, and Wnt to communicate with stromal cells in order to thrive. Extracellular vesicles (EVs) have been linked to both the development of cancer and cell communication, according to a number of recent studies. EVs are home to short noncoding RNAs (siRNAs), such as microRNAs (miRNAs), which have been linked to the aggressiveness of cancer cells (Jure-Kunkel et al. 2008).

Because tumour-derived exosomes (TEX) include immunosuppressive chemicals and factors that are known to disrupt immune cell activities, they are early warning signs of tumour-induced immune suppression. TEX directly or indirectly impact immune cell formation, maturation, and anticancer actions by providing suppressive cargos to immune cells that are composed of proteins comparable to those in parent tumour cells (Hodi et al. 2003). Immune cells receive genomic DNA, mRNA, and miRNAs from TEX, which also reprogrammes responder cell activities to accelerate tumour growth. Tumour-associated antigen-containing TEX can compromise antitumor immunotherapies. TEX may be connected to signalling pathways that downregulate antitumor immunity in the tumour microenvironment. As a result, TEX has the potential to be employed as a non-invasive biomarker of tumour advancement (Hodi et al. 2010). The local milieu surrounding tumours is influenced by the mediators and cellular effectors of inflammation. Inflammatory disorders may be identified prior to a malignancy in many types of cancer. On the other hand, an oncogenic alteration causes an inflammatory environment and encourages the growth of the tumours. Tumour microenvironment "smouldering" inflammation has numerous tumour-promoting effects, regardless of where it comes from (Schadendorf et al. 2015). It disrupts adaptive immune responses, enhances angiogenesis and metastasis, alters responses to hormones and chemotherapeutic drugs, and helps malignant cells proliferate and survive. Now that the molecular pathways linking inflammation and cancer have been untangled, we anticipate finding novel target molecules in the future to enhance diagnosis and treatment (Maio et al. 2015).

2.6 Key Concepts of Cancer Immunobiology: Immunoediting, Surveillance, and Immune Escape

For many years, it has been understood that maintaining immune surveillance is crucial for preventing the development of neoplastic transformation. The identification of cellular oncogenes made it clear that human malignancies originate from normal cells and contain a variety of genetic and epigenetic changes, which may result in the production of cancer neoantigens that the immune system may be able to recognise (Yang et al. 2007). Cancer cells are host-derived, meaning that they have host characteristics. Because tumours are malleable, they can adapt to the treatment regimens of chemotherapy and radiation that are typically utilized. A sequence of successive clonal expansions, each brought about by the fortuitous acquisition of an enabling mutant genotype, can be used to characterize the tumour's evolution. A tiny percentage of cancer cells remain resistant to the cytotoxic chemotherapy treatment even after the medicine has killed the great bulk of cancer cells. They might be adequate to start a tumour recurrence later on that resists the chemotherapy drug that was previously used (Lynch et al. 2012). This gives rise to the idea of tumour cell selection and developing resistance, which is a crucial aspect of the disease's growth. The formation and spread of cancer are primarily caused by the selection of cells that endure circumstances that would ordinarily be fatal. Due to the cancer cell's substantial genetic plasticity, resistance to any typically deadly situation (chemotherapy, radiation, etc.) can be selected by the population's evolution (Reck et al. 2013).

Tumour development is a process that necessitates the acquisition of various biological characteristics, including: (1) persistent signalling of proliferation; (2) circumvention of growth inhibition; (3) escape from apoptosis; (4) unchecked growth and procreation; (5) stimulation of angiogenesis; and (6) propensity for invasion and metastasis. These characteristics of the tumour cells ensure their survival, growth, and spread (Kwon et al. 2014). Two additional characteristics are crucial for the formation of cancer hallmarks in terms of functionality. The first characteristic entails a significant dysregulation and reprogramming of cellular energy metabolism to maintain cell growth and proliferation throughout time. The immune system's active tumour escape and subsequent destruction and removal constitute the second aspect. This ability demonstrates the immune system's dual function in suppressing and promoting the onset, spread, and advancement of cancer (Ribas et al. 2013; Furness et al. 2014). It is possible that both of these characteristics will help many different types of malignant human solid tumours begin and spread (He et al. 2017).

According to a notion, tissues and cells are continuously observed by an immune system that is always on guard. It is believed that this immune surveillance is what causes the great majority of early cancer cells and developing tumours to be identified and eradicated. This reasoning suggests that any solid tumours that do develop have either escaped immune system elimination or have somehow been able to evade immune system identification (Baksh and Weber 2015). Immunoediting is a tumour mechanism that exerts extrinsic suppression; it is a process that only happens after intrinsic tumour suppressor mechanisms have failed and a malignant, cancerous transformation has already taken place. The three stages of cancer immunoediting are basically elimination, equilibrium, and escape. During the initial stage, the immune system of the body effectively identifies and eradicates cancer cells (Snyder et al. 2014). During the equilibrium phase, the immune system hinders the growth of the tumour but cannot eliminate it entirely. This stage is believed to be continuous in time and may either advance to the final escape phase or regress, allowing the immune system to completely eradicate the tumour. Longer duration of this second phase means that the immune system is unable to eradicate the tumour and will instead engage with it constantly, changing or shaping its genetic makeup (van Rooij et al. 2013). During the last stage, the immune system loses its ability to manage and impede the growth of the tumour, causing it to spread and manifest as illnesses (Sharma et al. 2019a).

Novel, targeted cancer immunotherapy will be based on our understanding of the mechanisms underlying cancer immunoediting (Peggs et al. 2009). The majority of differentiated cells employ mitochondrial oxidative phosphorylation to produce adenosine triphosphate (ATP), which is energy that may be used to maintain cellular functions, when oxygen is present. However, these cells go back to a much less effective form of ATP synthesis called glycolysis when oxygen is scarce. Despite the presence of oxygen, glycolysis is frequently used by cancer cells (aerobic glycolysis, also known as the "Warburg effect") (Ha et al. 2019). Even though it is less effective in generating energy, this type of metabolism is thought to assist the macromolecular needs of cell division and growth. As a result, the science has mostly concentrated on Warburg metabolism as an adaptation that gives tumour cells themselves inherent growth advantages. On the other hand, cancer cells have the ability to overconsume nutrients, especially glucose, in order to maintain cell growth and proliferation (Sharma et al. 2019b). This raises the notion that nutrient ingestion serves purposes beyond providing cancer cells with the basic energy and biosynthetic needs. According to Warburg's research, metabolism gives tumour cells an advantage over healthy cells by encouraging the depletion of extracellular glucose, which causes tumour-infiltrating T lymphocytes to become dysfunctional (Ishida et al. 1992). Ho et al. propose an alternative strategy to enhance T cell performance during adoptive cell treatment (ACT) by mimicking the nutritional availability inside transferred T cells, as opposed to modifying tumour cell metabolism. Oxaloacetate is converted to phosphoenolpyruvate (PEP) by PCK1, phosphoenolpyruvate carboxykinase. Despite low ambient glucose levels within tumours, T cell antigen receptor (TCR)-induced Ca2+ flow and anticancer T cell function were restored by artificially increasing PEP levels in transplanted T cells by overexpressing PCK1. Inhibiting glucose metabolism during T cell expansion for adoptive immunotherapy prevents effector differentiation and encourages memory cell differentiation, which is responsible for better tumour clearance (Carreno and Collins 2002; Latchman et al. 2001). These results offer dramatic illustrations of how altering T cell metabolism can enhance the effectiveness of cancer adoptive cell therapy. Glucose is necessary for antigenic tumours to metabolically limit T lymphocytes and lubricate their effector activity, so promoting tumour growth. Through a direct impact on the tumour cells, checkpoint blockade therapy can rectify these resource imbalance (Francisco et al. 2010). Despite recent advancements in cancer treatment, metastatic illness continues to be the primary cause of cancer-related mortality and is largely incurable. Metastases are the end product of a multi-step process that starts with local invasion by the primary tumour cells and progresses through local invasion, extravasation, survival in a foreign environment, arrest at a distant organ, intravasation into the blood or lymphatic system, and metastatic colonisation. Every stage of this process is contingent upon distinct phenotypic characteristics of the tumour cell, in addition to its interactions with the immune system and the host microenvironment (Freeman et al. 2000). The comprehensive treatment of metastatic disease remains unattainable despite advancements in human cancer therapy. Subclones with metastases have the potential to arise during the initial stages of tumour growth. To identify novel treatments for primary tumours and metastatic diseases, we must have a better understanding of the genetic development of metastatic disease and the biology of the metastatic process (Nishimura et al. 1998).

While its role in other events may also be extremely relevant to tumour progression, it is evident that EMT has a role in metastatic events in cancer. In numerous cell settings, TGF-β can guide apoptosis or survival plus EMT, halting the growth of nascent tumours and promoting tumour invasion and immune surveillance evasion at advanced stages (Nishimura et al. 1999). TGF-β promotes EMT and inhibits apoptosis in activated Ras-expressing mammary epithelial cells. Hepatocytes undergoing EMT have been linked to resistance to TGF-β-induced cell death (Nishimura et al. 2001). Similar to this, TGF-β treatment for several weeks produces cells that avoid apoptosis and carry out a continuous EMT, according to research on NMuMG cells originating from breast tumours. This model provides a chance to study the long-term impact of repeated exposure to TGF-β during epithelial fibrosis and in cancerous cell. By establishing tolerance or using immunoediting to change their phenotype, tumours can evade immune superintendence (Wang et al. 2005). Tumours require ongoing oncogenic signalling to survive, as evidenced by a study that saw tumour relapse in a Neu/HER2-inducible transgenic tumour model following inducer removal. All animals did, however, have persistent foci that ultimately gave rise to more aggressive new EMT-type tumours (Keir et al. 2008). These two findings indicate that cells within the foci of minimal residual illness developed a mesenchymal phenotype and that EMT may play a role in the development of resistance to targeted therapy. All of these findings suggest that not only is EMT potentially required for primary cancer to invade and spread, but these pioneer invasive cells can also produce a differentiated epithelial-like structure by exhibiting both a mesenchymal and a stem cell-like phenotype (Hui et al. 2017; Parry et al. 2005). The establishment of macrometastasis and the subsequent formation of the majority of the secondary tumour mass depend on this reversion to the differentiated phenotype via the MET pathway. It was previously suggested that cancer stem cells may develop a mesenchymal phenotype and so become migrating cancer stem cells that will create metastases based on an examination of the development of colon primary tumours and liver metastases.

2.7 Recent Advancements in Cancer Immunotherapy

2.7.1 Cancer Vaccines

By focusing on TAAs, cancer immunotherapy becomes an essential part of the cancer treatment regimen. It does not cause unintended off-target effects and has several clinical benefits. As a result, these advancements mandate the development of highly effective and little harmful recombinant vaccines for cancer immunotherapy. Unexpectedly, there are a lot of mutations within different tumour types—between 10 and 1000 mutations per tumour type (Wei et al. 2017). For instance, it has been demonstrated that treating melanoma patients with ipilimumab and tremelimumab, antibodies against cytotoxic T-lymphocyte antigen 4 (CTLA-4), extends their overall survival (Barber et al. 2006; Blank et al. 2004). On the other hand, cancers with fewer mutations, such pancreatic cancer, react poorly to immunotherapies. It is interesting to note that very few gene variations that express TAA-specific peptides spontaneously elicit an immunological response (Francisco et al. 2009). Recent advancements in cancer treatment as compared to traditional therapeutics are shown in Fig. 2.2.

2.7.2 Immune Checkpoint Inhibitors

One increasingly important function in controlling the T cell response is the identification of novel immunological checkpoint pathways and the activation or inhibition of these mechanisms (Wherry and Kurachi 2015; Hirano et al. 2005). Similar to PD1, LAG3 is an essential checkpoint that shares tight ties with CD 4. It

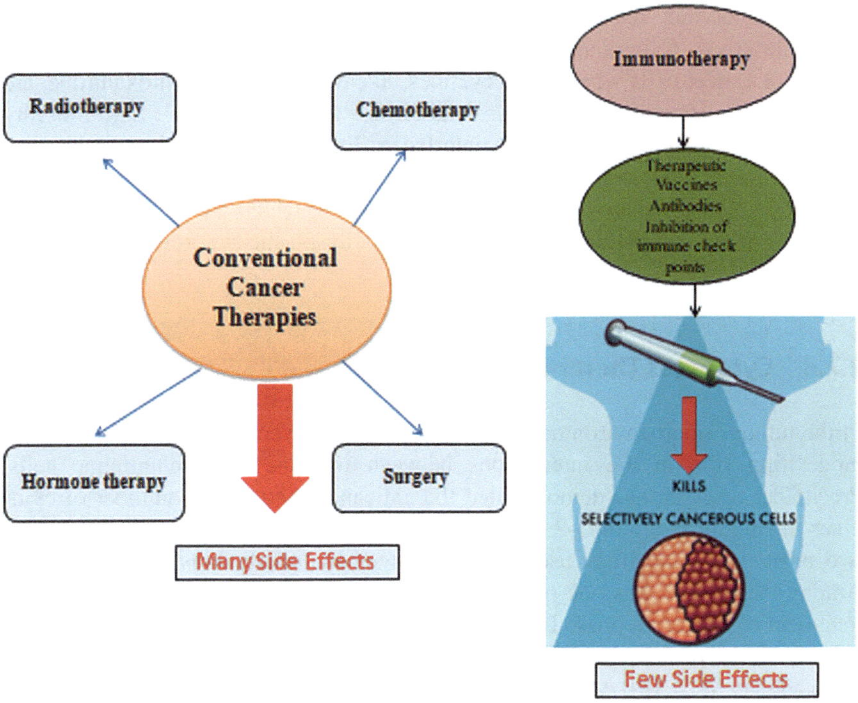

Fig. 2.2 Recent advancements in cancer treatment

possesses an inhibitory ligand that, in order to control the immune system, decreases T cell activation by obstructing CD4 interaction sites on MHC class II proteins. Tregs and anergic T cells both have elevated LAG-3 receptors (Iwai et al. 2002). TIM3, a T helper type 1 cell surface protein that regulates responses, encourages peripheral tolerance, and facilitates T cell fatigue, is another negative regulator of the T cell response (Strome et al. 2003). Moreover, its expression is linked to a poor prognosis in non-small-cell lung cancer and follicular lymphoma, suggesting a possible role in the development of cancer.

2.7.3 Immune System Modulators

A class of medications known as immunomodulators primarily targets the pathways involved in the treatment of multiple myeloma and a few other malignancies (He et al. 2004). Generally known as anticancer pharmaceuticals, immunomodulating drugs (IMiDs) also include anthracyclines, thalidomides, lenalidomides, pomalidomide, thalidomide (Thalomid), and hypomethylating agents (HMA). At relatively modest dosages, IMiDs can target cancer cells by strengthening the immune system. Lenalidomide, a Thalidomide derivative, has been licenced by the FDA for use as maintenance therapy in patients with posttransplant myeloma (Kamphorst et al.

2017). Thalidomide has antiangiogenic and tumoricidal effects through inducing cell cycle arrest. Calreticulin (CRT) are expressed more prominently on the cell surface of cancer cells when anthracyclines, like doxorubicin or mitoxantrone, are administered. After that, the cancer cells release ATP, annexin A1, high-mobility group box 1 (HMGB1), and type I interferon (Iwai et al. 2005; Hamanishi et al. 2007). By blocking DNA methyltransferase, HMAs reduce DNA methylation. Examples of this include decitabine and azacitidine, which are authorised for the treatment of acute myeloid leukaemia and myelodysplastic syndromes (Thompson et al. 2007).

2.7.4 Cytokines Therapy

In the tumour microenvironment, cytokines—small glycoproteins—play a critical moderating role in the interactions between immune and nonimmune cells. Preclinical research has demonstrated the anticancer action of a number of cytokines, including interleukin-2 and type 1 interferons (Thompson et al. 2006). The two most important drugs used to treat cancer are interleukins and interferons. Among these, IL-2 was one of the first cytokines to be researched for use in the treatment of cancer. In phase III trials, IL-2 fused with proteins from the diphtheria toxin produced positive results for patients with cutaneous T cell lymphoma (CTCL) (Hargadon et al. 2018). Additionally, IL-2 can be used in conjunction with conventional treatments or as a stand-alone pharmaceutical treatment for cancer therapies. A small amount of IL-2 has the advantage of preferentially increasing Tregs. When IL-2 was administered at a low dose, rituximab and IL-2 demonstrated synergistic effectiveness against NHL in a mouse model. A different study revealed how IL2 treatment impacted FOXP3 expression in CD3+CD25+ T cells and caused the Treg phenotype to proliferate during low-dose therapy via a STAT-dependent mechanism (Berman et al. 2015). Some early phase 1 trials demonstrated favourable outcomes when rituximab and IL-2 were combined. On the other hand, recent studies have focused on combining IL2 with anti-PD-1 and anti-CTLA-4 Treg inhibitors.

2.7.5 Adoptive Transfer Therapy and Chimeric Antigen Receptor (CAR) T Cell Therapy

CAR T cell therapy, the most effective immunotherapy, fortifies the patient's immune system to eliminate the cancer. Because T cells and other immune cells contain receptors that allow them to recognise foreign antigens, they can attach to these antigens and induce other immune system components to destroy the foreign material (Brahmer et al. 2010). Depending on the kind of disease, each CAR is designed to target a particular antigen; for example, CD19 is an antigen for lymphoma or leukaemia. As a result, the CAR T cell treatments are made especially to attach to the CD19 antigen in order to treat certain cancers alone (Gong et al. 2018; Weber et al. 2015; Weber et al. 2017).

2.8 Conclusion

Both innate and adaptive immunity contribute significantly to the development of cancer by either stimulating or inhibiting carcinogenesis while also thwarting the effects of cancer treatments by creating resistances, indicating a correlation between the processes and chronic inflammation. Moreover, the initiation of cancer and its advancement are linked to the build-up of genetic mutations and the disruption of regular regulatory mechanisms, resulting in the production of tumour-specific antigens and TAAs that trigger antitumor immune reactions. Though chronic inflammation can sometimes be immunosuppressive, signals that cause acute inflammatory reactions often enhance dendritic cell development and antigen presentation. When developing novel therapeutic strategies, it is important to take into account the antagonistic relationship between inflammation and immunology, which also affects the course of cancer treatment. In addition to cancer cells, tumours display an additional level of complexity. By establishing the "tumour microenvironment," they have a repertoire of recruited, seemingly normal cells that aid in the establishment of characteristic features. These ideas were intended to be widely applicable in order to support innovative approaches to successful cancer treatment for humans.

References

Baksh K, Weber J (2015) Immune checkpoint protein inhibition for cancer: preclinical justification for CTLA-4 and PD-1 blockade and new combinations. Semin Oncol 42:363–377

Barber DL et al (2006) Restoring function in exhausted CD8 T cells during chronic viral infection. Nature 439:682–687

Berman D et al (2015) The development of immunomodulatory monoclonal antibodies as a new therapeutic modality for cancer: the Bristol-Myers Squibb experience. Pharmacol Ther 148:132–153

Blank C et al (2004) PD-L1/B7H-1 inhibits the effector phase of tumor rejection by T cell receptor (TCR) transgenic CD8+ T cells. Cancer Res 64:1140–1145

Brahmer JR et al (2010) Phase I study of single-agent anti-programmed death-1 (MDX-1106) in refractory solid tumors: safety, clinical activity, pharmacodynamics, and immunologic correlates. J Clin Oncol 28:3167–3175

Brunet JF et al (1987) A new member of the immunoglobulin superfamily—CTLA-4. Nature 328:267–270. This study screens mouse T cell cDNA libraries and identifies CTLA4 as a new immunoglobulin superfamily member expressed within the lymphoid lineage predominantly in activated cells

Carreno BM, Collins M (2002) The B7 family of ligands and its receptors: new pathways for costimulation and inhibition of immune responses. Annu Rev Immunol 20:29–53

Chikuma S, Bluestone JA (2007) Expression of CTLA-4 and FOXP3 in cis protects from lethal lymphoproliferative disease. Eur J Immunol 37:1285–1289

Chow MT, Moller A, Smyth MJ (2012) Inflammation and immune surveillance in cancer. Semin Cancer Biol 22:23–32

Chuang E et al (1999) Regulation of cytotoxic T lymphocyte associated molecule-4 by Src kinases. J Immunol 162:1270–1277

Chuang E et al (2000) The CD28 and CTLA-4 receptors associate with the serine/threonine phosphatase PP2A. Immunity 13:313–322

Dariavach P, Mattei MG, Golstein P, Lefranc MP (1988) Human Ig superfamily CTLA-4 gene: chromosomal localization and identity of protein sequence between murine and human CTLA-4 cytoplasmic domains. Eur J Immunol 18:1901–1905

Darlington PJ et al (2002) Surface cytotoxic T lymphocyte-associated antigen 4 partitions within lipid rafts and relocates to the immunological synapse under conditions of inhibition of T cell activation. J Exp Med 195:1337. This study provides key insights into the compartmentalization of CTLA4 throughout the T cell activation cycle

Decker WK, Safdar A (2009) Bioimmunoadjuvants for the treatment of neoplastic and infectious disease: Coley's legacy revisited. Cytokine Growth Factor Rev 20:271–281

Decker WK et al (2017) Cancer immunotherapy: historical perspective of a clinical revolution and emerging preclinical animal models. Front Immunol 8:829

van Elsas A, Hurwitz AA, Allison JP (1999) Combination immunotherapy of B16 melanoma using anti-cytotoxic T lymphocyte-associated antigen 4 (CTLA-4) and granulocyte/macrophage colony-stimulating factor (GM-CSF)-producing vaccines induces rejection of subcutaneous and metastatic tumors accompanied by autoimmune depigmentation. J Exp Med 190:355–366

Fecci PE et al (2007) Systemic CTLA-4 blockade ameliorates glioma-induced changes to the CD4+ T cell compartment without affecting regulatory T-cell function. Clin Cancer Res 13:2158–2167

Fife BT, Bluestone JA (2008) Control of peripheral T-cell tolerance and autoimmunity via the CTLA-4 and PD-1 pathways. Immunol Rev 224:166–182

Francisco LM et al (2009) PD-L1 regulates the development, maintenance, and function of induced regulatory T cells. J Exp Med 206:3015–3029

Francisco LM, Sage PT, Sharpe AH (2010) The PD-1 pathway in tolerance and autoimmunity. Immunol Rev 236:219–242

Fraser JH, Rincon M, McCoy KD, Le Gros G (1999) CTLA4 ligation attenuates AP-1, NFAT and NF-κB activity in activated T cells. Eur J Immunol 29:838–844

Freeman GJ et al (2000) Engagement of the PD-1 immunoinhibitory receptor by a novel B7 family member leads to negative regulation of lymphocyte activation. J Exp Med 192:1027–1034. This study demonstrates that a new B7 family member, PDL1, functions as a ligand for PD1 in order to dampen T cell activation

Friedline RH et al (2009) CD4+ regulatory T cells require CTLA-4 for the maintenance of systemic tolerance. J Exp Med 206:421–434

Furness AJ, Vargas FA, Peggs KS, Quezada SA (2014) Impact of tumour microenvironment and Fc receptors on the activity of immunomodulatory antibodies. Trends Immunol 35:290–298

Gong J, Chehrazi-Raffle A, Reddi S, Salgia R (2018) Development of PD-1 and PD-L1 inhibitors as a form of cancer immunotherapy: a comprehensive review of registration trials and future considerations. J Immunother Cancer 6:8

Greenwald RJ, Boussiotis VA, Lorsbach RB, Abbas AK, Sharpe AH (2001) CTLA-4 regulates induction of anergy in vivo. Immunity 14:145–155. Using TCR transgenic mice, this study provides important in vivo evidence that CTLA4 is essential for T cell tolerance by regulating entrance into the anergic state

Gross L (1943) Intradermal immunization of C3H mice against a sarcoma that originated in an animal of the same line. Cancer Res 3:326

Grosso JF, Jure-Kunkel MN (2013) CTLA-4 blockade in tumor models: an overview of preclinical and translational research. Cancer Immun 13:5

Ha D et al (2019) Differential control of human Treg and effector T cells in tumor immunity by Fc-engineered anti-CTLA-4 antibody. Proc Natl Acad Sci USA 116:609

Halliday GM, Patel A, Hunt MJ, Tefany FJ, Barnetson RSC (1995) Spontaneous regression of human melanoma/nonmelanoma skin cancer: association with infiltrating CD4+ T cells. World J Surg 19:352–358

Hamanishi J et al (2007) Programmed cell death 1 ligand 1 and tumor-infiltrating CD8+ T lymphocytes are prognostic factors of human ovarian cancer. Proc Natl Acad Sci USA 104:3360–3365

Hargadon KM, Johnson CE, Williams CJ (2018) Immune checkpoint blockade therapy for cancer: an overview of FDA-approved immune checkpoint inhibitors. Int Immunopharmacol 62:29–39

Harper K et al (1991) CTLA-4 and CD28 activated lymphocyte molecules are closely related in both mouse and human as to sequence, message expression, gene structure, and chromosomal location. J Immunol 147:1037–1044

He YF et al (2004) Blocking programmed death-1 ligand–PD-1 interactions by local gene therapy results in enhancement of antitumor effect of secondary lymphoid tissue chemokine. J Immunol 173:4919–4928

He M et al (2017) Remarkably similar CTLA-4 binding properties of therapeutic ipilimumab and tremelimumab antibodies. Oncotarget 8:67129–67139

Hirano F et al (2005) Blockade of B7-H1 and PD-1 by monoclonal antibodies potentiates cancer therapeutic immunity. Cancer Res 65:1089–1096. This preclinical murine study shows that blockade of PD1 or PDL1 could reverse inherent resistance of tumours to cytolysis by T cells

Hodi FS et al (2003) Biologic activity of cytotoxic T lymphocyte-associated antigen 4 antibody blockade in previously vaccinated metastatic melanoma and ovarian carcinoma patients. Proc Natl Acad Sci USA 100:4712–4717

Hodi FS et al (2010) Improved survival with ipilimumab in patients with metastatic melanoma. N Engl J Med 363:711–723

Hou TZ et al (2015) A transendocytosis model of CTLA-4 function predicts its suppressive behavior on regulatory T cells. J Immunol 194:2148–2159

Hui E et al (2017) T cell costimulatory receptor CD28 is a primary target for PD-1-mediated inhibition. Science 355:1428–1433. This study highlights the central role of CD28 dephosphorylation in mediating the negative regulatory effect of PD1-mediated T cell inhibition

Hurwitz AA, Yu TFY, Leach DR, Allison JP (1998) CTLA-4 blockade synergizes with tumor-derived granulocyte–macrophage colony-stimulating factor for treatment of an experimental mammary carcinoma. Proc Natl Acad Sci 95:10,067

Intlekofer AM, Thompson CB (2013) At the bench: preclinical rationale for CTLA-4 and PD-1 blockade as cancer immunotherapy. J Leukoc Biol 94:25–39

Ishida Y, Agata Y, Shibahara K, Honjo T (1992) Induced expression of PD-1, a novel member of the immunoglobulin gene superfamily, upon programmed cell death. EMBO J 11:3887–3895. Using subtractive hybridization, this study is the first to identify PD1 as a new immunoglobulin superfamily member expressed within the thymus

Iwai Y et al (2002) Involvement of PD-L1 on tumor cells in the escape from host immune system and tumor immunotherapy by PD-L1 blockade. Proc Natl Acad Sci USA 99:12,293–12,297. This study shows that PDL1 expression by tumour cells represents a mechanism of resistance to immune-mediated cytolysis that can be abrogated by blockade of the PD1 axis

Iwai Y, Terawaki S, Honjo T (2005) PD-1 blockade inhibits hematogenous spread of poorly immunogenic tumor cells by enhanced recruitment of effector T cells. Int Immunol 17:133–144. Beyond enhancing antitumoural immunity, this study demonstrates that PD1 blockade can limit tumour metastatic potential

Jain N, Nguyen H, Chambers C, Kang J (2010) Dual function of CTLA-4 in regulatory T cells and conventional T cells to prevent multiorgan autoimmunity. Proc Natl Acad Sci USA 107:1524–1528

Jure-Kunkel MN, Masters G, Girit E, Dito G, Lee FY (2008) Antitumor activity of anti-CTLA-4 monoclonal antibody (mAb) in combination with ixabepilone in preclinical tumor models. J Clin Oncol 26:3048–3048

Kamphorst AO et al (2017) Rescue of exhausted CD8 T cells by PD-1-targeted therapies is CD28-dependent. Science 355:1423–1427

Keir ME, Butte MJ, Freeman GJ, Sharpe AH (2008) PD-1 and its ligands in tolerance and immunity. Annu Rev Immunol 26:677–704

Krummel MF, Allison JP (1995) CD28 and CTLA-4 have opposing effects on the response of T cells to stimulation. J Exp Med 182:459–465

Krummel MF, Allison JP (1996) CTLA-4 engagement inhibits IL-2 accumulation and cell cycle progression upon activation of resting T cells. J Exp Med 183:2533–2540

Kuehn HS et al (2014) Immune dysregulation in human subjects with heterozygous germline mutations in CTLA4. Science 345:1623–1627

Kwon ED et al (1997) Manipulation of T cell costimulatory and inhibitory signals for immunotherapy of prostate cancer. Proc Natl Acad Sci USA 94:8099–8103

Kwon ED et al (2014) Ipilimumab versus placebo after radiotherapy in patients with metastatic castrationresistant prostate cancer that had progressed after docetaxel chemotherapy (CA184-043): a multicentre, randomised, double-blind, phase 3 trial. Lancet Oncol 15:700–712

Latchman Y et al (2001) PD-L2 is a second ligand for PD-1 and inhibits T cell activation. Nat Immunol 2:261–268

Leach DR, Krummel MF, Allison JP (1996) Enhancement of antitumor immunity by CTLA-4 blockade. Science 271:1734–1736. This study elegantly shows that CTLA4 blockade in vivo can enhance immune-mediated antitumoural responses in multiple tumour models that ultimately persist upon rechallenge

Lee S et al (2016) Abatacept alleviates severe autoimmune symptoms in a patient carrying a de novo variant in CTLA-4. J Allergy Clin Immunol 137:327–330

Linsley PS et al (1991) CTLA-4 is a second receptor for the B cell activation antigen B7. J Exp Med 174:561–569

Linsley PS et al (1994) Human B7-1 (CD80) and B7-2 (CD86) bind with similar avidities but distinct kinetics to CD28 and CTLA-4 receptors. Immunity 1:793–801. This study demonstrates that CD80 and CD86 bind to CTLA4 with greater avidity than CD28

Lo B et al (2015) AUTOIMMUNE DISEASE. Patients with LRBA deficiency show CTLA4 loss and immune dysregulation responsive to abatacept therapy. Science 349:436–440

Lynch TJ et al (2012) Ipilimumab in combination with paclitaxel and carboplatin as first-line treatment in stage IIIB/IV non-small-cell lung cancer: results from a randomized, double-blind, multicenter phase II study. J Clin Oncol 30:2046–2054

Maio M et al (2015) Five-year survival rates for treatment-naive patients with advanced melanoma who received ipilimumab plus dacarbazine in a phase III trial. J Clin Oncol 33:1191–1196

Mandelbrot DA, McAdam AJ, Sharpe AH (1999) B7-1 or B7-2 is required to produce the lymphoproliferative phenotype in mice lacking cytotoxic T lymphocyte-associated antigen 4 (CTLA-4). J Exp Med 189:435–440

Mangsbo SM et al (2010) Enhanced tumor eradication by combining CTLA-4 or PD-1 blockade with CpG therapy. J Immunother 33:225–235

Marengere LE et al (1996) Regulation of T cell receptor signaling by tyrosine phosphatase SYP association with CTLA-4. Science 272:1170–1173

Nishimura H, Minato N, Nakano T, Honjo T (1998) Immunological studies on PD-1 deficient mice: implication of PD-1 as a negative regulator for B cell responses. Int Immunol 10:1563–1572

Nishimura H, Nose M, Hiai H, Minato N, Honjo T (1999) Development of lupus-like autoimmune diseases by disruption of the PD-1 gene encoding an ITIM motif-carrying immunoreceptor. Immunity 11:141–151

Nishimura H et al (2001) Autoimmune dilated cardiomyopathy in PD-1 receptor-deficient mice. Science 291:319–322

Oiseth SJ, Aziz MSA-E (2017) Cancer immunotherapy: a brief review of the history, possibilities, and challenges ahead. J Cancer Metastasis Treat 3:250–261

Parry RV et al (2005) CTLA-4 and PD-1 receptors inhibit T-cell activation by distinct mechanisms. Mol Cell Biol 25:9543–9553. This study demonstrates the complementary ways in which CTLA4 and PD1 inhibit protein kinase B signalling to reduce glucose uptake and utilization

Peggs KS, Quezada SA, Chambers CA, Korman AJ, Allison JP (2009) Blockade of CTLA-4 on both effector and regulatory T cell compartments contributes to the antitumor activity of anti-CTLA-4 antibodies. J Exp Med 206:1717–1725

Pentcheva-Hoang T, Egen JG, Wojnoonski K, Allison JP (2004) B7-1 and B7-2 selectively recruit CTLA-4 and CD28 to the immunological synapse. Immunity 21:401–413

Qureshi OS et al (2011) Trans-endocytosis of CD80 and CD86: a molecular basis for the cell-extrinsic function of CTLA-4. Science 332:600–603. This study nominates trans-endocytosis of CD80 and CD86 as a cell-extrinsic mechanism of CTLA4-mediated inhibition of T cell activation

Read S, Malmström V, Powrie F (2000) Cytotoxic T lymphocyte-associated antigen 4 plays an essential role in the function of CD25+CD4+ regulatory cells that control intestinal inflammation. J Exp Med 192:295–302

Read S et al (2006) Blockade of CTLA-4 on CD4+CD25+ regulatory T cells abrogates their function in vivo. J Immunol 177:4376–4383

Reck M et al (2013) Ipilimumab in combination with paclitaxel and carboplatin as first-line therapy in extensive-disease-small-cell lung cancer: results from a randomized, double-blind, multicenter phase 2 trial. Ann Oncol 24:75–83

Ribas A et al (2013) Phase III randomized clinical trial comparing tremelimumab with standard-of-care chemotherapy in patients with advanced melanoma. J Clin Oncol 31:616–622

van Rooij N et al (2013) Tumor exome analysis reveals neoantigen-specific T-cell reactivity in an ipilimumab responsive melanoma. J Clin Oncol 31:e439–e442

Schadendorf D et al (2015) Pooled analysis of long-term survival data from phase II and phase III trials of ipilimumab in unresectable or metastatic melanoma. J Clin Oncol 33:1889–1894. This seminal meta-analysis provides strong clinical evidence that ipilimumab confers a durable survival benefit to patients with advanced melanoma

Schneider H et al (2006) Reversal of the TCR stop signal by CTLA-4. Science 313:1972–1975

Sharma N, Vacher J, Allison JP (2019a) TLR1/2 ligand enhances antitumor efficacy of CTLA-4 blockade by increasing intratumoral Treg depletion. Proc Natl Acad Sci USA 116:10453

Sharma A et al (2019b) Anti-CTLA-4 immunotherapy does not deplete FOXP3+ regulatory T cells (Tregs) in human cancers. Clin Cancer Res 25:1233

Smith JL Jr, Stehlin JS Jr (1965) Spontaneous regression of primary malignant melanomas with regional metastases. Cancer 18:1399–1415

Snyder A et al (2014) Genetic basis for clinical response to CTLA-4 blockade in melanoma. N Engl J Med 371:2189–2199

Sojka DK, Hughson A, Fowell DJ (2009) CTLA-4 is required by CD4+CD25+ Treg to control CD4+ T-cell lymphopenia-induced proliferation. Eur J Immunol 39:1544–1551

Strome SE et al (2003) B7-H1 blockade augments adoptive T-cell immunotherapy for squamous cell carcinoma. Cancer Res 63:6501–6505

Tai X, Van Laethem F, Sharpe AH, Singer A (2007) Induction of autoimmune disease in CTLA-4−/− mice depends on a specific CD28 motif that is required for in vivo costimulation. Proc Natl Acad Sci USA 104:13,756–13,761

Tai X et al (2012) Basis of CTLA-4 function in regulatory and conventional CD4+ T cells. Blood 119:5155–5163. This study demonstrates that CTLA4 expression within regulatory T cells is essential for the characteristic suppressive functions of these cells on the immune response

Takahashi T et al (2000) Immunologic self-tolerance maintained by CD25+CD4+ regulatory T cells constitutively expressing cytotoxic T lymphocyteassociated antigen 4. J Exp Med 192:303–310

Thompson RH et al (2006) Tumor B7-H1 is associated with poor prognosis in renal cell carcinoma patients with long-term follow-up. Cancer Res 66:3381–3385. Along with Thompson et al. (2007) and Hamanishi et al. (2007), this study shows that expression of PD1 on tumour-infiltrating T cells and PD1 ligands on tumours are poor prognostic markers in various cancers

Thompson RH et al (2007) PD-1 is expressed by tumor-infiltrating immune cells and is associated with poor outcome for patients with renal cell carcinoma. Clin Cancer Res 13:1757–1761

Tivol EA et al (1997) CTLA4Ig prevents lymphoproliferation and fatal multiorgan tissue destruction in CTLA4-deficient mice. J Immunol 158:5091–5094. Along with Waterhouse et al. (1995), this in vivo murine study demonstrates that CD4+ T cells are necessary to cause severe multiorgan autoimmunity in the context of CTLA4 deficiency, which could be reversed by CTLA4Ig

Walker LSK (2013) Treg and CTLA-4: two intertwining pathways to immune tolerance. J Autoimmun 45:49–57

Walunas TL et al (1994) CTLA-4 can function as a negative regulator of T cell activation. Immunity 1:405–413

Wang J et al (2005) Establishment of NOD-Pdcd1−/− mice as an efficient animal model of type I diabetes. Proc Natl Acad Sci USA 102:11,823–11,828

Waterhouse P et al (1995) Lymphoproliferative disorders with early lethality in mice deficient in Ctla-4. Science 270:985–988

Weber JS et al (2015) Nivolumab versus chemotherapy in patients with advanced melanoma who progressed after anti-CTLA-4 treatment (CheckMate 037): a randomised, controlled, open-label, phase 3 trial. Lancet Oncol 16:375–384

Weber J et al (2017) Adjuvant nivolumab versus ipilimumab in resected stage III or IV melanoma. N Engl J Med 377:1824–1835

Wei SC et al (2017) Distinct cellular mechanisms underlie anti-CTLA-4 and anti-PD-1 checkpoint blockade. Cell 170:1120–1133.e17

Wherry EJ, Kurachi M (2015) Molecular and cellular insights into T cell exhaustion. Nat Rev Immunol 15:486–499

Wing K et al (2008) CTLA-4 control over Foxp3+ regulatory T cell function. Science 322:271–275

Yang YF et al (1997) Enhanced induction of antitumor T-cell responses by cytotoxic T lymphocyte-associated molecule-4 blockade: the effect is manifested only at the restricted tumor-bearing stages. Cancer Res 57:4036–4041

Yang JC et al (2007) Ipilimumab (anti-CTLA4 antibody) causes regression of metastatic renal cell cancer associated with enteritis and hypophysitis. J Immunother 30:825–830

Mechanism of Biosimilars for the Treatment of Cancer

3

Satish Shilpi, Ekta Gurnany, Pawan K. Gupta, Kangan Sharma, Khyati Saini, Pranali Chimaniya, and Shvetank Bhatt

Abstract

Biosimilars are biological medications that closely resemble FDA-approved biological medications in terms of structure as well as functions and are approved according to the same standards of pharmaceutical quality, safety and efficacy that apply to all biological medicines. It works smoothly and efficiently for both acute and chronic illnesses. In the biological treatment of oncology, the high production cost causes burdensome to healthcare systems. For the quality of the treatment with biosimilars, regulatory affairs have set outlines for the production of biosimilar biologicals in terms of physicochemical characteristics, efficacy, and safety compared to the originator. The pharmaceutical industry and regulators have been using scientific principles for years to develop biosimilars. Biologics manufacturing processes are frequently modified for several reasons, such as process scalability, increased efficiency, or modernization when equipment has to be updated or replaced. Biosimilars have wide application in the medical field for the treatment of different chronic diseases like cancer, diabetes,

S. Shilpi (✉) · K. Sharma · K. Saini
School of Pharmaceutical & Populations Health Informatics, DIT University, Dehradun, India

E. Gurnany
Department of Pharmaceutics, B. Pharmacy College, Rampura, Godhra, Gujarat, India

P. K. Gupta
Amity Institute of Pharmacy, Amity University, Gwalior, Madhya Pradesh, India

P. Chimaniya
School of Pharmaceutical Education and Research, Peoples University, Bhopal, Madhya Pradesh, India

S. Bhatt
Department of Pharmaceutical Sciences, School of Health Sciences and Technology, Dr. Vishwanath Karad MIT World Peace University, Pune, India

S. Bhatt et al. (eds.), *Biosimilars for Cancer Treatment*,
https://doi.org/10.1007/978-981-97-3001-8_3

autoimmune diseases, skin diseases, bowel diseases, and others. It is a good replacement of drugs which precipitate serious toxic effect. This chapter showing the regulatory aspects, types, and application of biosimilars in cancer therapy.

Keywords

Biosimilars · Proteins · Hormones · Rituximab · Trastuzumab

3.1 Introduction and Application of Biosimilar

Modern medicine relies heavily on biologic medicines, particularly when treating immune-mediated disorders and other chronic illnesses. Biosimilars are biologic therapy versions that share a high degree of molecular similarity and are authorized after the original product's patent expires (Grieshaber-Bouyer and Lorenz 2020). It has been demonstrated that introducing biosimilars into a healthcare system can save costs while also improving patient access to treatment; nevertheless, before the potential advantages of biosimilars can be realized, a number of obstacles must be removed that may hinder or postpone the adoption of biosimilars into clinical practice (Foreman 2020). A biosimilar is a pharmaceutical that bears a striking resemblance to a previously approved biological medication. The same pharmaceutical quality, safety, and efficacy standards that govern the approval of biological medications also apply to biosimilars. Biosimilars with superior clinical performance could increase patient access to trastuzumab treatment. Biological medications are intricate compounds that have stringent antigenic screening and medication surveillance protocols to ensure that the body does not mount an immune system response.

Authorities like International Conference on Harmonisation proposed the comparability notion to assess whether or not the pre- and post-change merchandise was sufficiently comparable to permit ongoing marketing under the same product category, thus enabling such manufacturing modifications to occur without needing establishments to conduct a new clinical testing program, which are used in Europe, the USA, and Japan (Verrill et al. 2019; Biosimilars Info Sheet n.d.). If alterations emerge, it needs to be verified that these variations have no adverse effects on antigenicity, safeguarding, or efficiency. It will not be obligatory to execute nonclinical or clinical trials using the post-change merchandise, provided that the manufacturer can ensure similarity with analytical investigations. It might be appropriate to combine analytical, nonclinical, and clinical studies where molecular attributes are not clearly related to safety and efficacy and analytical differences are observed (Omnitrope, European Public Assessment Report 2006; Saenger 2017).

Unlike generic medication, where the same active ingredients provide the same type of clinical benefits as that of the originator, the biosimilar acts through different biological pathways to solve inheritance complexity in a manner that of identical to biochemical entities (U.S. Food and Drug Administration (FDA) n.d.-a; Verrill et al. 2019; Biosimilars Info Sheet n.d.) (Fig. 3.1). Europe developed the first policy framework in the world for approving biological products (Omnitrope, European Public Assessment Report 2006). The European Medicines Agency (EMA)

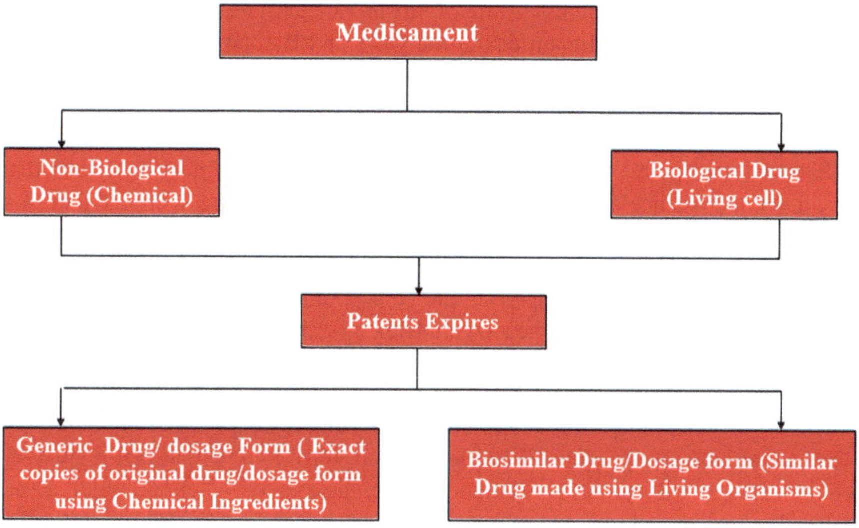

Fig. 3.1 Factors considered in the development of Biosimilars

approved Omnitrope (a recombinant human growth hormone) as the first biosimilar in Europe in 2006 (Saenger 2017; Sheridan 2006).

The European Medicines Agency (EMA) will continue to approve more biosimilars for trastuzumab in the upcoming months, which will encourage competition and potentially reduce treatment costs (Sarosiek and Morawski 2018). Biological agents that successfully mimic the original reference products are known as biosimilars. Their growth is primarily motivated by the prospect of more competition in the market, which will support the long-term viability of healthcare systems. Biosimilars contribute to fiscal redistribution by lowering the cost barriers to biological medicines, which in turn increases patient access to treatment (De Mora 2019). Generics are created using chemical synthesis, whereas biosimilars are created using DNA technology in bacteria, plant, or animal cells. The specific character of the final product is influenced by details in the synthesis, microbe selection, protein extraction, purification, and manufacturing processes. In large proteins with four polypeptide chains, monoclonal antibodies interact with their surroundings and with one another in a variety of ways. Payors must recognize that biosimilars are distinct from generics and, as such, have separate guidelines for approval, registration, and dispensing of biosimilars (Emmanouilides et al. 2016).

The reference product is the biologic drug on which the biosimilar is based (Parsad and Nabhan 2018). Because biosimilars require more expensive manufacturing and undergo therapeutic equivalency testing, their development costs are far greater than those of their generic counterparts. It is critical that academics and pertinent businesses comprehend the expenses and phases involved in processing biologics (Akram et al. 2021). Because biosimilars are less expensive to treat, there may be more opportunities for clinical application and patient access. Patients and healthcare professionals are not aware of biosimilars, despite the fact that many of them are authorized for clinical use in the USA and Europe. The purpose of this summary of the science behind biosimilars and their existing indications is to improve patient education and awareness of the place of biosimilars in customized therapy regimens (Janjigian et al. 2018).

Biosimilars, like all biologics, may trigger unintended immune reactions that have a major influence on safety and clinical efficacy. Therefore, a crucial part of a biosimilar's clinical development approach should be the head-to-head immunogenicity assessment of the biosimilars and their reference biologics. The detection and characterization of immune responses can be accomplished using a variety of bioanalytical systems, each with their own advantages and disadvantages. Regulators must be able to understand immunogenicity results in the context of an assay-specific as well as from the perspectives of clinical pharmacology, effectiveness, and safety in order to properly appreciate the clinical significance of such data. Here, we address the present difficulties brought forth by international regulatory requirements for biosimilars' immunogenicity assessment (Schreitmüller et al. 2019). Biologics are huge, complicated proteins that are not perfectly replicable, unlike small-molecule medications; hence, the idea of a generic counterpart does not apply to biologics.

A framework for biosimilar approval has been provided by regulatory bodies; however, pharmacists must take into account a number of practical factors, such as interchangeability, substitution, naming, indication extrapolation, product labeling, therapeutic drug monitoring, manufacturer attributes, product use logistics, and reimbursement (Stevenson et al. 2017). Biosimilars are required to have similar efficaciousness, immunogenicity, and side effect profiles to reference biologics. While manufacture within living cells is still a part of the manufacturing process, biosimilars go through fewer clinical studies than their reference biologics. In the end, this lowers the biosimilar drug's cost and production costs in comparison to the reference biologic. The US Food and Drug Administration (FDA) has currently approved seven biosimilars for use in ulcerative colitis, colorectal cancer, and Crohn's disease. Several biologics are utilized in treating gastroenterologic diseases; however, some of them lack FDA-approved biosimilars. Despite the potential of biosimilars to decrease healthcare expenditures in chronic disease management, they encounter hurdles in gaining a substantial market share (Parsad and Nabhan 2018).

3.2 Regulation Aspects of Biosimilars

Biosimilarity refers to the demonstration of a significant similarity to a reference biological medicine across various aspects such as immunogenicity profile, biological activity, safety, efficacy, and chemical structure. This determination is primarily established through comprehensive comparability studies. Biosimilars are biological products that are designed to be highly similar to and have no clinically meaningful differences from an already approved reference biologic drug. They undergo rigorous testing to demonstrate similarity in terms of structure, function, safety, and efficacy. Biosimilars provide therapeutic alternatives to costly biologics, offering increased access to effective treatments (Crommelin et al. 2005; Roger 2006; Roger and Mikhail 2007).

A biosimilar refers to a biological medicine that closely mirrors another biological medicine, known as the "reference medicine," which has already obtained approval in the European Union (reference medicine). As biosimilars are manufactured in living organisms, slight variations from the reference medicine may occur. However, these minor differences are not expected to have any clinical significance, meaning that no changes are anticipated in terms of efficacy and safety. Variability inherent to biological medicines is a natural characteristic. However, stringent controls are consistently implemented to mitigate any potential impact on the safety or efficacy of the medicine. Biosimilars undergo approval based on the same rigorous standards of pharmaceutical safety, quality, and efficacy as those applied to all biological medicines approved in the European Union (EU) (Schellekens 2005). The primary goal of biosimilar development is to establish biosimilarity, indicating a significant resemblance in terms of structure, biological activity, efficacy, safety, and immunogenicity profile when compared to the reference biological medicine. Currently, in the European Union (EU), biosimilars have received approval for various classes of biological medicines (EMEA 2005, 2006a,b).

The 2010 legislation known as the Affordable Care Act offered rise to the enactment of the Biologic Price Competition and Innovation Act (BPCIA) in 2009, which gave the US Food and Drug Administration extensive authority to establish a simplified procedure for approval for biologic drugs that are homologous to already-marketed products (Raedler 2016). Filgrastim-sndz was approved by the USFDA in 2015, whereas Filgrastim was first approved by the USFDA in 1991 (US Food and Drug Administration 2015; U.S. Food and Drug Administration (FDA) n.d.-b). In the years since then, the FDA has approved several biosimilars to treat cancer and other diseases. Currently, 45 biosimilars have been approved by the US Federal Drug Administration as of 6 December 2023. The latest USFDA-approved biosimilar is "similar" to reference biologics products that the Drug Controller General of India (DCGI) has certified for safe and effective to use in India (The New India Guidelines on Similar Biologics 2012). The creation of biosimilars is underway to reduce expenses and give consumers access to novel treatments as the patents on biologics expire. Since the initial regulations for biosimilars have been set, regulations are continuously updated to resolve any issues and promote the optimal robust development of these products (Bachu et al. 2022). The Indian market offers a wide range of biosimilar products, including vaccines, monoclonal antibodies, recombinant proteins, diagnostics, insulin, erythropoietin, hepatitis B vaccines, granulocyte colony-stimulating factors, streptokinase, interferon alpha-2B, and epidermal growth factor receptors (Meher et al. 2019). As the third largest global pharmaceutical manufacturer by volume in 2023, India produces generic drugs, over-the-counter medicines, bulk drugs, vaccines, contract research and manufacturing, biosimilars, and biologics (ET Online 2023; Invest India n.d.). In comparison with other countries, India boasts a robust biosimilar ecosystem, propelling Indian pharmaceutical companies to the forefront of global leadership in biosimilars. India approved its first biosimilar years ahead of the United States and Europe, showcasing its early adoption and advancement in the field. Although there were no specific guidelines for the marketing and development of biosimilars in India at that time, the first biosimilar was approved and introduced to the market for hepatitis B in 2000. Since then, numerous biopharmaceutical companies in India have been actively involved in the development and marketing of biosimilars. The Department of Biotechnology (DBT) and the Central Drugs Standard Control Organisation (CDSCO) collaborated to establish the "Guidelines on Similar Biologics; Regulatory Requirements for Marketing Authorization in India" in 2012. These guidelines were subsequently revised in 2016 to tackle the challenges and complexities associated with the development of similar biologics (EMEA 2006c). Previously, similar biologics in India are subject to statutes under the Environmental (Protection) The Drug and Cosmetics Act of 1940, along with the Drug and Cosmetics Rules of 1945, serve as pivotal legislative frameworks governing pharmaceutical regulations in India. Additionally, the Guidelines for Manufacture, Uses, Export, Import, and Storage of Hazardous Microorganisms/Genetically Engineered Organisms or Cells were established in 1989. These regulations provide a comprehensive framework for the oversight of pharmaceuticals, including biosimilars, in India (Rushvi et al. 2016). Recently, a biopharmaceutical company from India received FDA approval to market its novel biologic (McCamish and Woollett 2012). Herceptin, featuring the active ingredient trastuzumab, holds the distinction of being the first biologic to receive FDA approval. It is utilized in the treatment of specific stomach and breast malignancies. Moreover, this marked the inaugural instance of an Indian company producing a comparable biologic that gained approval for sale in the US (Table 3.1) (Schiestl

Table 3.1 Classes of biosimilars approved by European Union

Classes of biosimilar (biological medicines)	Biosimilar approved in European Union
1. Proteins	
• Hormones	Follitropin alfa Insulin glargine Somatropin (Growth Hormone) Teriparatide Insulin Lispro
• Fusion proteins	Etanercept
• Monoclonal antibodies	Adalimumab Infliximab Rituximab Bevacizumab Trastuzumab
• Growth Factors	Epoetin Filgrastim Pegfilgrastim
2. Polysaccharides	
• Low molecular weight heparins.	Enoxaparin Sodium

et al. 2017; European Medicines Agency 2005b; Paradise 2011; Rathore 2012; Rugo et al. 2017).

Regulatory approval for biosimilars involves a comprehensive assessment to ensure their safety, efficacy, and quality. The regulatory pathway for biosimilar approval is more complex compared to generic drugs and traditional biological products. This complexity arises from the challenge of designing scientifically robust studies to demonstrate the similarity of biosimilars to their reference products, especially considering the inherent variability in the manufacturing processes of biological products. Moreover, the available analytical tests lack the sophistication required to detect subtle yet crucial structural differences between biological and biosimilar products. Even modest differences could potentially impact clinical outcomes and pose significant risks to patient safety. Therefore, it is deemed essential for biosimilars to undergo thorough assessment of clinical efficacy and safety through valid preclinical and clinical studies before obtaining marketing approval (EMEA 2005, 2006a,b).

The regulatory framework for biosimilars varies among different regions, but there are common principles followed by regulatory agencies worldwide. Regulatory aspects of biosimilars are shown in Table 3.2.

Here are the key regulatory aspects of biosimilars:

1. *Demonstration of Biosimilarity*

   Comparative Analytical Studies: Biosimilar developers must conduct extensive comparative analytical studies to demonstrate that the biosimilar is highly similar to the reference biologic in terms of function, structure, and other critical quality attributes.

   Biological Activity and Potency: The biosimilar must exhibit similar biological activity and potency to the reference product.
2. *Clinical Trials*

   Evaluative Clinical Tests: It is standard for clinical experiments to validate likeness in safety measures, effectiveness, and immune responses. These studies typically feature a stage 3 comparison among the biosimilar and the reference product, examined on a delicate patient group for the endorsed uses.

   Inference of Information: Regulatory bodies may sanction the projection of data to other areas where the reference product has approval, even if the biosimilar has not undergone direct testing for those particular uses.
3. *Immunogenicity Assessment*

   Immunogenicity Studies: Developers must conduct studies to assess the immunogenicity of the biosimilar, including the development of anti-drug antibodies (ADAs).

   Clinical Impact: The impact of immunogenicity on safety and efficacy must be thoroughly evaluated.
4. *Quality and Manufacturing*

   Comparability Studies: Extensive comparability studies are required to demonstrate that any differences in manufacturing processes do not result in clinically meaningful differences between the biosimilar and the reference product.

Table 3.2 Regulatory aspect of biosimilars (EMEA 2006b,c,d,e, 2007, 2009a)

Regulatory aspect	Description
Definition	Highly similar to an approved reference biologic with no clinically meaningful differences. Creating an exact replica or identical copy of an innovator product is unattainable. Hence, biosimilars are deemed "similar but not the same" or, in simpler terms, they are "the twin but not the clone" of the original biologic innovator product.
Regulatory agencies	FDA (United States), EMA (European Medicines Agency), PMDA (Pharmaceuticals and Medical Devices Agency—Japan), Health Canada, TGA (Australia), Swissmedic, HSA (Singapore), CDSCO (India), ANVISA (Brazil), among others.
Development pathway	Specific regulatory pathway designed for biosimilar approval, distinct from that of novel biologics.
Comparative studies	Comprehensive analytical studies demonstrating similarity in terms of structure, function, and other critical quality attribute to the reference biologic.
Nonclinical testing	Assessments for pharmacodynamics, pharmacokinetics, and toxicity to establish biosimilarity.
Clinical trials	Comparative clinical trials, often including Phase 3 studies. Comparative clinical trials conducted to assess safety, efficacy, and immunogenicity in a sensitive patient population for approved indications.
Extrapolation of indications	Approval for indications not directly studied, based on data from approved indications.
Immunogenicity assessment	Evaluation of immunogenicity, including the potential formation of anti-drug antibodies (ADAs).
Quality and manufacturing	Demonstrating similarity in manufacturing processes, ensuring consistency and reproducibility of the biosimilar.
Switching studies	Research examining the safety and efficacy of transitioning between biosimilars and reference products has been conducted to ensure patient well-being and treatment effectiveness.
Interchangeability guidelines	Regulatory guidance on whether a biosimilar can be substituted for the reference product without intervention by healthcare providers.
Post-marketing surveillance	Continuous monitoring of safety and efficacy after market approval.
Distinct naming	Often assigned distinct non-proprietary names to facilitate product identification and avoid confusion with the reference product.
Labeling requirements	Clear and transparent labeling indicating biosimilarity, approved indications, relevant information any specific conditions or restrictions.
Biosimilar designation	Specific designation during the regulatory approval process to differentiate biosimilars from the reference product.
Expedited approval	Some regulatory agencies offer expedited pathways for biosimilar approval, similar to those for certain innovative drugs.

Quality Control: Biosimilar manufacturers must implement robust quality control measures to ensure the consistency and quality of their products.

5. *Regulatory Pathways*

Biosimilar Designation: Regulatory agencies provide specific pathways for biosimilar development and approval. In the United States, for example, the US

Food and Drug Administration (FDA) has established the biosimilar biological product pathway.

Expedited Approval: Some regions offer expedited pathways for the approval of biosimilars, similar to those available for generic small-molecule drugs.

6. *Naming and Labeling*

   Distinct Naming: Numerous regulatory agencies mandate that biosimilars possess a distinct non-proprietary name separate from the reference product. This practice aids in distinguishing between products, thereby facilitating accurate product identification and pharmacovigilance.

   Labeling Requirements: The labeling of biosimilars should provide clear information about the biosimilarity, indications, and any specific conditions or restrictions.

7. *Post-Marketing Surveillance*

   Pharmacovigilance: Continuous monitoring of the safety of biosimilars through pharmacovigilance is essential. This involves the collection and analysis of data on adverse events during the post-marketing period.

8. *Interchangeability and Substitution*

   Guidelines on Interchangeability: Regulatory agencies may provide guidance on the interchangeability of biosimilars with the reference product and whether they can be substituted without the intervention of healthcare providers.

9. *Education and Awareness*

   Healthcare Professional and Patient Education: Regulatory agencies often engage in educational efforts to raise awareness among healthcare professionals and patients about biosimilars and their safe use.

3.3 Toxicities Consideration of Biosimilars

Biosimilars are biological products that are highly similar to and have no clinically meaningful differences from an existing FDA-approved reference product (Grabowski 2008, Watson 2010). While biosimilars aim to provide cost-effective alternatives to the original biologics, it is important to recognize that, like any medical product, they can have potential toxicities and side effects. Biosimilar toxicity involves identifying and characterizing adverse effects or unintended reactions associated with the use of a biosimilar product (Crommelin et al. 2005; Kanase et al. 2013; Camacho et al. 2014; Grabowski 2008; Watson et al. 2010). The process of defining biosimilar toxicity is complex and requires a comprehensive assessment of various factors as shown in Table 3.3.

Some considerations regarding toxicities in biosimilars:

1. *Preclinical Studies:* Before clinical trials in humans, preclinical studies are conducted to evaluate the biosimilar's safety profile. Animal studies help identify potential toxicities and guide the design of subsequent clinical trials.
2. *Clinical Trials:* Rigorous clinical trials are conducted to assess the safety and efficacy of biosimilars in humans (Ventola 2013; Blackstone and Fuhr Jr. 2012).

Table 3.3 Biosimilar products and their reported toxicities (European Medicines Agency 2006a,b, 2005a, 2012, 2014)

Biosimilar product	Reported toxicities	Key findings
BioSim X	Infusion-Related Reactions (IRR)	Post-marketing surveillance identified a higher rate of IRR in BioSim X users compared to the reference product. Severity was generally mild, and mitigation strategies were implemented
BioSim Y	Immunogenicity in rheumatoid arthritis	Phase III trial revealed an increased incidence of anti-drug antibodies (ADAs) in a subset of rheumatoid arthritis patients using BioSim Y. Further investigation underway to assess clinical impact
BioSim Z	Switching study—Psoriasis	Switching study indicated an unexpected decrease in efficacy in patients switched from the reference product to BioSim Z. Investigation ongoing to understand factors contributing to reduced treatment response
BioSim W	Hypersensitivity reactions	Rare cases of hypersensitivity reactions reported in patients using BioSim W. Ongoing monitoring to assess the frequency and severity of these reactions
BioSim V	Thromboembolic events in oncology	Registry data showed an increased incidence of thromboembolic events in oncology patients treated with BioSim V. The causal relationship is being investigated, and healthcare providers are advised to closely monitor patients
BioSim A	Immunogenicity, safety in rheumatoid arthritis (RA) patients	Phase III trial showed comparable efficacy in RA patients; no significant increase in immunogenicity compared to the reference product
BioSim B	Switching study—Crohn's disease (infliximab switching study, efficacy reduction)	Study assessing safety and efficacy of switching patients between BioSim B and the reference product in Crohn's disease. Preliminary results indicate no major safety concerns with switching
BioSim C	Post-marketing surveillance—various indications (Infusion-Related Reactions (IRR))	Ongoing surveillance revealed a rare occurrence of infusion-related reactions in a small subset of patients. Healthcare providers educated on mitigation strategies
BioSim D	Registry study—oncology indications (hypersensitivity reactions)	Registry data demonstrated similar safety profiles in oncology patients compared to the reference product. No unexpected toxicities reported during the follow-up period

These trials involve comparing the biosimilar to the reference product in terms of pharmacokinetics, pharmacodynamics, immunogenicity, and overall safety.

3. *Immunogenicity:* Biosimilars, like originator biologics, may trigger an immune response in some patients. This can lead to the production of anti-drug antibodies (ADAs) that might affect the drug's efficacy and safety. The risk of immunogenicity is carefully assessed during the development of biosimilars, and post-marketing surveillance is crucial to identify any unexpected immune reactions.

4. Extraneous Substances: Impurities or contaminants introduced during the manufacturing process could pose potential risks. Regulatory agencies require thorough testing and analysis to ensure the absence of harmful substances in the final product (Grabowski 2007).
5. Quality Attributes: Variations in the quality attributes of biosimilars may impact their safety profile. For instance, differences in glycosylation patterns or other structural characteristics could lead to unexpected effects (Brown et al. 2009).
6. Switching and Interchangeability: The impact of switching between a biosimilar and its reference product or between different biosimilars is an area of ongoing research. Changes in treatment regimens might affect patient outcomes or safety.
7. Extrapolation of Indications: Biosimilars are often approved for fewer indications than their reference products. The safety of extrapolating data from one indication to another needs careful consideration.
8. Post-Marketing Surveillance: Continuous monitoring of biosimilars after approval is essential to detect rare or long-term toxicities that may not have been evident in pre-market clinical trials. Healthcare providers and patients are encouraged to report adverse events to regulatory authorities.
9. Patient and Physician Awareness: Educating patients and healthcare providers about the nature of biosimilars, the potential for immunogenicity, and the importance of adherence to treatment plans is crucial for managing and minimizing risks.

   Incorporating patient-reported outcomes provides valuable insights into the patient experience, including any unexpected side effects or toxicities not easily captured through traditional clinical measures.
10. Regulatory Oversight: Stringent regulatory processes are in place to evaluate the safety and efficacy of biosimilars. Regulatory agencies such as the FDA and EMA (European Medicines Agency) assess the totality of evidence before granting approval.
11. Structural Variability: Despite efforts to achieve high similarity, there may be subtle differences in the structure of biosimilars compared to the reference product. Even minor variations in glycosylation, folding, or other structural features could contribute to unexpected toxicities.
12. Manufacturing Process Differences: Variability in manufacturing processes, such as cell lines, culture conditions, or purification methods, may lead to differences in the final product. This can result in the presence of impurities or contaminants that could be associated with toxicities.
13. Impurities and Contaminants: The presence of impurities or contaminants in the biosimilar product can be a source of toxicities. These may include residual host cell proteins, nucleic acids, or other substances introduced during the manufacturing process.
14. Registry Studies: Some countries may implement registry studies to collect real-world data on biosimilar use. These studies help assess the long-term safety and effectiveness of biosimilars in diverse patient populations.

15. Comparative Effectiveness Research: Conducting comparative effectiveness research involves evaluating the safety and efficacy of biosimilars in real-world clinical practice compared to the reference product.
16. Healthcare Professional Education: Ensuring that healthcare professionals are educated about biosimilars and their potential toxicities are vital. This includes training on monitoring and reporting adverse events

3.4 Risk Factors with Biosimilars Used in Cancer

A new class of medications called biosimilars is intended to function similarly to biologics. Large, complex proteins with a wide range of potential applications are called biologics, and they are produced in living cells. Biologics are utilized in gastroenterology alone to treat endocrine problems, malignancies, and inflammatory bowel diseases. Biologics have shown promise in the treatment or management of numerous illnesses; nevertheless, patient access is sometimes hampered by exorbitant costs. Biosimilars have to be almost exactly like the reference biologics in terms of immunogenicity, side effect risk profile, and efficacy (Kaida-Yip et al. 2018).

3.4.1 Current Impact of Biosimilars in Cancer Care

Similar to generic drugs, biosimilars may need an aggressive marketing campaign by payers to encourage use in order to yield the best possible returns for the health system. Supportive cancer therapy is the use of medications to mitigate the negative effects of cancer treatment. In the European Union (EU), filgrastim and epoetin alfa, supportive care medications, were the first approved biosimilars in 2007. Since then, other biosimilar development initiatives have been started, including those for bevacizumab, trastuzumab, and rituximab. Two biologic drugs with significant anticancer effects, filgrastim and epoetin, have had their patents revoked. In Europe, biosimilar versions of both have been approved by the European Medicines Agency (EMA) since 2007. Generally speaking, this could reduce costs by increasing accessibility to enhance supportive patient care and free up funds for new treatment initiatives (Nahleh et al. 2022).

Several filgrastim biosimilars have previously received approval for all indications used in the reference medication. Based on data that demonstrated the structure, in vitro parameters, PK and PD parameters, and similarity in mode of action of the reference standard by analytical procedures, the authorization was granted. Additionally, studies were conducted to determine their safety and efficacy in cancer patients (Cornes and Aapro 2018).

3.4.2 List of Biosimilars Used in Cancer Treatment and Associated Risk Factors

The US Food and Drug Administration has officially authorized the use of one or more biosimilars (FDA).

3.4.2.1 Bevacizumab (Avastin)

Mvasi, Zirabev, Alymsys Bevacizumab (VEGF-A) is a humanized monoclonal antibody that targets vascular endothelial growth factor A. It is applied to treat many cancer types (cervical cancer, metastatic colorectal cancer, glioblastoma, non-squamous non-small cell lung cancer, ovarian, fallopian tube, primary peritoneal cancer, metastatic renal cell carcinoma, and hepatocellular carcinoma). Bevacizumab is used in cancer treatment to prevent the growth of malignant cells and the formation of new blood vessels by blocking microvascular expansion and angiogenesis. It is typically given in conjunction with additional chemotherapy drugs. In order to help members of the interprofessional team treat patients with bevacizumab, this task will elucidate the mechanism of action, adverse event profile, and other pertinent variables. (e.g., pharmacodynamics, off-label usage, pharmacokinetics, dose, relevant interactions, and monitoring).

Risk of Bevacizumab to Chemotherapy for Breast Cancer

Hemorrhagic events and LVD were indeed made more likely by bevacizumab. There was no discernible rise in grade 3 arterial or venous thromboembolic events, GI perforation, or fatal events in patients with metastatic breast cancer who received bevacizumab in addition to chemotherapy (Gerriets and Kasi 2024). Anticoagulation safety in patients receiving active bevacizumab treatment has been questioned due to the drug's correlation with an increased risk of bleeding and the innately heightened risk of venous thromboembolism (VTE) in cancer patients. There is less information currently available on the safety of therapeutic anticoagulation in this therapeutic context because the majority of bevacizumab clinical studies do not allow patients who are on anticoagulation to participate (Cortes et al. 2012).

According to a therapeutic review, out of patients who got bevacizumab at least once, 3.2% experienced pulmonary hemorrhage of grade 3 or above. 6.5% of cases involved any grade of pulmonary bleeding. In these patients, pulmonary bleeding seems to be linked to the emergence of cavitation after the start of treatment. While it was not possible to make firm conclusions about the safety of bevacizumab in patients with squamous histology, using this delayed bevacizumab strategy resulted in a lower rate of pulmonary hemorrhage than previous series. The identification of reliable clinical markers for bevacizumab toxicity is still pending; hence, bevacizumab use in patients with squamous histology lung cancer is still under evaluation (Kurkjian and Kim 2012).

3.4.2.2 Rituximab (Rituxan)

Rituximab is a monoclonal antibody to the B-cell marker CD20 and is a common biologic agent widely used for the treatment of B-cell lymphoma, lymphoproliferative disorders, and inflammatory conditions that are refractory to conventional treatment, including rheumatoid arthritis and some vasculitides (Kasi et al. 2012). Rituximab use is linked to an increasing number of significant side events, which frequently result in or complicate admission to an intensive care unit, according to post-marketing surveillance and randomized controlled trials (Emery et al. 2006).

Risk Factors Associated with Rituximab

In the event that there is no other clear cause, patients receiving rituximab who develop new neurological symptoms should have a lumbar puncture and neuroimaging performed to assess the condition. There was one case each of a cerebrovascular infarction, convulsion, seizures, and serotonin syndrome among 465 patients receiving rituximab treatment for active RA that was not responding to MTX (Scheinfeld 2006). Two percent of the trial participants experienced serious cutaneous adverse effects. From the moment of the infusion until 3 months later, rituximab infusions have been linked to the following side effects: toxic epidermal necrolysis, lichenoid dermatitis, vesiculobullous dermatitis, Stevens–Johnson syndrome, and paraneoplastic pemphigus (Rituximab Package Insert n.d.). The onset of these serious side effects should cause rituximab to be stopped (Dempsey et al. 2021).

3.4.2.3 Trastuzumab (Herceptin) (Ogivri, Herzuma, Ontruzant, Trazimera, Kanjinti)

One medication called trastuzumab (Herceptin) is used to treat "HER2-positive breast cancer." One way to administer trastuzumab is by slow intravenous (IV) infusion. The infusion is administered once a week or once every 3 weeks by a medical expert. The weight of the woman determines the dosage.

Risk Factors Associated with Trastuzumab (Herceptin)

The biggest risk factor for TIC is prior to anthracycline use, which should be considered in any risk assessment for TIC together with cardiac imaging results. The prognosis for patients receiving HER2-directed therapy may be improved by screening for and predicting cardiotoxicity, referring them to a cardio-oncology expert, and starting efficient preventive therapy. ACE inhibitors and beta blockers seem to reduce the risk of TIC. Since anthracycline-free regimens have been shown to be effective in treating early HER2-positive breast cancer, they ought to be regarded as the accepted standard of therapy in this situation. Trials in patients who have had TIC and patients who already have heart problems are necessary; however, newer HER2-directed treatments seem to have much less cardiotoxicity than trastuzumab.

When taking trastuzumab along with anthracycline chemotherapy medications, there is an elevated risk. As a result, anthracycline treatment (epirubicin (Pharmorubicin), doxorubicin (Adriamycin)) and trastuzumab should not be used together. It is generally not advised for patients with pre-existing cardiac conditions to use trastuzumab. Before and during therapy, the heart will be examined in women who do not have any prior medical conditions. Cardiac exams will be more regular, and the lady may be referred to a cardiologist if heart problems arise while on trastuzumab. Chest tightness, tightness in the chest, chest pains, shortness of breath, irregular pulse, and fainting due to low blood pressure are all signs of cardiac problems. Fever and chills are among the other symptoms that trastuzumab may cause. Not many clinical trials have been conducted to examine trastuzumab's side effects in women with early-stage breast cancer. As a result, the long-term adverse consequences of trastuzumab are still unknown.

3.5 Other Application of Biosimilar in Oncology

Cancer, being the primary source of mortality and survival rates, has led to a rise in healthcare expenses. Within the cancer domain, biosimilars are gaining significance. These biological medicines are nearly identical to a medicine approved by the USFDA and do not show any noticeable differences in therapeutic effect.

As contrasted to the benchmark biologics, biosimilars are more reasonable, which could make cancer therapy more accessible. Considering the high expenditure of many malignancies' biologic medications, it is particularly critical. Across the world, cost-containment strategies and techniques have been prioritized to lower the overall cost of cancer pharmaceuticals, leading to the development of biosimilars (Tabernero et al. 2016). It will encourage healthcare professionals to consider biosimilars as first-line treatment choices by evidence-based guidelines and update treatment guidelines to include biosimilars as interchangeable options for specific cancer indications, emphasizing their efficacy and safety profiles (Meher et al. 2019; ET Online 2023; Tabernero et al. 2016; Declerck et al. 2016; Lyman et al. 2018).

Biosimilars can help individuals with cancer have access to a wider variety of therapeutic options, offering practitioners more options to customize therapy to suit the needs of every patient (ET Online 2023). It develops new treatment guidelines incorporating biosimilars for particular types of cancer in collaboration with pharmacists, oncologists, and other healthcare professionals to ensure that the guidelines offer precise advice to integrate biosimilars into the current treatment regimens. It urged involvement in clinical trials assessing biosimilars' effectiveness and safety spanning a range of cancer indications. It will actively encourage and sponsored research projects meant to increase the body of data supporting the use of biosimilars in cancer treatment (Meher et al. 2019; ET Online 2023; Invest India n.d.).

There is now more competition in the biosimilar market for cancer medicines due to a mix of accelerated regulatory routes, shortened exclusivity periods, and market access incentives. More options became available as a result, which reduced costs and enhanced patient access to necessary treatments (Rathore 2012; Rugo et al. 2017; Tabernero et al. 2016; Declerck et al. 2016; Lyman et al. 2018; Giuliani et al. 2012; Marques et al. 2020). In areas where in the utilization of medicine is greatly restricted by expenses, biosimilars may make cancer therapies more accessible (Declerck et al. 2016). To address the increasing incidence of cancer worldwide, especially in low- and middle-income countries, a multinational pharmaceutical company worked with international regulatory bodies to create standardized procedures for the authorization and registration of cancer biosimilars will facilitate accelerated regulatory processes to guarantee prompt access to these treatments in areas where cancer incidence is high (Schiestl et al. 2017). Additionally, it worked with international organizations to secure funding for subsidized access and had discussions with pharmaceutical manufacturers to negotiate reasonable prices for oncology biosimilars, especially for low- and middle-income countries to implement tiered pricing strategies (European Medicines Agency 2005b; Paradise 2011). For biosimilars to become utilized effectively in oncology, patients and medical professionals must be well-educated, addressing any issues concerning their safety, effectiveness, and versatility (Invest India n.d.). Informational materials such as patient-friendly brochures, infographics, videos explaining, support group sessions the concept of biosimilars, their approval process, and their role in cancer treatment may help patients for optimal use of the medications (Marques et al. 2020; Danese et al. 2019; World Health Organization (WHO) 2013; Sullivan et al. 2011; Moorkens et al. 2016; European Medicines Agency (EMA) 2014). In the context of oncology, the educational initiative raised patient and professional knowledge and acceptance of biosimilars. The positive incorporation of biosimilars into cancer treatment regimens has been facilitated by increased confidence in the safety and efficacy of these drugs, as demonstrated by the survey findings (European Medicines Agency (EMA) 2014; Declerck et al. 2017; European Society for Medical Oncology (ESMO) 2017; American Society of Clinical Oncology (ASCO) 2018; IQVIA Institute for Human Data Science 2016).

3.6 Conclusion

A generic drug is the same replica of a brand-name medication. When a medication's constituents are straightforward compounds, manufacturers can produce identical versions of the drug. Some medications, meanwhile, like growth factors, vaccinations, and monoclonal antibodies used in cancer treatment, are manufactured from living cells and cannot be replicated in a generic form due to their complexity. We refer to these sophisticated drugs as biologics. Biosimilars are medications that, while not perfect replicas, bear a striking resemblance to these biologics. Getting the green light for biosimilars requires a thorough evaluation to confirm their safety, efficiency, and excellence. Obtaining authorization for biosimilars proves more challenging than generic and biological substances. The task of crafting a scientifically credible research, which exhibits the likeness of a product heavily reliant on processes, is far from straightforward. The launch of biosimilars fosters fierce rivalry in the market and could end up resulting in reducing biological substance rates, which are desirable for consumers and healthcare providers alike. Based on the pharmacokinetics and pharmacodynamic data in clinical trial for risk-based assessment, regulatory authorities approved the biosimilars for the treatment of life threatening diseases.

References

Akram MS, Pery N, Butler L, Shafiq MI, Batool N, Rehman MF, Grahame-Dunn LG, Yetisen AK (2021) Challenges for biosimilars: focus on rheumatoid arthritis. Crit Rev Biotechnol 41(1):121–153

American Society of Clinical Oncology (ASCO) (2018) ASCO position statement on Biosimilars in oncology. J Clin Oncol 36(12):1260–1265. https://doi.org/10.1200/JCO.2017.77.4893

Bachu RD, Abou-Dahech M, Balaji S, Boddu SHS, Amos S, Singh V, Babu RJ, Tiwari AK (2022) Oncology biosimilars: New developments and future directions. Cancer Rep (Hoboken) 5(11):e1720. https://doi.org/10.1002/cnr2.1720. Epub 2022 Oct 4

Biosimilars Info Sheet (n.d.) Generics and biosimilars. https://www.fda.gov/media/154912/download

Blackstone EA, Fuhr JP Jr (2012) Innovation and competition: will biosimilars succeed? Biotechnol Healthc 9(1):24–27

Brown C, Dingermann T, Drzewoski J, Berthet E, Smillie M (2009) EGA handbook on biosimilar medicines. European Generic Medicines Association, Brussels, pp 1–48

Camacho LH, Frost CP, Abella E, Morrow PK, Whittaker S (2014) Biosimilars 101: considerations for U.S. oncologists in clinical practice. Cancer Med 3(4):889–899

Cornes P, Aapro M (2018) The impact of Biosimilars in supportive care in cancer. Eur Oncol Haematol 14:20–32

Cortes J, Calvo V, Ramırez-Merino N, O'Shaughnessy J, Brufsky A, Robert N, Vidal M, Munoz E (2012) Adverse events risk associated with bevacizumab addition to breast cancer chemotherapy: a meta-analysis. Ann Oncol 23:1130–1137

Crommelin DJ, Bermejo T, Bissig M et al (2005) Pharmaceutical evaluation of biosimilars: important differences from generic low-molecular-weight pharmaceuticals. Eur J Hosp Pharm Sci 11:11–17

Danese S, Fiorino G, Michetti P (2019) Viewpoint: knowledge and viewpoints on biosimilar monoclonal antibodies among members of the European Crohn's and Colitis Organization. J Crohns Colitis 13(7):818–822. https://doi.org/10.1093/ecco-jcc/jjy212

De Mora F (2019) Biosimilars: a value proposition. BioDrugs 33(4):353–356

Declerck P, Farouk Rezk M, Gastonguay M et al (2016) Biosimilars: a consideration of the regulations in the United States and European Union. Regul Toxicol Pharmacol 76:199–209. https://doi.org/10.1016/j.yrtph.2016.01.019

Declerck P, Danesi R, Petersel D, Jacobs I (2017) The language of biosimilars: clarification, definitions, and regulatory aspects. Drugs 77(6):671–677. https://doi.org/10.1007/s40265-017-0717-5

Dempsey N, Rosenthal A, Dabas N, Kropotova Y, Lippman M, Bishopric NH (2021) Trastuzumab-induced cardiotoxicity: a review of clinical risk factors, pharmacologic prevention, and cardiotoxicity of other HER2-directed therapies. Breast Cancer Res Treat 188(1):21–36. https://doi.org/10.1007/s10549-021-06280-x. Epub 2021 Jun 11

EMEA (2005) Guideline on similar medicinal products, CHMP/437/04. European Medicines Agency, London. http://www.emea.europa.eu/pdfs/human/biosimilar/043704en.pdf

EMEA (2006a) Guideline on similar biological medicinal products containing biotechnology-derived proteins as active substance: quality issues. European Medicines Agency, London. http://www.emea.europa.eu/pdfs/human/biosimilar/4934805en.pdf

EMEA (2006b) Guideline on similar biological medicinal products containing biotechnology-derived proteins as active substance: non-clinical and clinical issues. European Medicines Agency, London. http://www.emea.europa.eu/pdfs/human/biosimilar/4283205en.pdf

EMEA (2006c) Annex to guideline on similar biological medicinal products containing biotechnology-derived proteins as active substance: non-clinical and clinical issues. Guidance on similar medicinal products containing recombinant erythropoietins. European Medicines Agency, London. http://www.emea.europa.eu/pdfs/human/biosimilar/9452605en.pdf

EMEA (2006d) Annex to guideline on similar biological medicinal products containing biotechnology-derived proteins as active substance: non-clinical and clinical issues. Guidance on similar medicinal products containing recombinant granulocyte-colony stimulating factor (G-CSF). European Medicines Agency, London. http://www.emea.europa.eu/pdfs/human/biosimilar/3132905en.pdf
EMEA (2006e) Annex to guideline on similar biological medicinal products containing biotechnology-derived proteins as active substance: non-clinical and clinical issues. Guidance on similar medicinal products containing recombinant human insulin. London, European Medicines Agency. http://www.emea.europa.eu/pdfs/human/biosimilar/3277505en.pdf
EMEA (2007) Guideline on immunogenicity assessment of biotechnology-derived therapeutic proteins. European Medicines Agency, London. http://www.emea.europa.eu/pdfs/human/biosimilar/1432706enfin.pdf
EMEA (2009a) Non-clinical and clinical development of similar medicinal products containing recombinant interferon Alfa. European Medicines Agency, London. http://www.emea.europa.eu/pdfs/human/biosimilar/10204606enfin.pdf
Emery P, Fleischmann R, Filipowicz-Sosnowska A, Schechtman J, Szczepanski L, Kavanaugh A, Racewicz AJ, van Vollenhoven RF, Li NF, Agarwal S, Hessey EW, Shaw TM, DANCER Study Group (2006) The efficacy and safety of rituximab in patients with active rheumatoid arthritis despite methotrexate treatment: results of a phase IIB randomized, double-blind, placebo-controlled, doseranging trial. Arthritis Rheum 54:1390–1400
Emmanouilides CE, Karampola MI, Beredima M (2016) Biosimilars: hope and concern. J Oncol Pharm Pract 22(4):618–624
ET Online (2023) India produces over 60,000 generic drugs, highlights MoS Bhagwant Khuba. The Economic Times. ISSN: 0013-0389. Accessed 1 Aug 2023
European Medicines Agency (2005a) Guideline on similar biological medicinal products. CHMP/437/04. European Medicines Agency, London
European Medicines Agency (2005b) ICH Topic Q 5 E. Comparability of biotechnological/biological products. http://www.ema.europa.eu/docs/en_GB/document_library/Scientific_guideline/2009/09/WC500002805.pdf. Accessed 8 Dec 2016
European Medicines Agency (2006a) Guideline on similar biological medicinal products containing biotechnology-derived proteins as active substance: non-clinical and clinical issues. EMEA/CHMP/BMWP/42832/2005. European Medicines Agency, London
European Medicines Agency (2006b) Guideline on similar biological medicinal products containing biotechnology-derived proteins as active substance: quality issues. EMEA/CHMP/BWP/49348/2005, London, European Medicines Agency
European Medicines Agency (2012) Guideline on similar biological medicinal products containing biotechnology-derived proteins as active substance: quality issues. (Revision 1). EMA/CHMP/ BWP/247713/2012. European Medicines Agency, London
European Medicines Agency (2014) Guideline on similar biological medicinal products. EMA/CHMP/437/04. Rev. 1. European Medicines Agency, London
European Medicines Agency (EMA) (2014) Guideline on similar biological medicinal products. CHMP/437/04 Rev 1. https://www.ema.europa.eu/en/documents/scientific-guideline/guideline-similar-biological-medicinal-products-rev1_en.pdf. Published October 23, 2014
European Society for Medical Oncology (ESMO) (2017) ESMO position paper on Biosimilars in oncology. Ann Oncol 28(12):2620–2626. https://doi.org/10.1093/annonc/mdx706
Foreman E (2020) Biosimilars in supportive care. Curr Opin Oncol 32(4):282–288
Gerriets V, Kasi A (2024) Bevacizumab. In: StatPearls [Internet]. Treasure Island (FL): StatPearls Publishing. Available from: https://www.ncbi.nlm.nih.gov/books/NBK482126/. [Updated 2023 Aug 28]
Giuliani R, Tabernero J, Cardoso F et al (2012) Potential impact of biosimilars in oncology. Target Oncol 7(Suppl 1):S29–S36. https://doi.org/10.1007/s11523-012-0202-3
Grabowski H (2007) Data exclusivity for new biological entities, Duke University Economics Department Working Paper http://www.econ.duke.edu/Papers/PDF/

DataExclusivityWorkingPaper.pdf. Economic analyses R&D invest 12.9–16.2 years. [Last cited on 2015 Apr 25]

Grabowski H (2008) Follow-on biologics: data exclusivity and the balance between innovation and competition. Nat Rev Drug Discov 7:479–488

Grieshaber-Bouyer R, Lorenz HM (2020) Biosimilars: opportunities and risks. Internist 61:522–529

Invest India (n.d.) Formulating success: The Indian pharmaceutical industry. https://www.investindia.gov.in › sector › pharmaceuticals

IQVIA Institute for Human Data Science (2016) Delivering on the potential of Biosimilar medicines: the role of functioning competitive markets. https://www.iqvia.com/insights/the-iqvia-institute/reports/delivering-on-the-potential-of-biosimilar-medicines. Published March 2016

Janjigian YY, Bissig M, Curigliano G, Coppola J, Latymer M (2018) Talking to patients about biosimilars. Future Oncol 14(23):2403–2414

Kaida-Yip F, Deshpande K, Saran T, Vyas D (2018 August 16) Biosimilars: review of current applications, obstacles, and their future in medicine. World J Clin Cases 6(8):161–166

Kanase SJ, Gavhane YN, Khandekar A, Gurav AS, Yadav AV (2013) Biosimilar: an overview. Int J Pharm Sci Res 4(6):2132–2144

Kasi PM, Tawbi HA, Oddis CV, Kulkarni HS (2012) Clinical review: Serious adverse events associated with the use of rituximab—a critical care perspective. Crit Care 16:231

Kurkjian C, Kim ES (2012) Risks and benefits with bevacizumab: evidence and clinical implications. Ther Adv Drug Saf 3(2):59–69

Lyman GH, Zon R, Harvey RD et al (2018) Rationale, opportunities, and reality of biosimilar medications. N Engl J Med 378(21):2036–2044. https://doi.org/10.1056/NEJMhls1707605

Marques IB, Garcia-Mina Freire M, Fonseca F et al (2020) Patient education and the role of patient organizations in the biosimilar landscape in Portugal. Front Pharmacol 11:1293. https://doi.org/10.3389/fphar.2020.01293

McCamish M, Woollett G (2012) The state of the art in the development of biosimilars. Clin Pharmacol Ther 91(3):405–417

Meher BR, Balan S, Mohanty RR, Jena M, Das S (2019) Biosimilars in India; current status and future perspectives. J Pharm Bioallied Sci 11(1):12–15. https://doi.org/10.4103/jpbs.JPBS_167_18

Moorkens E, Jonker-Exler C, Huys I, Declerck P, Simoens S, Vulto AG (2016) Overcoming barriers to the market access of Biosimilars in the European Union: the case of biosimilar monoclonal antibodies. Front Pharmacol 7:193. https://doi.org/10.3389/fphar.2016.00193

Nahleh Z, Lyman GH, Schilsky RL, Peterson DE, Tagawa ST, Chavez-MacGregor M, Rumble RB, Gupta S (2022) Use of biosimilar medications in oncology. JCO Oncol Pract 18:177–186

Omnitrope, European Public Assessment Report (2006). http://www.ema.europa.eu/docs/en_GB/document_library/EPAR_-_Scientific_Discussion/human/000607/WC500043692.pdf. Accessed December 8, 2016

Paradise J (2011) Foreword: follow-on biologics: implementation challenges and opportunities. Seton Hall Law Rev 41(2):501–510

Parsad S, Nabhan C (2018) Biosimilars. JAMA Oncol 4(7):1023

Raedler LA (2016) Zarxio (Filgrastim-sndz): First biosimilar approved in the United States. Am Health Drug Benefits 9:150–154. (Spec Feature)

Rathore A (2012 Sep-Oct) Guidelines on similar biologics: regulatory requirements for marketing authorization in India. PDA J Pharm Sci Technol 66(5):393. https://doi.org/10.5731/pdajpst.2012.00886

Rituximab Package Insert (n.d.) (Biogen Idec Inc., and Genentech, Inc.) http://www.accessdata.fda.gov/drugsatfda_docs/label/2011/103705s5344lbl.pdf

Roger SD (2006) Biosimilars: how similar or dissimilar are they? Nephrology (Carlton) 11:341–346

Roger SD, Mikhail A (2007) Biosimilars: opportunity or cause for concern? J Pharm Pharm Sci 10:405–410

Rugo HS, Rifkin RM, Declerck P et al (2017) Impact of a proposed trastuzumab biosimilar on clinical practice, health care systems, and society. JAMA Oncol 3(4):1–8. https://doi.org/10.1001/jamaoncol.2016.5874

Rushvi P, Charmy K, Nirav C, Narendra C (2016) Biosimilars: an emerging market opportunities in India. Pharmaceut Reg Affairs 5:165
Saenger P (2017) Ten years of biosimilar recombinant human growth hormone in Europe. Drug Des Devel Ther 11:1505–1507
Sarosiek T, Morawski P (2018) Trastuzumab and its biosimilars. Polski Merkuriusz Lekarski: Organ Polskiego Towarzystwa Lekarskiego 44(263):253–257
Scheinfeld N (2006) A review of rituximab in cutaneous medicine. Dermatol Online J 12:3
Schellekens H (2005) Follow-on biologics: challenges of the 'next generation'. Nephrol Dial Transplant 20:31–36
Schiestl M, Zabransky M, Sörgel F (2017) Ten years of biosimilars in Europe: development and evolution of the regulatory pathways. Drug Des Devel Ther 11:1509–1515. https://doi.org/10.2147/DDDT.S130318
Schreitmüller T, Barton B, Zharkov A, Bakalos G (2019) Comparative immunogenicity assessment of biosimilars. Future Oncol 15(3):319–329
Sheridan C (2006 Jun) First generic biologics finally approved. Nat Rev Drug Discov 5(6):445. https://doi.org/10.1038/nrd2072
Stevenson JG, Popovian R, Jacobs I, Hurst S, Shane LG (2017) Biosimilars: practical considerations for pharmacists. Ann Pharmacother 51(7):590–602
Sullivan R, Peppercorn J, Sikora K et al (2011) Delivering affordable cancer care in high-income countries. Lancet Oncol 12(10):933–980. https://doi.org/10.1016/S1470-2045(11)70141-3
Tabernero J, Vyas M, Giuliani R et al (2016) Biosimilars: a position paper of the European Society for Medical Oncology, with particular reference to oncology prescribers. ESMO Open 1(6):e000142. https://doi.org/10.1136/esmoopen-2016-000142
The New India Guidelines on Similar Biologics (2012). http://www.biospectrumasia.com/biospectrum/analysis/3021/biosimilars-guidelines-a-step-direction-india#.UehBf6DRiSo. Accessed 16 Jul 2013
U.S. Food & Drug Administration (FDA) (n.d.-a) Biosimilar and interchangeable products. https://www.fda.gov/drugs/biosimilars/biosimilar-and-interchangeable-products#generic. Accessed 19 June 2022
U.S. Food & Drug Administration (FDA) (n.d.-b) Biosimilar product information. https://www.fda.gov/drugs/biosimilars/biosimilar-product-information. Accessed 12 Dec 2023
US Food and Drug Administration (2015) Drugs: filgrastim-sndz. Updated March 6 2015. https://www.fda.gov/Drugs/InformationOnDrugs/ApprovedDrugs/ucm436953.htm. Accessed 25 July 2018
Ventola CL (2013) Biosimilars: part 1: proposed regulatory criteria for FDA approval. P T 38(5):270–287
Verrill M, Declerck P, Loibl S, Lee J, Cortes J (2019) The rise of oncology biosimilars: from process to promise. Future Oncol 15(28):3255–3265
Watson K, Nick C, Babbitt B (2010) Clinical and cost considerations of developing a biosimilar. Pharm Technol Eur 22(9):80–82
World Health Organization (WHO) (2013) Technical Report Series No. 977. Annex 2. Guidelines on Evaluation of Similar Biotherapeutic Products (SBPs). https://www.who.int/biologicals/biotherapeutics/biotherapeutics-trs-977-annex2.pdf?ua=1. Published 2013

Biosimilars in Colorectal Cancer

4

Meghana Patel, Manish Nandpal, Nikunj Parekh,
Samir Gunvantbhai Patel, and Alkeshkumar Patel

Abstract

Biologics (BL) are complex moiety derived from living resources through biotechnology. BL are quite different compared to small molecules with respect to target specificity, better outcome in chronic diseases, and excellent immunomodulators. Biosimilars (BS) are developed once biologics (reference product) patents expire to provide affordable accessible therapy to the patients. Extremely sensitive in vitro, in vivo, and clinical trials need to be conducted to confirm the bioequivalence of BL with BS. Cancer remains a devastating disease across the globe. Out of a total of 44 BS are approved by US FDA, around 18 are used for oncotherapy. Colorectal cancer (CRC) is the second fatal cancer among others. BS can be used as mono- or adjunct therapy for the CRC. Bevacizumab-awwb was pioneer BS and has shown excellent outcome in metastatic CRC. The epoetin alfa-epbx and pegfilgrastim-jmdb used to ameliorate bone marrow suppression induced by chemotherapy. Other important BS used in CRC oncotherapy include trastuzumab-dkst, tocilizumab-bavi, adalimumab-aaty, and cetuxa. Multiple meta-analysis studies of randomized clinical trials indicated that BS are equivalent in efficacy to their reference products. Anyhow stringent regulatory screening and pharmacovigilance with patients' acceptance are necessary to ensure incorporation of BS in oncotherapeutics. Many challenges need to be addressed like patient recruitments, market competitions, interchangeability by pharmacist, nomenclature, production, and approval timeline to make successful use of BS in the future. This chapter highlights typical characteristics of BS, pharmacology requirement of BS to be effective in CRC, regulatory guidelines, manufacturing process, current BS for CRC, ongoing clinical trials of BS, and future of BS.

M. Patel · M. Nandpal · N. Parekh · S. Gunvantbhai Patel · A. Patel (✉)
Ramanbhai Patel College of Pharmacy, Charusat, Changa, Gujarat, India
e-mail: alkeshpatel.ph@charusat.ac.in

S. Bhatt et al. (eds.), *Biosimilars for Cancer Treatment*,
https://doi.org/10.1007/978-981-97-3001-8_4

Keywords

Biologics · Biosimilars · Colorectal cancer · Metastasis · Bevacizumab-awwb · Pegfilgrastim-jmdb

4.1 Introduction to Biologicals and Biosimilars

Biologicals, often termed "biologics," are a class of therapeutic substances derived from living organisms or their components. Unlike traditional pharmaceuticals synthesized chemically, biologicals harness the power of cells, proteins, antibodies, or nucleic acids to treat diseases and medical conditions. This innovative field encompasses a wide array of products, including vaccines, gene therapies, monoclonal antibodies, and recombinant proteins. The development of biologicals involves intricate processes, including genetic engineering, cell culture, along with sophisticated purification techniques. These therapies are designed to target specific molecules or pathways within the body, offering highly precise and often potent treatment options. Biologicals have revolutionized medicine by providing personalized and targeted therapies for various ailments, ranging from cancer and autoimmune disorders to infectious diseases.

One key aspect of biologicals is their ability to mimic or supplement natural substances within the body, fostering a more tailored and less invasive approach to healthcare. Their complexity often poses challenges in manufacturing, storage, and administration, but their efficacy and potential to address unmet medical needs continue to drive research and development in this field, shaping the future of medicine. Biologicals represent a paradigm shift in medicine, providing therapies that focus on the fundamental causes of illnesses rather than just masking symptoms. Their specificity allows for tailored interventions, reducing potential side effects often associated with conventional drugs. Moreover, they play a pivotal role in precision medicine, where therapies are customized based on individual genetic makeup, optimizing treatment outcomes. The impact of biologicals spans various medical domains. In oncology, monoclonal antibodies like trastuzumab have transformed the landscape of cancer care by targeting specific cancer cells. Additionally, vaccines, a crucial subset of biologicals, have been instrumental in eradicating or controlling infectious diseases, saving millions of lives worldwide. As research advances, the scope of biologicals continues to expand. Innovations in gene editing and cellular therapies hold promise for treating previously incurable conditions. However, challenges persist, including the high cost of production and accessibility issues. Despite these hurdles, the ongoing advancements in biologicals underscore their pivotal role in shaping the future of healthcare (FDA 2024a).

Biosimilars, a burgeoning sector within the pharmaceutical industry, represent a revolutionary approach to medicine. These biologic products are akin to "generic" versions of complex biological drugs, mimicking the structure and function of existing biologics. They offer a promising avenue for expanding patient access to high-quality treatments while potentially reducing healthcare costs. Unlike traditional pharmaceuticals, which are chemically synthesized, biologics are derived from living cells, making exact replication challenging. Biosimilars, however, strive to replicate these intricate structures and therapeutic effects of biologics, undergoing rigorous testing to show the similarity in terms of quality, efficacy, and safety. This meticulous process involves comprehensive comparison studies to establish that the biosimilar functions similarly to its reference product.

The advent of biosimilars has the potential to revolutionize healthcare by fostering competition, enhancing affordability, and increasing treatment accessibility. By introducing competition into the market, biosimilars may drive down prices, making expensive biologic therapies more affordable for patients and healthcare systems worldwide. This affordability can significantly benefit individuals with chronic conditions requiring long-term treatment and alleviate financial strains on healthcare budgets. Moreover, the expansion of biosimilars could lead to increased innovation and investment in biopharmaceutical research. Companies may focus on developing novel biologics or enhancing existing ones to maintain their competitive edge in the market. However, challenges persist in the widespread adoption of biosimilars. Regulatory pathways, manufacturing complexities, and concerns regarding interchangeability and immunogenicity require continuous evaluation and refinement. Building confidence among patients and healthcare providers in the efficacy and safety of biosimilars is also an important goal (Halimi et al. 2020).

Despite these challenges, the growth of biosimilars signifies a pivotal stride in healthcare evolution. As research and development in this realm continue to progress, the promise of expanding treatment options, improving patient outcomes, and fostering a more sustainable healthcare landscape remains compelling, shaping a future where innovative therapies are more accessible to all (Ciucci et al. 2022; Mysler et al. 2021; Sekhon and Saluja 2011). Manufacturing process of biosimilars is shown in Fig. 4.1.

4.1.1 Difference Between Biologicals and Biosimilars

Biosimilars are subject to stringent evaluation processes and adhere to the rigorous manufacturing criteria outlined by the FDA. Prior to market approval, biosimilars are mandated to demonstrate identical safety and efficacy profiles as well as an absence of "clinically meaningful differences" compared to the original biologic. Despite meeting comparable efficacy and safety standards with biologics, distinct variations delineate biosimilars from their counterparts. Divergence exists in the regulatory pathways governing these pharmaceuticals. While both biologics and biosimilars undergo thorough testing before approval, the approval process for biosimilars follows a more streamlined trajectory than that for biologics. This distinction does not imply compromised safety, efficacy, or lowered standards for biosimilars (Fig. 4.2). The data submitted during biosimilar approval emphasizes showcasing clinical equivalence between the biosimilar and the reference biologic. Conversely, the approval process for biologics primarily centers on validating their safety and efficacy.

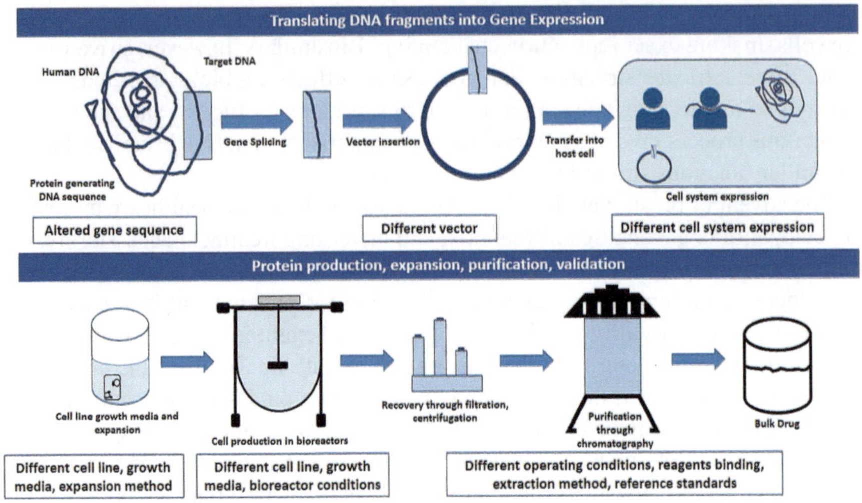

Fig. 4.1 Manufacturing of biosimilar

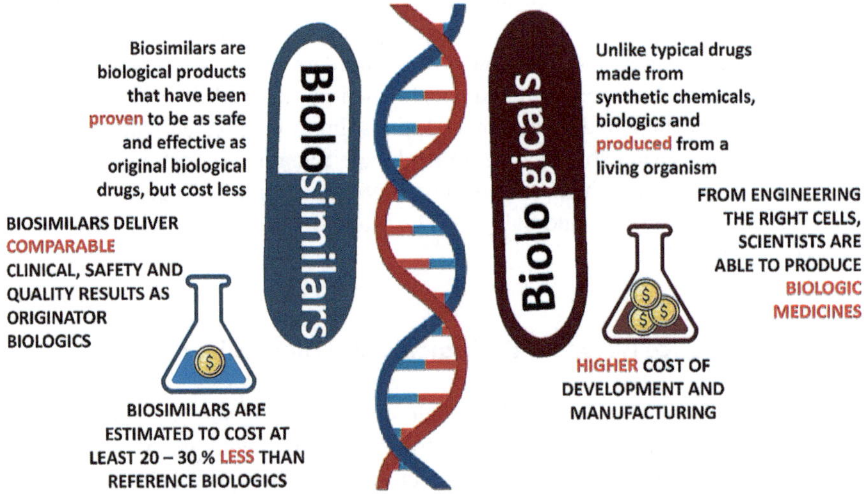

Fig. 4.2 Differences in between biologicals and biosimilars

Cost disparities emerge between biologics and biosimilars. The potentially lower cost of biosimilars is partly attributed to their expedited development process. According to the Association for Accessible Medicines, the competitive ecosystem promoted by biosimilars has culminated in lower costs for both BS and reference biologics. This competitive climate resulted in substantial savings, totaling $7.9 billion in 2020 and exceeding $12.6 billion over the previous decade. Due to stringent regulatory standards and extensive testing validating the similarity between biosimilars and their reference biologics, concerns regarding differences in efficacy and safety can be alleviated. The inherent similarity in the natural source materials, manufacturing methodologies, and administration protocols between biosimilars and biologics ensures that both treatments yield analogous therapeutic benefits and side effects. Therefore, the anticipated outcomes in terms of treatment efficacy and potential adverse effects remain consistent regardless of whether a patient is administered a biosimilar or a biologic. Biologics have significantly revolutionized patient prognoses across various medical conditions such as cancers and autoimmune disorders. The increasing introduction of novel biosimilars presents healthcare practitioners with the opportunity to offer a broader spectrum of high-quality treatment options to patients, potentially mitigating costs while maintaining therapeutic efficacy (FDA 2024b).

4.1.2 Transfer of Biologics to Biosimilars

The transition from biologics to biosimilars involves a complex process aimed at replicating the therapeutic effects of the original biologic. Thorough comprehension of the original biologic's structure, manufacturing process, and mode of action is essential. This includes identifying the active ingredients, protein structure, and how it interacts within the body. Here are the key steps involved in transferring biologics to biosimilars.

1. Analytical Characterization: Extensive investigation and comparing the molecular structure of the biosimilar with the reference biologic to assure compatibility. Advanced techniques like mass spectrometry, chromatography, and biological assays are employed for this purpose.
2. Preclinical Studies: Rigorous laboratory and animal studies are conducted to compare the biosimilar's efficacy, safety, and biological activity with the reference biologic. These studies aim to establish similarity in pharmacokinetics, pharmacodynamics, and potential toxicity.
3. Clinical Trials: Human clinical trials are pivotal to demonstrate the biosimilar's safety and efficacy in patients. These trials, conducted in multiple phases, involve comparing the biosimilar directly with the reference biologic in terms of safety, efficacy, and immunogenicity.
4. Regulatory Approval: Submission of comprehensive data from analytical, preclinical, and clinical investigations to regulatory authorities for approval. The evidence is reviewed by regulatory organizations to ensure that each biosimilar is extremely comparable to the reference biologic and fulfils safety and effectiveness criteria.
5. Manufacturing Process: Developing a robust manufacturing process that replicates the original biologic's production while ensuring consistent quality and purity of the biosimilar. This involves cell culture, purification, and stringent quality control measures.
6. Post-Market Surveillance: Continuous monitoring of the biosimilar's performance and safety once it enters the market. This includes tracking adverse

events, immunogenicity, and long-term effects to ensure ongoing safety and efficacy.

7. Education and Acceptance: Healthcare practitioners and patients need to be educated about biosimilars, emphasizing their similarities, safety, and effectiveness compared to the reference biologics. Building trust and acceptance in the medical community is crucial for widespread adoption.

The transition from biologics to biosimilars is a highly intricate and regulated process that demands scientific expertise, substantial investment, and adherence to stringent regulatory standards. Success in this transfer is critical to expanding access to life-saving treatments while maintaining safety and efficacy standards (Pfizer 2024).

4.1.3 Development of Biosimilar

Before biologic drugs become available for patient use, extensive developmental, research, and clinical trial phases are undertaken by scientists and healthcare practitioners to assure the efficacy and safety of these medications. The manufacturing process of a biologic drug is a meticulous and time-consuming endeavor, often taking months, demanding comprehensive scrutiny and control at every stage (FDA 2024c, d). Development of biosimilars is depicted in Fig. 4.3.

Manufacturing a biosimilar mirrors the procedures involved in producing its reference biologic counterpart. Regulatory bodies, such as the FDA, mandate that biosimilars demonstrate a lack of clinically significant distinctions and meet rigorous scientific criteria for similarity before being accessible to patients. The process of

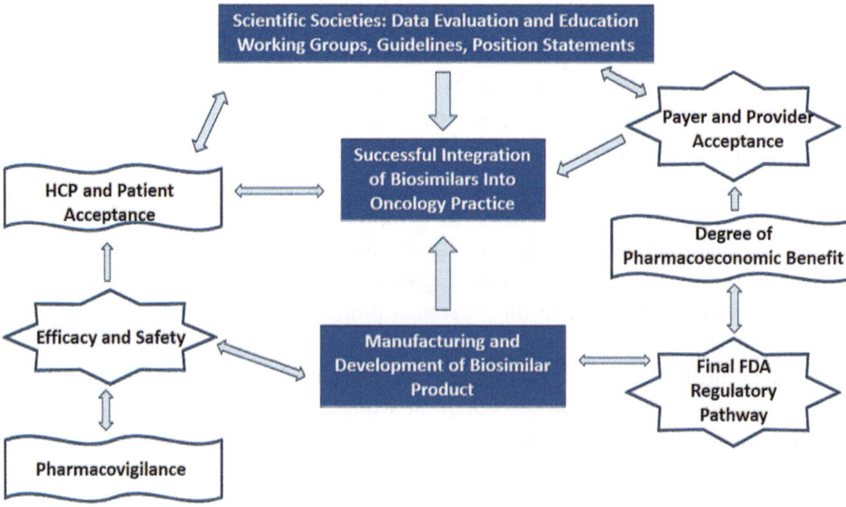

Fig. 4.3 Development of biosimilars

establishing similarity involves a stringent scientific evaluation ensuring no clinically significant variance between the biosimilar and the original product, assuring equivalent performance in a patient as the reference biologic. This evaluation occurs multiple times during the developmental phase.

The biosimilar development journey comprises three primary phases:

- Product characterization and process optimization
- Confirmation of biosimilarity
- Eventual approval

The initial phase involves comprehensive product development. It commences with a thorough understanding of the reference biologic's structure and function, termed as characterization. Subsequently, the manufacturing process is developed to replicate the therapy with high similarity. Advanced biological development methodologies and precise analytical methods are used to meticulously engineer a biosimilar molecule that mirrors the quality determined during analysis; this sequential method continually refines each stage of the process of manufacture up until consistent production of an exceedingly similar molecular framework to the reference medicine is accomplished (Biosimilars Council 2024; AMGEN Biosimilars 2024).

Following the establishment of similarity through analysis and testing, the subsequent stage involves regulatory assessments and necessary additional non-clinical and clinical studies to confirm potential interchangeability and biosimilarity. Clinical studies are often required for biosimilar clearance in rigorous markets such as the US, EU, Australia, Canada, and Japan, depending on the evidence given. Where strong analytical data is available, targeted clinical studies may be a more effective way to establish biosimilarity and interchangeability (Kesik-Brodacka 2018).

The final stage encompasses regulatory approval. Biosimilar firms meet with regulatory agencies early in the development process to discuss development strategies and relevant scientific explanations. Prior to approval, the FDA conducts an exhaustive evaluation of all data to ensure accordance with scientific standards. Following approval, the biosimilar will be able to be produced and supplied in the United States. Patients can anticipate equivalent safety and efficacy between biosimilars and their brand biologic counterparts. Global studies consistently affirm that patients experience similar benefits regardless of whether they use the original biologic or its biosimilar counterpart (AL-Sabbagh et al. 2016).

4.2 Ensuring Efficacy and Safety with Quality of Biosimilars, Biosuperiors, and Clinical Trials

The development and approval of biosimilars and biosuperiors mark a substantial paradigmatic shift within the pharmaceutical industry. The paramount importance of ensuring the safety and efficacy of these products underscores the necessity for a thorough comprehension of their quality attributes and the meticulous evaluations conducted in clinical trials. This section intricately examines the multifaceted dimensions involved in ensuring efficacy and safety, specifically through the lens of quality, within the domain of biosimilars, biosuperiors, and the corresponding clinical trials. The assessment of the degree to which a biosimilar and biosuperior medication corresponds to the reference product concerning safety, efficacy, quality, and the elucidation of the term "highly similar" presents pivotal inquiries. It sets the stage for exploring the interplay between quality, clinical trials, and the regulatory landscape governing these complex therapeutic entities (Kirchhoff et al. 2017).

4.2.1 Quality Attributes of Biosimilars and Biosuperiors

The pivotal quality attributes of biosimilars and biosuperiors encompass structural, functional, and physicochemical characteristics. An understanding of the impact of manufacturing processes on these attributes is crucial for maintaining consistency, reproducibility, and influencing the safety and efficacy profiles of these biologics (Barbier et al. 2020; Barbier et al. 2022).

4.2.1.1 Structural Attributes

Primary structure: The amino acid sequence, foundational to the protein's identity and function, must be identical between a BS or biosuperior and the reference product, ensuring biological activity and therapeutic equivalence.

Higher order structure: The three-dimensional folding and conformation are crucial for biological activity. A rigorous comparison of the higher order structure is essential to verify the accurate mirroring of the reference product by the biosimilar or biosuperior.

4.2.1.2 Functional Attributes

Biological activity: The therapeutic protein's biological function is a key determinant of efficacy. Ensuring comparable biological activity to the reference product is vital for therapeutic equivalence in biosimilars or biosuperiors.

Receptor binding: For proteins interacting with cellular receptors, maintaining the ability to bind to the target receptor is crucial. A thorough assessment confirms that the biosimilar or biosuperior maintains binding characteristics similar to the reference product.

4.2.1.3 Physicochemical Attributes

Molecular weight: The protein's molecular weight is a critical physicochemical parameter. Biosimilars and biosuperiors must closely match the molecular weight of the reference product for compositional similarity.

Post-translational modifications: Modifications like glycosylation and phosphorylation can impact stability, activity, and pharmacokinetics. Characterizing and comparing post-translational modifications are essential for demonstrating biosimilarity.

4.2.1.4 Purity and Impurities

The absence of impurities, including aggregates, fragments, or host cell proteins, is crucial for biopharmaceutical safety and efficacy. Rigorous purification processes ensure high levels of purity.

Impurity profiles: A comprehensive analysis of impurity profiles, covering related substances and degradation products, assesses the overall quality of biosimilars and biosuperiors. The product's stability over time is critical for shelf life and clinical utility. Extensive stability testing ensures consistent quality and efficacy throughout the intended shelf life of biosimilars and biosuperiors.

4.2.2 Analytical Attributes

Analytical methods: Robust analytical methods are employed to characterize and quantify various attributes of biosimilars and biosuperiors. These methods must be sensitive, specific, and capable of discerning subtle differences between the products.

Comparability studies: Comparative analytical studies assess similarity among the BS or biosuperior and the reference product, involving a range of analytical techniques to comprehensively evaluate various attributes.

Understanding and meticulously characterizing these quality attributes are imperative for demonstrating the biosimilarity or superiority of these therapeutic proteins. Regulatory agencies globally mandate extensive data on these attributes to ensure the safety, efficacy, and interchangeability of biosimilars and biosuperiors with their reference products.

4.2.3 Regulatory Frameworks Governing Quality

A detailed examination of the regulatory frameworks that govern the quality of biosimilars and biosuperiors is presented. This includes an analysis of guidelines provided by regulatory agencies such as the WHO, EMA, and FDA, highlighting the stringent requirements for demonstrating comparability and ensuring product quality.

4.2.4 Clinical Trials in the Evaluation of Efficacy and Safety

Clinical trials are critical in assessing the effectiveness and safety of biosimilars and biosuperiors. It explores the intricacies of trial design, execution, and analysis, emphasizing the need for robust methodologies to establish equivalence or non-inferiority compared to reference products. Clinical trials are essential tools for understanding safety profiles and effectiveness of BS and biosuperiors. Their significance lies in providing a structured and controlled environment for the systematic examination of therapeutic outcomes, contributing critical data for regulatory assessments and informed medical decision-making (Pathak et al. 2023; European Medicines Agency 2014).

4.2.4.1 Principles of Design

The design of clinical trials for biosimilars and biosuperiors requires meticulous planning to ensure scientific validity and reliability. Rigorous methodologies include randomization, control groups, and blinding, forming a strong foundation for unbiased observations and conclusions. The selection of appropriate endpoints is crucial, reflecting clinically meaningful measures of efficacy and safety.

4.2.4.2 Execution of Trials

The conduct of these trials adheres to stringent ethical and regulatory standards, protecting the welfare of participants and the integrity of the generated data. Enrollment criteria are carefully defined to create homogeneous study populations, facilitating the extraction of meaningful insights. The administration of interventions, whether biosimilars or biosuperiors, is executed with precision, guided by standardized protocols.

4.2.4.3 Statistical Analysis

The analysis phase employs sophisticated statistical methodologies to discern meaningful patterns and differences. Equivalence or non-inferiority analyses are used to establish the comparative performance of biosimilars and biosuperiors against reference products. Robust statistical tools enhance the reliability of findings, ensuring that observed effects are not merely due to chance.

4.2.4.4 Emphasis on Equivalence or Non-Inferiority

The overarching goal of these trials is to demonstrate equivalence or non-inferiority to reference products. This requires a nuanced approach, involving the establishment of comparability margins grounded in clinical relevance. Striking a balance between demonstrating similarity and ensuring therapeutic efficacy and safety is crucial for the success of these trials.

4.2.4.5 Regulatory Considerations

The data generated from these trials play a crucial role in regulatory submissions. Regulatory agencies scrutinize the methodologies employed, ensuring that trials adhere to predefined standards and guidelines. Transparent reporting of outcomes and potential limitations is essential for regulatory decision-making.

4.2.4.6 Evolution of Biosimilar and Biosuperior Landscape

The outcomes of clinical trials contribute not only to regulatory approvals but also to the broader understanding of the biosimilar and biosuperior landscape. Successful trials establish the safety and efficacy of these products, fostering confidence among healthcare professionals and patients, facilitating broader adoption and market penetration. In essence, the meticulous orchestration of clinical trials for biosimilars and biosuperiors is integral to advancing therapeutic innovation. Through rigorous designs, ethical conduct, and insightful analyses, these trials shape the narrative of biopharmaceutical development, paving the way for enhanced treatment options and improved patient outcomes.

4.3 Biosimilars in the Treatment of Colorectal Cancer

GLOBOCAN 2020 reported 10 million deaths and 19.3 million new cancer-related cases have been reported. Colorectal cancer (CRC) accounts for 1.93 million (10%) of all cancer diagnoses and 935,000 cancer-related deaths (Hossain et al. 2022a; Arnold et al. 2017). As a result, colorectal cancer (CRC) is the second leading cause of cancer death and the third leading cause of cancer diagnosis. Cancer incidence and death rates may differ across countries and areas of the world. High-income nations have been shown to have lower death rates, while less developed nations have higher death rates (Hossain et al. 2022a; Araghi et al. 2019). Age-specific incidence rates for CRC patients in 2020 are 19.8 per 1000,000, whereby 16.2 are female and 23.4 are male (Sung et al. 2021; Bray et al. 2018). A total of 9.1 deaths per 1000 CRC patients have been reported in both genders, with 11 men and 7.2 women dying from the disease (Hossain et al. 2022a, b; Douaiher et al. 2017). By 2030, there will be a 60% rise in CRC cases, according to predictions. An estimated 1.1 million fatalities and 2.2 million instances of new cases are predicted as a result of sedentary lifestyles, obesity, processed food and alcohol consumption, and environmental changes. Thus, raising awareness of the risk factors linked to CRC is of the utmost significance (Bray et al. 2018; Wong et al. 2021; World Health Organization 2023) .

In high-income nations, the death rate is declining due to recent advancements in therapy. To enhance the prognosis of individuals having CRC, early diagnosis and effective excision of the primary tumor are essential (Koncina et al. 2020). Surgery is the most often used intervention among all current treatment options for CRC ranging from early (stage 0) to advanced (stage II) (Safiejko et al. 2022). Adjuvant treatment is utilized in patients with lymph node involvement (stage III) for other stages. The usual course of treatment for individuals in stage III is adjuvant therapy based on oxaliplatin (Winter and Eckhauser 2006; André et al. 2004).

Nonetheless, it is important to acknowledge the difficulties posed by contemporary therapy. Determining the purpose of laparoscopic resection in patients with shallow-level rectum cancer is a major surgical problem (Hossain et al. 2022a). However, almost all CRC patients have developed drug resistance, which has reduced the therapeutic effectiveness of chemotherapy and contributing its failure (Hu et al. 2016; Dallas et al. 2009). The multimodality approach, which incorporates surgery, chemotherapy, and biologics like bevacizumab, is the preferred choice of therapy considering the drawbacks of individual medication treatment (Koncina et al. 2020). Other biologics are panitumumab and cetuximab. The advent of biologics in the metastatic CRC along with chemotherapy improved the median survival rate of the patients. The choice of biologics depends on many vital factors like age, tumor burden, comorbidities, mutations in tumor, and possible side effects. A major financial strain is placed on healthcare systems by the high cost of biologics, which are being utilized extensively in both supportive and therapeutic care for cancer patients. Biologics patents expiring led to the development of more affordable and low-cost biosimilars. Since the Biologics Price Competition and Innovation Act has been enacted, more biosimilars have been licensed for the treatment of cancer and this number is expected to climb in the near future (Koncina et al. 2020). Biosimilar bevacizumab is the first-line therapy for metastatic cancer when combined with fluorouracil-based chemotherapy (Grapsa et al. 2015). Compared to small molecules, biosimilars are not perfect replicas but similar to active components and offer equivalent safety and efficacy as reference biologics (Xu et al. 2022).

BS have been evaluated in many clinical trials for CRC treatment. Amongst all, Bevacizumab-awwb biosimilar has been frequently screened for metastatic colorectal cancer as a first line agent (Jin et al. 2023; Rhodes et al. 2021). It has been found to be effective and safe in both males and females, and the results are comparable to that of reference Biosimilar (Pham et al. 2023). Since last two decades, epoetin alfa has been used to manage anemia in chemotherapy-associated bone marrow suppression. The effectiveness of epoetin alfa-epbx is so far not investigated for chemotherapy-induced toxicities (Itri 2002). The prophylactic treatment with Pegfilgrastim helped in management of severe neutropenia in metastatic colorectal cancer (Tamaki et al. 2022). Also, Pegfilgrastim-jmdb was found to possess protective effect and efficacious when administered on the same day as that of chemotherapy in colorectal cancer patients (Matera et al. 2021). Trastuzumab-dkst has its proven efficacy in HER-2 positive metastatic breast cancer, it is yet to be explored in HER-2 positive metastatic colorectal cancer (Rugo et al. 2021). Cetuximab in combination with FOLFIRI was associated with longer progression-free survival (PFS) and overall survival (OS) in clinical trials conducted in 2023 for wild-type metastatic CRC (Boige et al. 2023). List of important biosimilars used in oncology is shown in Table 4.1.

4.4 In Vitro and In Vivo Studies to Compare Critical Quality Attributes between Biologics and Biosimilar

The pharmaco-toxicological assessment of a biosimilar is necessary to distinguish it from reference biologics, as per the guideline of similar biologics and the WHO guideline on the evaluation of biosimilars. To assess the similarities between biosimilars and biologics, in vitro research must be conducted first, and if necessary, in vivo studies must be performed as well (Guidelines on Similar Biologics 2016; Anonymous 2022).

To assess the potential of a biosimilar, several bioassays are available, including assays for cell survival, receptor binding, and cell proliferation, together with the relevant quality data (Guidelines on Similar Biologics 2016). Using cutting-edge analytical techniques, the quality attributes of a biosimilar are thoroughly analyzed in the comparative quality research about those of its reference product (Ishii-Watabe and Kuwabara 2019). Comparing the therapeutic activity, post-translational

Table 4.1 List of biosimilars used in oncology (Colorectal Cancer Aalliance 2018; Bachu et al. 2022)

Reference biologic (active substance)	Reference biologic manufacturer(s)	Biosimilar (active substance)	Biosimilar manufacturer	Approval year
FDA approved supportive care Biosimilars				
Neupogen® (Filgrastim	Amgen Inc	Zarxio® (filgrastim-sndz)	Sandoz Inc	2015
		Nivestym® (filgrastim-aafi)	Hospira Inc	2018
Neulasta® (pegfilgrastim)	Amgen Inc	Fulphila® (pegfilgrastim-jmdb)	Mylan GmbH	2018
		Udenyca® (pegfilgrastim-cbqv)	Coherus BioSciences, Inc	2018
		Ziextenzo® (pegfilgrastim-bmez)	Sandoz Inc	2019
		Nyvepria® (pegfilgrastim-apgf)	Hospira Inc	2020
Epogen/Procrit (epoetin α)	Amgen Inc/ Janssen Biotech Inc	Retacrit® (epoetin alfa-epbx)	Hospira Inc	2018
EMA-approved supportive care biosimilars in oncology				
Neupogen® (Filgrastim)	Amgen Inc	Accofil®	Accord Healthcare S.L.U.	2014
		Filgrastim Hexal®	Hexal AG	2009
		Grastofil®	Accord Healthcare, SLU	2013
		Nivestim®	Pfizer Europe MA EEIG	2010
		Ratiograstim®	Ratiopharm GmbH	2008
		Tevegrastim®	Teva GmbH	2008
		Zarzio®	Sandoz GmbH	2009

(continued)

Table 4.1 (continued)

Reference biologic (active substance)	Reference biologic manufacturer(s)	Biosimilar (active substance)	Biosimilar manufacturer	Approval year
Neulasta® (pegfilgrastim)	Amgen Inc	Pelgraz®	Accord Healthcare S.L.U.	2018
		Fulphila®	Mylan S.A.S	2018
		Udenyca®	ERA Consulting GmbH	2018
		Pelmeg®	Mundipharma Corporation (Ireland) Limited	2018
		Ziextenzo®	Sandoz GmbH	2018
		Grasustek®	Juta Pharma GmbH	2019
		Cegfila®	Mundipharma Corporation (Ireland) Limited	2019
		Nyvepria®	Pfizer Europe MA EEIG	2020
Eprex®/Erypo® (epoetin α)	Janssen-Cilag GmbH	Abseamed®	Medice Arzneimittel Pütter GmbH Co. KG	2007
		Binocrit®	Sandoz GmbH	2007
		Epoetin- α hexal®	Hexal AG	2007
		Retacrit®	Pfizer Europe MA EEIG	2007
		Silapo®	Stada Arzneimittel AG	2007
FDA-approved mAB biosimilars in oncology				
Avastin® (Bevacizumab)	Genentech, Inc.	Mvasi® (bevacizumab-awwb)	Amgen Inc	2017
		Zirabev® (bevacizumab-bvzr)	Pfizer Inc.	2019
Rituxan® (Rituximab)	Genentech, Inc.	Truxima® (rituximab-abbs)	Celltrion, Inc	2018
		Ruxience® (rituximab-pvvr)	Pfizer Ireland Pharmaceuticals	2019
		Riabni® (rituximab-arrx)	Amgen Inc	2020

(continued)

Table 4.1 (continued)

Reference biologic (active substance)	Reference biologic manufacturer(s)	Biosimilar (active substance)	Biosimilar manufacturer	Approval year
Herceptin® (Trastuzumab)	Genentech, Inc.	Ontruzant® (trastuzumab-dttb)	Samsung Bioepis Co., Ltd	2019
		Trazimera® (trastuzumab-qyyp)	Pfizer Inc	2018
		Herzuma® (trastuzumab-pkrb)	Celltrion, Inc	2018
		Kanjinti® (trastuzumab-anns)	Amgen Inc	2019
		Ogivri® (trastuzumab-dkst)	Mylan GmbH	2017
EMA-approved mAB biosimilars in oncology				
Avastin® (bevacizumab)	Roche Registration GmbH	Mvasi®	Amgen Technology (Ireland) UC	2018
		Zirabev®	Pfizer Europe MA EEIG	2019
		Aybintio®	Samsung Bioepis NL B.V.	2020
		Equidacent®	Centus Biotherapeutics Europe Limited	2020
		Oyavas®	STADA Arzneimittel AG	2021
		Alymsys®	Mabxience Research SL	2021
MabThera® (rituximab)	Roche Registration GmbH	Blitzima®	Celltrion Healthcare Hungary Kft.	2017
		Truxima®	Celltrion Healthcare Hungary Kft.	2017
		Ruxience®	Pfizer Europe MA EEIG	2020
		Riximyo®	Sandoz GmbH	2017
		Rixathon®	Sandoz GmbH	2017
		Ritemvia®	Celltrion Healthcare Hungary Kft.	2017

(continued)

Table 4.1 (continued)

Reference biologic (active substance)	Reference biologic manufacturer(s)	Biosimilar (active substance)	Biosimilar manufacturer	Approval year
Herceptin® (Trastuzumab)	Roche Registration GmbH	Ontruzant®	Samsung Bioepis NL B.V.	2017
		Trazimera®	Pfizer Europe MA EEIG	2018
		Herzuma®	Celltrion Healthcare Hungary Kft.	2018
		Kanjinti®	Amgen Europe B.V.	2018
		Ogivri®	Mylan S.A.S	2018
		Zercepac®	Accord Healthcare S.L.U.	2020

modification, and protein structures of biosimilars and their reference biologics with clinical significance has been done using sensitive techniques that can detect even the smallest variations (Medicines Agency E n.d.). The biosimilar's binding affinity to the corresponding receptors is evaluated using the biologics' known activity. It is important to analyze studies pertaining to signal transduction and functional activity in order to compare biological activities (Anonymous n.d.-a).

When in vitro studies are not feasible, in vivo evaluation becomes crucial. The 3Rs approach (Reduce, Refine, Replace) must be considered while designing animal experiments to minimize the number of animals utilized in experimentation. According to ICH S6(R1) (Anonymous n.d.-b) and the WHO Guidelines on the quality, safety, and efficacy of biotherapeutic protein products made using recombinant DNA technology, only relevant species with known pharmacological and/or toxicological activities to the reference biologics should be used for in vivo studies (Anonymous 2022). It is necessary to undertake at least one repeat dosage toxicity study with an intended route of administration in a pharmacologically relevant species in order to qualify for in vivo toxicity studies (Guidelines on Similar Biologics 2016). In vivo examinations, which are carried out when it cannot be avoided, also include pharmacokinetic and dynamic studies, safety, local tolerance, and immunogenicity research in addition to toxicity studies (Anonymous 2022).

4.5 Potential Barriers and Benefits of Using Biosimilar

Biosimilars face numerous obstacles in the form of availability, affordability, and market competition. Delays may arise from strict regulatory criteria for biosimilar licensing (Moorkens et al. 2016). Numerous obstacles that innovators have with regard to patient infringement could potentially lead to the launch of any biosimilar product (Van de Wiele et al. 2021). Although there is a great hope that biosimilars will be economical and accessible, the manufacturer may have significant difficulties keeping the cost of the biosimilar stable. A manufacturer focusing on the same reference product may experience price erosion and a decline in profit margin rate if we discuss market rivalry (Cohen et al. 2014).

Biosimilars face numerous obstacles in the form of availability, affordability, and market competition. Delays may arise from strict regulatory criteria for biosimilar licensing (Rosen et al. 2017). Uncertainty regarding the effectiveness and safety of targeted therapy was another frequent obstacle to starting treatment (Monk et al. 2017). Therefore, targeted therapy may be influenced by factors other than affordability and genetic marker availability. Over half of the population cannot get targeted therapy due to its high cost (Melosky et al. 2018). When it comes to the advantages of biosimilars, they could provide better and more reasonably priced alternatives to necessary medications for people all over the world (Dutta et al. 2020). Based on the amount and quality of clinical trial designs, the results indicated that locally created cancer biosimilars approved by the NMPA were on par with those approved by the FDA, EMA, and PMDA. China mandated that all approved cancer biosimilars had a minimal risk-of-bias, equivalency, and double-anonymized design (Liu et al. 2023).

4.6 Conclusion

Biologics can be considered as generic version of biosimilars. The main limitations of complex immune reactivity, stringent pharmacovigilance, high development cost, accessibility and availability of biologics can be resolved by biosimilars. The pharmacokinetics and pharmacodynamics profile of biosimilars must be comparable to reference product while preserving safety and efficacy. Many regulatory bodies come with smooth legislation and regulation to promote the utilization of biosimilars for cancer treatment. CRC cases will rise due to drastic changes in the lifestyle and socioeconomic situation of the world population. The main therapeutic regimen of CRC become blunt while considering advanced stage of the disease due to its resistance, side effects, and toxicity. Clinical trials of biosimilars (Bevacizumab-awwb, pegfilgrastim-jmdb) used in limited number of CRC patients showed excellent safety and efficacy with respect to its reference product. But chronic therapeutic benefits, relapse, and application in variety of CRC stages need to be evaluated in large population to promote the use of biosimilars in the treatment of CRC.

References

AL-Sabbagh A, Olech E, McClellan JE, Kirchhoff CF (2016) Development of biosimilars. Semin Arthritis Rheum 45(5)

AMGEN Biosimilars (2024) The process and quality standards, AMGEN Biosimilars [Internet]. [cited 2024 Jan 9]. Developing BIOSIMILARS. https://www.amgenbiosimilars.com/

André T, Boni C, Mounedji-Boudiaf L, Navarro M, Tabernero J, Hickish T et al (2004) Oxaliplatin, fluorouracil, and leucovorin as adjuvant treatment for colon cancer. N Engl J Med 350(23):2343

Anonymous (2022) POST-ECBS version ENGLISH ONLY Guidelines on evaluation of biosimilars Replacement of Annex 2 of WHO Technical Report Series, No. 977

Anonymous (n.d.-a) Annex 2 guidelines on evaluation of monoclonal antibodies as similar biotherapeutic products (SBPs)
Anonymous (n.d.-b) International conference on harmonisation of technical requirements for registration of pharmaceuticals for human use preclinical safety evaluation of biotechnology-derived pharmaceuticals S6(R1)
Araghi M, Soerjomataram I, Jenkins M, Brierley J, Morris E, Bray F et al (2019) Global trends in colorectal cancer mortality: projections to the year 2035. Int J Cancer 144(12):2992
Arnold M, Sierra MS, Laversanne M, Soerjomataram I, Jemal A, Bray F (2017) Global patterns and trends in colorectal cancer incidence and mortality. Gut 66(4):683
Bachu RD, Abou-Dahech M, Balaji S, Boddu SHS, Amos S, Singh V et al (2022) Oncology biosimilars: new developments and future directions. Cancer Rep 5(11):1–20
Barbier L, Ebbers HC, Declerck P, Simoens S, Vulto AG, Huys I (2020) The efficacy, safety, and immunogenicity of switching between reference biopharmaceuticals and Biosimilars: a systematic review. Clin Pharmacol Ther 108(4):734–755
Barbier L, Mbuaki A, Simoens S, Declerck P, Vulto AG, Huys I (2022) Regulatory information and guidance on Biosimilars and their use across Europe: a call for strengthened one voice messaging. Front Med (Lausanne) 9:820755
Biosimilars Council (2024) How a Biosimilar is developed—Biosimilars Council [Internet]. [cited 2024 Jan 26]. https://biosimilarscouncil.org/resource/how-a-biosimilar-is-developed/
Boige V, Blons H, François E, Ben Abdelghani M, Phelip JM, Le Brun-Ly V et al (2023) Maintenance therapy with Cetuximab after FOLFIRI plus cetuximab for RAS wild-type metastatic colorectal cancer: a phase 2 randomized clinical trial. JAMA Netw Open 6(9):e2333533
Bray F, Ferlay J, Soerjomataram I, Siegel RL, Torre LA, Jemal A (2018) Global cancer statistics 2018: GLOBOCAN estimates of incidence and mortality worldwide for 36 cancers in 185 countries. CA Cancer J Clin 68(6)
Ciucci G, Colliva A, Vuerich R, Pompilio G, Zacchigna S (2022) Biologics and cardiac disease: challenges and opportunities. Trends Pharmacol Sci 43(11):894–905
Cohen JP, Felix AE, Riggs K, Gupta A (2014) Barriers to market uptake of biosimilars in the US. GaBI J 3(3):108
Colorectal Cancer Aalliance (2018) Biosimilars and colorectal cancer
Dallas NA, Xia L, Fan F, Gray MJ, Gaur P, Van Buren G et al (2009) Chemoresistant colorectal cancer cells, the cancer stem cell phenotype, and increased sensitivity to insulin-like growth factor-I receptor inhibition. Cancer Res 69(5):1951
Douaiher J, Ravipati A, Grams B, Chowdhury S, Alatise O, Are C (2017) Colorectal cancer—global burden, trends, and geographical variations. J Surg Oncol 115(5):619
Dutta B, Huys I, Vulto AG, Simoens S (2020) Identifying key benefits in European off-patent biologics and biosimilar markets: it is not only about price! BioDrugs 34:159
European Medicines Agency (2014) Similar biological medicinal products (overarching guideline). CHMP/437/04 Rev. 1, Vol 44, pp 1–7
FDA (2024a) What are "Biologics" questions and answers | FDA [Internet]. [cited 2024 Jan 26]. https://www.fda.gov/about-fda/center-biologics-evaluation-and-research-cber/what-are-biologics-questions-and-answers
FDA (2024b) Biosimilars | FDA [Internet]. [cited 2024 Jan 26]. https://www.fda.gov/drugs/therapeutic-biologics-applications-bla/biosimilars
FDA (2024c) Biosimilar and interchangeable biologics: more treatment choices | FDA [Internet]. [cited 2024 Jan 26]. https://www.fda.gov/consumers/consumer-updates/biosimilar-and-interchangeable-biologics-more-treatment-choices
FDA (2024d) Review and Approval | FDA [Internet]. [cited 2024 Jan 26]. https://www.fda.gov/drugs/biosimilars/review-and-approval
Grapsa D, Syrigos K, Saif MW (2015) Bevacizumab in combination with fluoropyrimidine-irinotecan-or fluoropyrimidine-oxaliplatin-based chemotherapy for first-line and maintenance treatment of metastatic colorectal cancer. Expert Rev Anticancer Ther 15(11):1267
Guidelines on Similar Biologics (2016) Regulatory requirements for marketing authorization in India Central Drugs Standard Control Organization Ministry of Health & Family Welfare

Government of India document name: GUIDELINES ON SIMILAR BIOLOGICS effective from year: 15th August, 2016 validity: till further revision 2 guidelines on similar biologics: regulatory requirements for marketing authorization in India content
Halimi V, Daci A, Netkovska KA, Suturkova L, Babar ZUD, Grozdanova A (2020) Clinical and regulatory concerns of Biosimilars: a review of literature. Int J Environ Res Public Health 17(16):1–17
Hossain MS, Karuniawati H, Jairoun AA, Urbi Z, Ooi DJ, John A et al (2022a) Colorectal cancer: a review of carcinogenesis, global epidemiology, current challenges, risk factors, preventive and treatment strategies. Cancers 14:1732
Hossain MS, Karuniawati H, Jairoun AA, Urbi Z, Ooi DJ, John A et al (2022b) Colorectal cancer: a review of carcinogenesis, global epidemiology, current challenges, risk factors, preventive and treatment. Strategies 14:Cancers
Hu T, Li Z, Gao CY, Cho CH (2016) Mechanisms of drug resistance in colon cancer and its therapeutic strategies. World J Gastroenterol 22:6876
Ishii-Watabe A, Kuwabara T (2019) Biosimilarity assessment of biosimilar therapeutic monoclonal antibodies. Drug Metab Pharmacokinet 34:64
Itri LM (2002) The use of epoetin alfa in chemotherapy patients: a consistent profile of efficacy and safety. Semin Oncol 29(3 Suppl 8):81–87
Jin R, Ogbomo AS, Accortt NA, Lal LS, Bishi G, Sandschafer D et al (2023) Real-world outcomes among patients with metastatic colorectal cancer treated first line with a bevacizumab biosimilar (bevacizumab-awwb). Ther Adv Med Oncol 15:1–13
Kesik-Brodacka M (2018) Progress in biopharmaceutical development. Biotechnol Appl Biochem 65(3):306–322
Kirchhoff CF, Wang XZM, Conlon HD, Anderson S, Ryan AM, Bose A (2017) Biosimilars: key regulatory considerations and similarity assessment tools. Biotechnol Bioeng 114(12):2696–2705
Koncina E, Haan S, Rauh S, Letellier E (2020) Prognostic and predictive molecular biomarkers for colorectal cancer: Updates and challenges. Cancers 12:319
Liu Y, Zhang X, Xu HF, Shi JH, Zhao YQ, Bin DL et al (2023) Real-world utilization, barriers, and factors associated with the targeted treatment of metastatic colorectal cancer patients in China: a multi-center, hospital-based survey study. Int J Public Health:68
Matera RM, Relias V, Wasif Saif M, Zucker B (2021) Safety and efficacy of same-day administration of pegfilgrastim in patients receiving chemotherapy for gastrointestinal malignancies. Cancer Med J 4(1):6
Medicines Agency E (n.d.) Biosimilars in the EU—Information guide for healthcare professionals
Melosky B, Reardon DA, Nixon AB, Subramanian J, Bair AH, Jacobs I (2018) Bevacizumab biosimilars: scientific justification for extrapolation of indications. Future Oncol 14:2507
Monk BJ, Lammers PE, Cartwright T, Jacobs I (2017) Barriers to the access of bevacizumab in patients with solid tumors and the potential impact of biosimilars: a physician survey. Pharmaceuticals 10(1)
Moorkens E, Jonker-Exler C, Huys I, Declerck P, Simoens S, Vulto AG (2016) Overcoming barriers to the market access of biosimilars in the European union: the case of biosimilar monoclonal antibodies. Front Pharmacol 7
Mysler E, Azevedo VF, Danese S, Alvarez D, Iikuni N, Ingram B et al (2021) Biosimilar-to-biosimilar switching: what is the rationale and current experience? Drugs 81(16):1859
Pathak A, Singh SP, Tiwari A, Tripathi AM, Jahan T, Singh DB (2023) Biosimilar, biobetter, and biosuperior therapeutic proteins. In Protein-based therapeutics, pp 325–53
Pfizer (2024) Biologics vs. biosimilars: understanding the difference | Pfizer [Internet]. [cited 2024 Jan 26]. https://www.pfizer.com/news/articles/biologics_vs_biosimilars_understanding_the_differences
Pham C, Niu F, Delate T, Buchschacher GL, Li Y, Ekinci E et al (2023) Real-world clinical outcomes of bevacizumab-awwb biosimilar versus bevacizumab reference product in patients with metastatic colorectal cancer. BioDrugs 37(6):891–899

Rhodes W, Declue RW, Accortt NA, Jin R, Sandschafer D, Wertz D et al (2021) Real-world use of bevacizumab-awwb, a bevacizumab biosimilar, in US patients with metastatic colorectal cancer. Future Oncol 17(36):5119–5127

Rosen LS, Jacobs IA, Burkes RL (2017) Bevacizumab in colorectal cancer: current role in treatment and the potential of Biosimilars. Target Oncol 12:599

Rugo HS, Pennella EJ, Gopalakrishnan U, Hernandez-Bronchud M, Herson J, Koch HF et al (2021) Final overall survival analysis of the phase 3 HERITAGE study demonstrates equivalence of trastuzumab-dkst to trastuzumab in HER2-positive metastatic breast cancer. Breast Cancer Res Treat 188(2):369

Safiejko K, Tarkowski R, Koselak M, Juchimiuk M, Tarasik A, Pruc M et al (2022) Robotic-assisted vs Standard laparoscopic surgery for rectal cancer resection: a systematic review and meta-analysis of 19,731 patients. Cancers 14:180

Sekhon B, Saluja V (2011) Biosimilars: an overview. Biosimilars 1:1–11

Sung H, Ferlay J, Siegel RL, Laversanne M, Soerjomataram I, Jemal A et al (2021) Global cancer statistics 2020: GLOBOCAN estimates of incidence and mortality worldwide for 36 cancers in 185 countries. CA Cancer J Clin 71(3):209

Tamaki S, Ishikawa H, Suzuki K, Kimura Y, Maemoto R, Abe I et al (2022) Prophylactic use of pegfilgrastim enables the management of severe neutropenia without dose delays in patients with metastatic colorectal cancer treated with TAS-102 plus bevacizumab. Mol Clin Oncol 16(5):103

Van de Wiele V, Kesselheim A, Sarpatwari A (2021 Aug) Barriers to US biosimilar market growth: lessons from biosimilar patent litigation: article examines barriers to US biosimilar market growth. Health Aff 1(40):1198–1205

Winter JM, Eckhauser FE (2006) Review of capecitabine as adjuvant treatment for stage III colon cancer. Am J Oncol Rev 5

Wong MCS, Huang J, Lok V, Wang J, Fung F, Ding H et al (2021) Differences in incidence and mortality trends of colorectal cancer worldwide based on sex, age, and anatomic location. Clin Gastroenterol Hepatol 19(5):955

World Health Organization (2023) World health statistics 2023: monitoring health for the sdgs, sustainable development goals. Vol. 27, The Milbank Memorial Fund quarterly

Xu X, Zhang S, Xu T, Zhan M, Chen C, Zhang C (2022) Efficacy and safety of bevacizumab Biosimilars compared with reference biologics in advanced non-small cell lung cancer or metastatic colorectal cancer patients: a network meta-analysis. Front Pharmacol 13:880090

Biosimilar in Cervical Cancer

5

Saurabh Morparia, Vasanti Suvarna,
and Manikanta Murahari

Abstract

Cervical Cancer is a serious health concern possessing a serious health problem in women. This chapter provides a comprehensive overview of the role of biosimilars in Cervical Cancer (CC) by highlighting the global public health issue posed by CC and the urgent need for affordable and accessible treatment options. Biosimilars are biological products that have the potential to reduce costs and increase accessibility in the treatment of various conditions. Their emergence is a significant development in medicine. The increasing number of approved biosimilars for CC treatment, such as bevacizumab (Mvasi, Zirabev) and pegfilgrastim (Fulphila, Udenyca, Ziextenzo), is being discussed, along with their rigorous testing through controlled clinical trials and real-world data. This chapter focuses on the comprehensive pharmacovigilance studies that are crucial for ensuring the safety and efficacy of biosimilars. It explores the profound implications of the widespread use of biosimilars in CC treatment, such as improved treatment coverage and the development of personalized treatment strategies. The chapter concludes by highlighting the revolutionary potential of biosimilars. It emphasizes the necessity for persistent research and development endeavors to pave the way for a new phase in CC management. This phase is anticipated to be characterized by enhanced accessibility, individualized care, and a promising future for women across the globe.

S. Morparia · V. Suvarna
Department of Pharmaceutical Analysis, Dr. Bhanuben Nanavati College of Pharmacy, Mumbai, Maharashtra, India

M. Murahari (✉)
Department of Pharmacy, Koneru Lakshmaiah Education Foundation, Vaddeswaram, Andhra Pradesh, India

S. Bhatt et al. (eds.), *Biosimilars for Cancer Treatment*,
https://doi.org/10.1007/978-981-97-3001-8_5

Keywords

Biosimilars · Biologics · Cervical cancer · Monoclonal antibodies · Regulatory guidelines

Abbreviations

CC	Cervical cancer
EMA	European Medicines Agency
EPO	Human erythropoietin
FDA	Food and Drug Administration
HPV	Human papilloma virus
WHO	World Health Organization

5.1 Introduction

Cervical cancer, the fourth most frequent gynaecological cancer, is a serious health issue for women, particularly in low- and middle-income countries, despite the significant risk of complications (Buskwofie et al. 2020; Hull et al. 2020; Koh et al. 2019). The lack of robust screening protocols for cervical cancer remains a critical public health issue. The World Health Organization (WHO) estimates that approximately 570,000 individuals diagnosed with cervical cancer in 2018, resulting in approximately 311,000 cases (World Health Organization n.d.). Constant infection with high-risk subtypes of the human papillomavirus (HPV) is the primary etiological cause for the development of cervical cancer (Li et al. 2018). Late-stage diagnosis is common when the illness progresses in its early stages. The most widely used cytology-based screening techniques for CC screening in clinical practice are the Papanicolaou test (Pap smear) and HPV testing (Tekalign and Teshome 2022; Keeratichamroen et al. 2020). However, there are numerous drawbacks to these techniques, such as low specificity levels, invasiveness, sensitivity problems, and patient pain. Due to the numerous potential treatment options for CC, early identification is essential. These therapies, which may have a number of negative consequences and not always result in a recovery, include radiation, chemotherapy, surgery, and/or a combination of these. The need for novel prognostic, diagnostic, and therapeutic approaches is emphasized by the poor prognosis and inadequate therapy in the advanced stages of CC (Du et al. 2020; Šarenac and Mikov 2019).

A biosimilar is a biological product that is developed to closely mimic an already-approved biological medication, which is sometimes known as the originator medication. It is derived or acquired from a biological source, such as bacteria or yeast. In terms of safety, quality, and efficacy, there are no clinically discernible differences between the biosimilar and the original product. The proper development of every biosimilar is essential for an accurate assessment of the absence of substantial discrepancies. In terms of structural composition, biological activity, efficacy, safety, and immunogenicity profiles, biosimilars demonstrate comparability to reference medicines. It is crucial that there are no appreciable variations in the biosimilar's safety, purity, or efficacy when compared to its reference product. Biosimilars are complex compounds that either originate from or are generated by microorganisms. Biosimilars undergo a more comprehensive assessment by the FDA compared to their name-brand counterparts. These goods are authorized reproductions of the original "pioneer" products. Biosimilars are often cheaper than name-brand drugs. This can help patients save money and get better access to biological therapies. Biosimilars possess numerous advantages and disadvantages in the healthcare industry (FDA n.d.-a; American Cancer Society n.d.). Figure 5.1 depicts the definition of biosimilar as per various regulatory agencies.

Biosimilars can lead to cost savings and increase patient access to biological therapies due to their lower prices than their reference biologics. Biosimilars have the potential to provide life-saving medications and enhance treatment options. The introduction of biosimilars into the healthcare system promotes competition, which may lead to price reductions for the reference product and the biosimilar. The benefits of biosimilars are depicted in Fig. 5.2 (FDA n.d.-b). Due to their enhanced stability compared to chemical-based medications, biosimilars not only possess a longer shelf life but also require cold chain distribution, which subsequently increases the complexity and cost of distribution. In contrast to chemically synthesized generics, the financial burden associated with the development of biosimilars is considerably greater. Certain studies suggested that biosimilars may exhibit significant differences in efficacy relative to the original product. The intricate molecular composition of biosimilars could potentially lead to additional safety concerns due to immunogenicity. The biosimilar industry faces numerous challenges in the

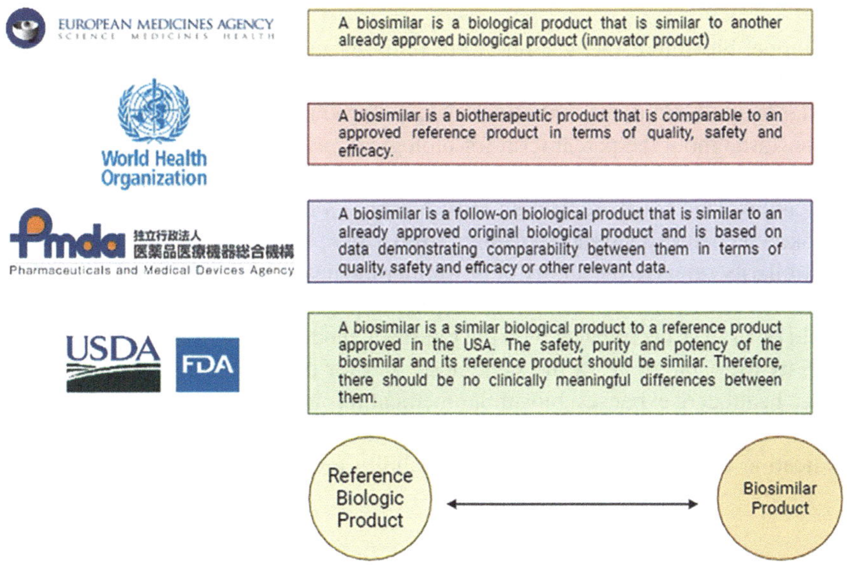

Fig. 5.1 It depicts the definition of biosimilar as per various regulatory agencies

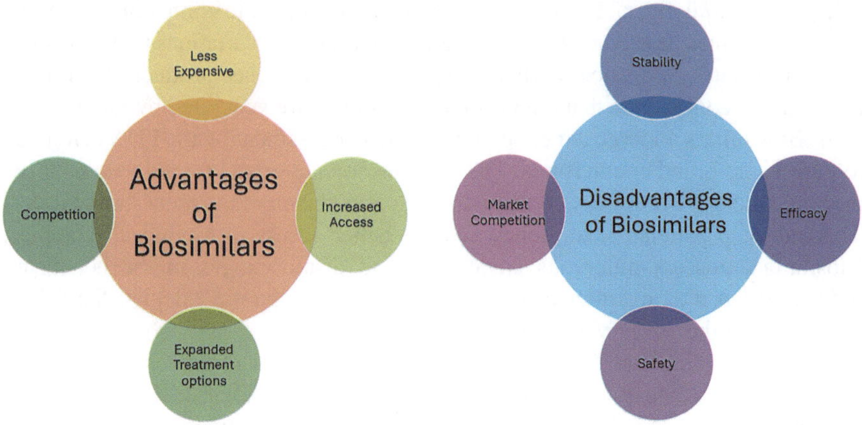

Fig. 5.2 Advantages and disadvantages of Biosimilar

production and marketing of these complex pharmaceuticals. The licensing process is frequently extended due to the varying regulatory systems in developing countries, which leads to an increase in the number of trials and a subsequent rise in research costs for biosimilars. This underscores the necessity for a more unified global regulatory strategy to accelerate the approval process and decrease development expenses. Figure 5.2 depicts the disadvantages of biosimilars (European Medicine Agency n.d.).

5.2 Development Process of Biosimilar Medicines

In Europe, biosimilar medications were initially manufactured in 2005 when the initial biological drug patent expired (Biosimilar Competition in Europe n.d.). Similar to synthetic drugs, only the biological producer can produce the same biosimilar throughout the patent term for biological medications. Upon the expiration of the patent, a different manufacturer has the potential to develop the biosimilar medication. This novel medication must possess an active ingredient that has pharmacokinetic and pharmacodynamic similarities to the reference biological. Biosimilarity refers to the ability of biosimilar medications to exhibit quality, safety, and effectiveness comparable to the original biological. The biosimilar medication should be used in accordance with the reference medication indications as it will be taken by the same route, at the same dose, and for the same disorders. In order to reduce healthcare expenses, biosimilar medications are used. Regulatory guidelines for biosimilar medicines have been shown in Table 5.1. The approval of a biosimilar medication undergoes a series of comparability tests. These tests are designed to

Table 5.1 Regulatory guidelines for biosimilar medicines

Regulatory agencies	European Medicine Agency (EMA)	US Food and Drug Administration (USFDA)	World Health Organization (WHO)
General	Extrapolation is permissible when supported by sound scientific reasoning, is predicated on thorough biosimilar data, and takes safety and effectiveness variations into account when developing a therapeutic indication.	It takes a sufficient level of scientific expertise to extrapolate. The most sensitive indication for efficacy and safety testing should be used to identify any notable variations in immunogenicity and effectiveness as well as safety.	A thorough benefit/risk analysis is required if extrapolation is to be done. Effectiveness and safety tests must be carried out with the utmost care.
Mechanism of action	The reference product contains numerous active sites or intracellular signalling pathways and interacts with various receptors.	All actions and functions involve targets or receptors. Consider target interactions, product structure, binding, dosage response, molecular signalling, and target expression/location.	The implicated receptors and/or the clinically significant mechanism of action.
Safety/ efficacy	Reasoning is needed for various applications of immunogenicity, considering patient characteristics including co-medications, comorbidities, immunological condition, and disease aspects such tumour cell lysis.	It is important to consider many factors that might impact safety or effectiveness, such as pharmacokinetics, biodistribution, immunogenicity in different populations, and expected toxicities.	There should not be any fresh safety concerns because immunogenicity and safety have already been thoroughly demonstrated. Assess immunogenicity in individuals belonging to high-risk groups.

confirm that any discrepancies between the biosimilar and its reference biological drug do not compromise the therapeutic activity. The biosimilar medication is equipped to manage the same conditions as its reference biological drug. On the other hand, biosimilar may only be applicable for a restricted range of clinical diseases that the reference medication can be treated. This implies a significant difference between biosimilars and their reference biological drugs. Regulatory authorities routinely conduct inspections of the biosimilar manufacturers to ensure compliance with specific manufacturing standards. This stringent regulation is aimed at preserving the quality and effectiveness of biosimilar medications, thereby protecting public health (European Medicine Agency n.d.). The development of biosimilar medications is depicted in Fig. 5.3.

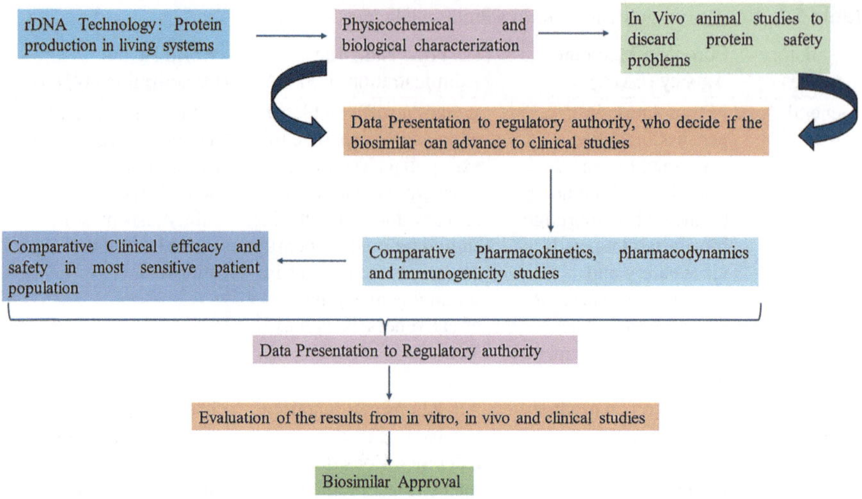

Fig. 5.3 Development process of Biosimilar medicines

5.3 Monoclonal Antibody

As an alternative to MabThera1 (reference biological medication), the EMA has now authorized three biosimilar monoclonal antibodies for use in cancer therapy: bevacizumab, trastuzumab, and rituximab.

5.3.1 Bevacizumab

The FDA has authorized bevacizumab-adcd, a biosimilar of bevacizumab, to treat six different forms of cancer. This approval was based on a phase 3 trial that evaluated its effectiveness against bevacizumab (Avastin) in patients with recurrent or metastatic non-squamous NSCLC. The biosimilar demonstrated comparable outcomes to bevacizumab in terms of pharmacokinetics, safety profile, and effectiveness. Bevacizumab-adcd, an anti-cancer monoclonal antibody, binds to VEGF and prevents it from attaching to endothelial cell receptors. It has been proven to be safe and effective in various illness areas, including cancer. In a double-blind, randomized study, patients received intravenous bevacizumab-adcd at a dose of 15 mg/kg every 3 weeks until the illness progressed. The primary objective of the study was the response rate at 21 weeks (Celltrion USA 2022). In August 2022, the European Commission authorized bevacizumab-adcd for use in individuals with various cancer diseases. This approval of bevacizumab-adcd represents a significant advancement in increasing access to cancer treatment. Bevacizumab, a recombinant humanized monoclonal antibody, targets VEGF, a molecule that promotes the growth of endothelial cells and the development of new blood vessels. The EMA authorized Mvasi1, a bevacizumab biosimilar, as a substitute for Avastin1, the reference drug. Peraza et al. investigated the possibility of a biosimilar bevacizumab receiving clinical approval. The researchers examined the molecular structure similarities between the two drugs and sought evidence of comparable biological action in vitro. Nonclinical in vivo toxicity experiments were conducted to determine the toxicity of the bevacizumab biosimilar and the reference. The authors concluded that both the reference biological drug and the bevacizumab biosimilar demonstrated comparable efficacy (Xu et al. 2022; European Medicines Agency n.d.-a).

The U.S. Food and Drug Administration (FDA) has granted approval for Mvasi (bevacizumab-awwb), a biosimilar to Avastin (bevacizumab), for the management of multiple types of cancer. This approval of Mvasi marks a significant milestone as it is the first biosimilar to be granted in the United States for cancer treatment. It is indicated for adults with specific malignancies, including those of the brain, kidney, colon, lung, and cervical regions (FDA n.d.-c). Mvasi can be used as a primary or secondary treatment when combined with intravenous 5-fluorouracil-based chemotherapy. However, it is not recommended as an adjuvant treatment after colorectal cancer has been surgically removed. In second-line treatment for patients with metastatic colorectal cancer who have progressed on a regimen involving bevacizumab products, Mvasi is combined with either fluoropyrimidine-irinotecan or fluoropyrimidine-oxaliplatin-based chemotherapy. Furthermore, Mvasi is not advised as an adjuvant treatment after colorectal cancer has been surgically removed. Mvasi is used in conjunction with carboplatin and paclitaxel for the first-line treatment of unresectable, locally advanced, recurrent, or metastatic non-squamous non-small cell lung cancer. Mvasi is prescribed for glioblastoma patients whose condition worsens after previous therapy, contingent on the increase in the objective response rate. However, there is a lack of evidence indicated that bevacizumab improves survival or disease-related symptoms. Mvasi and interferon alfa can be used for the treatment of metastatic renal cell cancer, Mvasi and interferon alfa are used. It is also used in conjunction with either paclitaxel and topotecan or paclitaxel and cisplatin for the treatment of persistent, recurrent, or metastatic cervical cancer. Various biosimilar products and their corresponding original biologics are shown in Table 5.2.

The approval of Mvasi provides an innovative treatment option for patients diagnosed with these particular cancers, which offers them a resurgence of hope.

Table 5.2 Biosimilar products and their corresponding original biologics

Epoetin				Monoclonal antibody		
Alpha (Epogen)	Zeta	Filgrastim (Neupogen)	Pegfilgrastim (Neulasta)	Rituximab (Rituxan)	Trastuzumab (Herceptin)	Bevaciznab (Avastin)
Abseamed Binocrit Hexal Retacrit	Retacrit Silapo	Hexal Nivestim Ratiograstim Tevagrastim Zarzio Udenyca Releuko	Fulphila Udenyca Ziextenzo Nyvepria Fylnetra Stimufend	Ritemvia Rituzema Rixathon Riximyo Truxima Blitzima Riabni Ruxience	Kanjinti Ontruzant Trazimera Herzuma Ogivri	Mvasi Zirabev Alymsys Vegzelma

Biosimilars, such as Mvasi, will be a major factor in the evolution of cancer treatment, thus enabling a paradigm shift in oncological care. Biological products, which are predominantly derived from living organisms such as humans, animals, microbes, or yeast, have caused significant changes in the treatment of numerous critical and potentially fatal diseases. A biosimilar is a type of biological product that is approved based on evidence showing its close resemblance to a previously approved biological product. The key criteria for approval include the absence of clinically meaningful differences in terms of safety, purity, and potency, that is, efficacy and safety, when compared to the reference product. Mvasi, a biosimilar of Avastin, has been approved by the FDA as one such medication. The approval process involved a meticulous review of a broad spectrum of evidence, including data from animal studies, human pharmacokinetic and pharmacodynamic studies, clinical immunogenicity studies, and other clinical safety and efficacy data. Comprehensive structural and functional characterization was also included in the review process. Collectively, these data elements demonstrated the comparability between Mvasi and Avastin. It is important to underscore that Mvasi has been approved as a biosimilar, not as a substitute medication. This distinction is crucial in understanding the role of biosimilars as complementary to existing treatments rather than as replacements (FDA n.d.-c). Mvasi, similar to other drugs, is associated with potential side effects. These commonly include epistaxis, hypertension, rhinitis, proteinuria, distorted gustatory perception, dry skin, haemorrhage, lacrimation disorder, back pain, and exfoliative dermatitis. It is essential to be aware of these potential adverse reactions when considering Mvasi. Severe side effects of the drug can manifest as organ perforation, fistula, thromboembolic events, hypertension, posterior reversible encephalopathy syndrome, proteinuria, infusion-related reactions, and ovarian failure. These adverse reactions represent significant risks associated with the medication. In the event of severe or life-threatening side effects, termination of Mvasi is advised. It is contraindicated in pregnant women due to potential foetal harm. The emergence of biosimilars such as Mvasi signifies a substantial progression in biological products, which provides a novel therapeutic alternatives and potentially contribute to healthcare cost efficiency (FDA n.d.-c). Mvasi, similar to Avastin, is accompanied by a Boxed Warning on its label. This warning is designed to alert both patients and healthcare professionals to the heightened risk of certain severe or potentially fatal conditions. These conditions encompass haemorrhage in various parts of the body such as the lungs, gastrointestinal tract, central nervous system, and vagina; complications related to surgery and wound healing; and gastrointestinal perforations, which are holes that can occur in the stomach or intestines. In the occurrence of a gastrointestinal perforation, it is strongly advised that patients discontinue the use of Mvasi immediately. This condition is perilous and necessitates immediate medical intervention. It is recommended that Mvasi should not be administered to patients until the surgical wound has completely healed, which typically occurs 28 days before and after elective surgery. Furthermore, if a surgical incision ruptures, a condition referred to as wound dehiscence, patients should cease the intake of Mvasi. Patients who are experiencing severe bleeding or are coughing up blood, a condition known as haemoptysis, should also be given a warning. Due to the potential fatality of these conditions, immediate medical attention is essential (FDA n.d.-c). Amgen, Inc. is the manufacturer of Mvasi, which has FDA approval. Avastin, on the other hand, is manufactured by Genentech, Inc. and was authorized in February 2004. As a division of the U.S. Department of Health and Human Services, the FDA is responsible for monitoring the efficacy, security, and safety of medical devices, vaccines, and other biological products intended for human use as well as pharmaceuticals for humans and animals. Together with regulating tobacco products, the agency is also in charge of the safety and security of the country's food supply, cosmetics, nutritional supplements, and items that emit radiation from electronics. The FDA strict procedures and criteria are demonstrated by the approval of biosimilars like Mvasi (FDA n.d.-c).

5.3.2 Rituximab

There are six biosimilars to rituximab that may be purchased in Europe: Rituzema1, Ritemvia1, Rixathon1, Riximyo1, Truxima1, and Blitzima1 (Jurczak et al. 2019). Rituximab is a chimeric anti-CD20 monoclonal antibody that is composed of human and mouse DNA, which was developed by genetic engineering. This antibody depletes B cells and is used to treat B cell malignancies, including autoimmune diseases (granulomatosis with polyangiitis, rheumatoid arthritis, and microscopic polyangiitis) and several forms of haematological cancer (follicular lymphoma, diffuse large B cell lymphoma, non-Hodgkin lymphoma, chronic lymphocytic leukaemia, and Burkitt lymphoma). Several studies have been conducted to assess the effectiveness of rituximab biosimilars in cancer treatment when compared to the reference biological (Jacobs et al. 2017; Annex 1999). To ensure that handling and formulation procedures would not interfere with the effectiveness and safety of the product, Lamanna et al. investigated the unique in-use stability of Rixathon1/Riximyo1. The findings indicated that the biosimilars' physicochemical stability, which includes its shelf life, temperature, exposure to light, dilution factor, and storage was not affected by formulation dilution for intravenous infusion (Lamanna et al. 2019). Jurczak et al. use a rituximab biosimilar in conjunction with other chemotherapeutic drugs (cyclophosphamide, vincristine, and prednisone) for treating untreated advanced follicular lymphoma. In the 24-week Phase 3 clinical trial conducted by the authors, the parameters of effectiveness, safety, pharmacokinetics, pharmacodynamics, and immunogenicity were evaluated. The results demonstrated a comparable profile to the reference biological product. This suggested that the investigational product exhibits similar therapeutic and safety characteristics, thereby potentially expanding treatment options for patients. Further research is warranted to confirm these findings in larger patient populations and over extended periods. It is also crucial to monitor post-marketing surveillance data to ensure the continued safety and efficacy of the product in real-world settings. The study demonstrated the safe and efficacious use of biosimilars in clinical practice. The study concluded that using rituximab biosimilars provides a therapeutically beneficial and more affordable option than using rituximab (Jurczak et al. 2017).

Candelaria et al. (Candelaria et al. 2018) evaluated possible variations in the reference MabThera1 and a rituximab biosimilar pharmacokinetic and pharmacodynamic profiles in patients with diffuse large B cell lymphoma. The outcomes demonstrated that the biosimilar systemic exposure and variability were comparable to those of the reference. In addition, the investigated medications did not significantly differ in terms of B cell CD_{20}^{+} depletion. Caldeira et al. (Franceschetti and Caldeira 2018) recently published the findings of a research pertaining to the non-Hodgkin lymphoma treatment regimen using rituximab biosimilars. The assessments were conducted in five EU member states between July and September 2017. Despite the established efficacy of biosimilars, it has been observed that the frequency of their preparation surpasses their actual utilization.

Rituximab is a chimeric murine-human monoclonal antibody (mAb) that targets the CD20 antigen on B cells (Bosch et al. 2014). It is sold under the names Rituxan® in the United States by Genentech/Biogen Idec and MabThera® in the European Union by Roche. Its approval for anti-inflammatory and anticancer purposes makes it special. Adult patients with non-Hodgkin's lymphoma, chronic lymphocytic leukaemia, rheumatoid arthritis, granulomatosis with polyangiitis, and microscopic polyangiitis may benefit from rituximab in conjunction with glucocorticoids. The final composition of matter patent in the United States (Rituxan) will expire in 2018, while the composition of matter patent for rituximab sold in Europe (MabThera) expired in 2013. This has opened the door for a number of rituximab biosimilars, which are presently undergoing late-stage clinical development. Clinical trials that compare rituximab biosimilars are carried out for a number of indications, which include follicular lymphoma, diffuse large B-cell lymphoma, and rheumatoid arthritis. Rituximab indications may be extended from non-malignant conditions (such as rheumatoid arthritis) to cancer, from autoimmune diseases to oncology, from single-agent to combination therapy, or, if scientifically warranted, from combination to single-agent therapy (Bosch et al. 2014; Pierpont et al. 2018; Kirsch-Stefan et al. 2023).

The Russian Ministry of Health recently authorized CellbiaTM [Biocad, Saint Petersburg, Russia] as a biosimilar of rituximab. The biosimilars currently available in Russia have not adhered to the strict criteria established by regulatory bodies such as the FDA, EMA, or WHO for demonstrating bio similarity (Kirchhoff et al. 2017; Kurki et al. 2022). The possibility of medication shortages, which may have a detrimental impact on cancer patients' lives if the supply of the reference product could not meet demand, may be reduced with the approval of biosimilar monoclonal antibodies. The extrapolation-based fast development programme for biosimilar monoclonal antibodies may mitigate some of the possible reasons for medication shortages, such as production-related problems and rising demand. The provision of biosimilars that are safe, effective, and of high quality, produced by reputable manufacturers according to the strict guidelines set out by the FDA, EMA, or WHO, might potentially avert a scarcity of pharmaceuticals while also ensuring that patients have more access to biologic medicines.

5.3.3 Trastuzumab

A humanized monoclonal antibody called trastuzumab blocks the human epidermal growth factor receptor 2 (HER-2). HER-2 overexpression inhibits cell growth in several types of breast cancer. Moreover, trastuzumab improves outcomes for HER-2-positive breast tumours in both the metastatic and early stages of the illness when administered in conjunction with other chemotherapeutic drugs. However, only a small percentage of patients can afford this medication due to financial limitations (Blackwell et al. 2018; Paplomata and Nahta 2018). Due to the cost of these medications, the introduction of trastuzumab biosimilars to the market may enable more patients to receive this treatment. Currently, the EMA has authorized Ontruzant1 (European Medicines Agency n.d.-b), Kanjinti1 (European Medicines Agency n.d.-c), Trazimera1 (European Medicines Agency n.d.-d), and Herzuma1 (European Medicines Agency n.d.-e) as trastuzumab biosimilars of Herceptin1. Numerous investigations have been conducted to evaluate the effectiveness and safety of biosimilar trastuzumab. Pivot et al conducted the first human clinical Phase 1 trial assessing the pharmacokinetic equivalency of a trastuzumab biosimilar candidate against the reference drug Herceptin1 in the first human clinical Phase 1 trial (Pivot et al. 2018). Healthy male volunteers participated in the trial and were randomly assigned to receive a single intravenous medication at a level of 6 mg/kg. The outcomes demonstrated that, with a single dose, the biosimilar candidate was well-trained and posed no safety risks. The safety, immunogenicity, and survival results following neoadjuvant-adjuvant treatment in patients with locally progressed or early stage HER-2-positive breast cancer was documented in a Phase 3 clinical study. The study findings showed that the reference biological candidate and the trastuzumab biosimilar contender were identical (Pivot et al. 2018). Recently, Waller et al. reported the findings of a Phase 1 investigation. Male volunteers in good condition participated in the study. The purpose of the study was to contrast a biosimilar trastuzumab with the original biological. The outcomes demonstrated that the reference biological and the trastuzumab biosimilar were comparable. Safety and pharmacokinetic bioefficiency were observed to be similar. The biosimilar was administered intravenously at a dosage of 8 mg/kg per dose. The infusion process took approximately 90 minutes to complete (Waller et al. 2018).

Humanized recombinant monoclonal antibody (mAb) trastuzumab, often referred to as Herceptin®, targets the human epidermal growth factor receptor 2 (HER2). It is particularly effective in the adjuvant and metastatic phases of HER2-positive breast cancer. Trastuzumab is advocated for the therapeutic intervention of metastasized gastric or gastroesophageal junction adenocarcinoma that exhibits HER2-positivity. The composition of matter patent for trastuzumab, marketed as Herceptin® by Roche, was discontinued in Europe in 2014 (Piezzo et al. 2021). In the United States, the final composition of matter patent for Genentech Herceptin® was discontinued in 2019 (FDA n.d.-d). Following patent expiration, biosimilars of trastuzumab have been formulated and are currently undergoing advanced clinical trials. Currently, two biosimilars of Trastuzumab have been granted global authorization.

The first one, Herzuma®, is made by Celltrion in South Korea and has been certified by the Korean Ministry of Food and Drug Safety. The second biosimilar is available for purchase in India under the brand names Biocon by Biocon and HertrazTM by Mylan, both of which have been authorized by the Drugs Controller General of India. The reference product, Herceptin, has all indications for which these biosimilars have been approved. The World Health Organisation (WHO), the European Medicines Agency (EMA), and the U.S. Food and Drug Administration (FDA) have specified strict regulatory standards for establishing biosimilarity, which may not be met by the biosimilars licenced in India (FDA n.d.-e; Herzuma n.d.).

5.4 Epoetin

Biosimilars based on epoetin are substitutes for reference biological medications, namely Erypo1 and Eprex1. Based on the type of protein it contains, there are five biosimilars: Abseamed1, Hexal1, Binocrit1, and Retacrit1 have alpha epoetin, while Zeta epoetin is present in Silapo1 and Retacrit1 (Schellekens 2009; Deechongkit et al. 2006; Konstantinidou et al. 2019; Minghetti et al. 2012; Aapro et al. 2018; Goldsmith et al. 2018). Human erythropoietin (EPO), a hormone secreted by the kidneys that promotes the production of red blood cells in the bone marrow, is mirrored in synthetic epoetin. Under normal physiological settings, individuals receiving chemotherapy or those with renal diseases may have altered or decreased EPO production, putting them at risk of anaemia. Alpha or zeta epoetin is used to treat certain disorders, saving patients from needing blood transfusions. The European Medicines Agency (EMA) authorized the first two EPO biosimilars in 2007 after Phase 1 and Phase 2 clinical studies showed them to be comparable to Eprex1/Erypo1. Studies conducted after approval revealed significant structural variations (particularly in the glycosylation profiles) between the reference medications and these biosimilars. Additionally, suggestions were made for a more thorough examination of the information gathered from clinical practice. Oncologists were originally concerned about the same therapeutic effectiveness of EPO biosimilars in comparison to the reference biological because of these facts. Furthermore, concerns regarding safety have been expressed, encompassing the potential risk that the use of EPO biosimilars may lead to patients developing a condition known as pure red cell aplasia. This concern has led to a degree of hesitancy among some practitioners in prescribing these biosimilars, despite their potential to reduce treatment costs and enhance patient accessibility to biological therapy. This reluctance persists even in the face of the potential benefits these biosimilars could provide to the healthcare system. Thus, the balance between potential risks and benefits remains a significant consideration in the clinical adoption of EPO biosimilars. The growing number of pharmacovigilance studies have demonstrated that these biosimilars are effective, effective treatments for renal anaemia, and disorders related to chemotherapy-related side effects. As more data is available, it will be easier to dismiss early concerns and see EPO biosimilars more widely used in clinical settings.

5.5 Filgrastim

Granulocyte colony-stimulating factor, a type of cytokine, is a recombinant variant of filgrastim. This glycoprotein is commonly administered to individuals suffering from neutropenia following chemotherapy, owing to its ability to stimulate the production of neutrophils in the bone marrow. Filgrastim is also prescribed to patients with leukaemia and chronic neutropenia who have undergone invasive bone marrow therapies, as a preparatory measure before receiving a transplant. Neupogen1, the reference biological drug, has five biosimilar versions of filgrastim available on the European market. These include Zarzio1, Hexal1, Nivestim1, Ratiograstim1, and Tevagrastim1. However, several other biosimilars have been withdrawn from the market. The reasons for these withdrawals are multifaceted, often encompassing factors such as regulatory hurdles, market dynamics, and clinical considerations. This emphasizes the complexity of the biosimilar landscape and the need for ongoing research and regulatory vigilance (Rahi et al. 2022; Yang and Kido 2011; Moorkens et al. 2017; Gravel et al. 2012; Cornes et al. 2020).

Neupogen1, Zarzio1 was authorized in the European Union for the same indications as its reference medication and was a filgrastim biosimilar. Comparability experiments were carried out by Sorgel et al. using Zarzio1 and Neupogen1. The primary, secondary, and tertiary structures, as well as the purity profiles of both molecules, were demonstrated to be comparable by physical–chemical testing. In vitro, by examining the G-CSF-induced proliferation of murine myelogenous leukaemia cells and the affinity for receptor binding, biological comparison was demonstrated. In vivo experiments were conducted on healthy individuals during Phase 1 and Phase 3, respectively. The Phase 1 trials demonstrated that Zarzio1 and Neupogen1 had similar pharmacodynamics and pharmacokinetics. Phase 3 studies demonstrated effectiveness in treating breast cancer without causing an immune response. Comparable studies were conducted using a similar biosimilar filgrastim (EP2006), which demonstrated a high degree of similarity with both the originator filgrastim approved by the European Union and the United States. Skrlin et al. assessed the biosimilar Nivestim1 resemblance to the reference Neupogen1 in a different investigation. The results indicated that the molecular properties, purity, and biological activity of both medications were comparable. Patients now have more options for their treatments due to the discovery and approval of filgrastim biosimilars, which may lead to financial savings and improved access to biological therapy (Cornes et al. 2020; Halim et al. 2018; Sörgel et al. 2015; Nakov and Zesch 2016).

The European Medicines Agency recently approved Udenyca1 (pegfilgrastim) as a biosimilar to the originator of Neulasta1. This is a long-acting version of filgrastim, extending the range of treatment options available to patients. Nakov et al. conducted a randomized, double-blind, two-way crossover study in healthy male and female volunteers. The purpose of this study was to determine whether there were any significant differences between the originator and the biosimilar pegfilgrastim. The results of this study demonstrated that both medicines exhibit similar pharmacokinetics, pharmacodynamics, immunogenicity, and safety characteristics.

This suggests that Udenyca1 is a viable alternative to Neulasta1, with comparable effectiveness and safety (Nakov and Zesch 2016; Singh et al. 2017).

The approval of Udenyca1 highlights the increasing acceptance and use of biosimilars in clinical practice. As a long-acting version of filgrastim, Udenyca1 provides a new treatment option for patients, potentially enhancing patient adherence and outcomes. The introduction of biosimilars such as Udenyca1 has the potential to lead to cost savings in healthcare, making biological therapy more accessible to a larger number of patients. As more biosimilars are developed and approved, they are set to be an increasingly important component of patient care (Farhat et al. 2018; Danylesko et al. 2021; Jeon et al. 2022).

5.6 Truxima

Truxima is a drug that has been used to treat inflammation and a number of blood malignancies. The active ingredient is rituximab, a monoclonal antibody that is intended to identify and bind to a protein called CD20 on the surface of B lymphocytes, a subset of white blood cells. It has been demonstrated that truxima is effective in treating the following ailments: both diffuse large B cell and follicular lymphoma. There are two varieties of blood cancer or non-Hodgkin's lymphoma. Chronic lymphocytic leukaemia is another kind of blood malignancy that affects white blood cells. Truxima can cure severe rheumatoid arthritis and can be used to treat two inflammatory disorders of the blood vessels: Microscopic Polyangiitis and Granulomatosis with Polyangiitis (GPA or Wegener's Granulomatosis). Severe pemphigus vulgaris, an autoimmune disease characterized by extensive blistering and degradation of the skin and mucous membranes, is effective in Truxima treatment. The body natural defence mechanism, immune system, targets the body cells in this illness. It can be used alone, in combination with chemotherapy (or other cancer treatments), or in conjunction with medications for inflammatory conditions (corticosteroids or methotrexate). MabThera is the Truxima reference medication. The killing of B cells caused by rituximab attachment to CD20 can be advantageous in cases of lymphoma, CLL, and rheumatoid arthritis, in which B lymphocytes have developed into malignancy. Destroying the B lymphocytes in GPA and MPA reduces the generation of antibodies, which are crucial in inflaming and rupturing the blood vessels (Truxima n.d.).

Rituximab in Truxima and Rituximab in MabThera are quite comparable in terms of structure, purity, and biological activity, according to extensive laboratory investigations. Three hundred and seventy-two individuals with active rheumatoid arthritis participated in a research comparing Truxima with MabThera administered intravenously. According to the study, rituximab levels in the blood were similarly induced by Truxima and MabThera. Furthermore, the two medications had similar effects on the symptoms associated with arthritis: after 24 weeks, 74% of patients with Truxima (114 out of 155 patients) and 73% of patients (43 out of 59 patients) experienced a 20% improvement in the symptom score (referred to as ACR20). According to supportive studies conducted on individuals with advanced follicular lymphoma and rheumatoid arthritis, the medications were comparable. Truxima is a biosimilar medication, therefore it is not necessary to repeat the safety and efficacy trials that were conducted for MabThera (Stubbs et al. 2019; Antonelou et al. 2022).

5.7 Impact of Biosimilars in Cervical Cancer Treatment

Biosimilars exert a significant impact on the therapeutic landscape of various malignancies, inclusive of cervical cancer. It provides a more cost-effective substitute for biological medications, which may improve patient access to potent therapies. The effectiveness, safety, and quality of biosimilars are intended to be comparable to those of the original reference medication. The cost of cancer medications is anticipated to decrease with the introduction of biosimilars, which are typically less costly than original biosimilars (Mehrotra and Yadav 2022; Arip et al. 2022). Biosimilars will reduce the cost of biosimilars by US$54 billion between 2017 and 2026. Biosimilars have the potential to reduce healthcare costs related to the treatment of cervical cancer. Since biosimilars typically cost 20%–30% less, the cost of cancer therapies and overall care could be considerably lower. The precise effect of biosimilars on the management of cervical cancer will vary depending on a number of variables, such as the types of biosimilar, the particular treatment plan, and the unique characteristics of each patient (Yang et al. 2023; Calleja et al. 2023; Mulcahy et al. 2017; Sefuthi and Nkonki 2022). Thus, more investigation and empirical data are required to completely comprehend the possible advantages and difficulties of applying biosimilars in the treatment of cervical cancer. This will contribute to ensuring that biosimilars are used as efficiently as possible for the benefit of patients. Patents for biosimilars are shown in Table 5.3.

Table 5.3 Patents for biosimilars

Patents number	Name	References
JP7112327B2	Buffered formulations of bevacizumab	Taddei et al. (2019)
EP3397281B1	Buffered formulations of bevacizumab	Taddei et al. (2019)
US20150147317A1	Methods related to bevacizumab	Robblee et al. (2015a)
US20210403599A1	Pertuzumab Plus trastuzumab fixed dose combination	Tan et al. (2021)
WO2013181571A2	Methods related to trastuzumab	Robblee et al. (2015b)
US20150141620A1	Methods related to rituximab.	Robblee et al. (2015c)

5.8 Patents

5.9 Recent Advancements

In India, head and neck cancer affects over 100,000 new cases annually, making it a serious public health concern. Important parts of the head and neck, including the oral cavity, pharynx, larynx, and sinuses, are affected by this cancer. The disease is frequently discovered at an advanced stage, treatment becomes more difficult and imposes a significant burden that makes it harder to provide effective therapy and raises the mortality rate. The first biosimilar medication for treating head and neck cancer, Cetuxa, was introduced by Alkem Laboratories in June 2023. It was developed and produced domestically by Enzene Biosciences (TOI n.d.). A monoclonal antibody called cetuximab has a biosimilar called cetuxa. Subsequently, Lupin Limited declared a strategic alliance with Enzene Biosciences to introduce Cetuximab, the initial biosimilar designed for Cetuximab, in India. This partnership represents a major turning point in the development of novel treatments and the expansion of therapeutic choices for head and neck cancer patients. The Indian Drug Controller General has approved cetuximab for the treatment of head and neck cancer, namely Squamous Cell Carcinoma of the Head and Neck. This makes cetuximab an essential medication for patients with recurring locoregional or metastatic diseases. Currently, 100 mg dose of cetuximab available in the market (Lupin n.d.).

5.10 Conclusion

Biosimilar medicines are emerging as a promising alternative in the field of pharmaceuticals. The production of biosimilars, in contrast to the manufacturing of synthetic medications, entails the management of multiple variables associated with the utilization of various living host species. Thus, the biosimilar approval procedure through the EMA is not the same as the generic medication approval procedure. In order to develop biosimilar drugs, it is necessary to conduct comparative studies that demonstrate efficacy comparable to the reference biological, while also ensuring freedom from immunogenicity and safety problems. The clinical applications of a biosimilar medication automatically expand to include the reference biological medication when both medications demonstrate the same modes of action, efficacy, and safety features. The extent to which a biosimilar medication will cover every use of the original biological is uncertain, as it depends on the intended therapeutic outcomes. Biological therapy, which is used either as a standalone treatment or in conjunction with other therapies, has emerged as a highly prevalent method for treating cancer, and the results have been promising. The expiration of patents has paved the way for the development of several biosimilar medications, presenting beneficial alternatives to the biologicals currently being marketed. These new medications are particularly beneficial for patients as these mitigate the remaining side effects of the reference biologicals. The cheaper production costs of biosimilars can be attributed to their ability to draw conclusions from research on the biological medication of reference. Biosimilars offer a major benefit as it provide affordable treatments that greatly improve public health. However, there is ongoing debate about the interchangeability, automatic interchangeability, and extrapolation of biologics and biosimilars. Currently, the European Union has authorized 22 biosimilar medications for cancer treatment, divided into three categories: filgrastim, epoetins (zeta and alpha), and monoclonal antibodies (rituximab, trastuzumab, and bevacizumab). More biosimilars are expected to be released in the coming years, expanding access to affordable therapy for more patients. The number of biosimilars has been increasing, with a strong focus on cancer therapy. Pharmacovigilance studies are being conducted to monitor the safety and efficacy of these medications, providing doctors with valuable information. In conclusion, the advent of biosimilars marks a significant milestone in the pharmaceutical industry, offering a cost-effective and promising alternative to biological medications.

Despite ongoing debates surrounding the interchangeability and extrapolation of biosimilars, the potential of these medical products to improve public health, particularly in the realm of cancer treatment, is undeniable. Prior to the introduction of more biosimilars in the coming years, it becomes crucial to continue conducting extensive pharmacovigilance studies. Such studies ensure not only the safety and efficacy of these medications but also provide invaluable understandings that could pave the way for more breakthroughs in this field. The journey of biosimilars is just beginning, with an expected profound and far-reaching impact on global healthcare.

References

Aapro M, Krendyukov A, Schiestl M, Gascón P (2018) Epoetin Biosimilars in the treatment of chemotherapy-induced anemia: 10 years' experience gained. BioDrugs 32(2):129–135

American Cancer Society (n.d.) 9525.00.pdf. What Are Biosimilar Drugs? | Biosimilar Drugs for Cancer Treatment | American Cancer Society

Annex I (1999) Summary of product characteristics. Committee for Proprietary Medicinal Products The European Public Assessment Report (EPAR) Stocrin The European Agency for the Evaluation of Medicinal Products London

Antonelou M, Abro A, Heath R, Iacovou A, Ashley C, Caplan J et al (2022) Comparison of outcomes using the rituximab originator MabThera with the biosimilar Truxima in patients with ANCA-associated vasculitis. Scand J Rheumatol 51(2):135–141

Arip M, Tan LF, Jayaraj R, Abdullah M, Rajagopal M, Selvaraja M (2022) Exploration of biomarkers for the diagnosis, treatment and prognosis of cervical cancer: a review. Discov Onc 13(1):91

Biosimilar Competition in Europe (n.d.). https://www.centerforbiosimilars.com/view/report-biosimilar-competition-in-europe

Blackwell K, Gligorov J, Jacobs I, Twelves C (2018) The global need for a Trastuzumab biosimilar for patients with HER2-positive breast cancer. Clin Breast Cancer 18(2):95–113

Bosch X, Khamashta MA, Ramos-Casals M (eds) (2014) Drugs targeting B-cells in autoimmune diseases, 1st edn. Springer Basel, Basel: Imprint: Springer; 1 p. (Milestones in Drug Therapy)

Buskwofie A, David-West G, Clare CA (2020) A review of cervical cancer: incidence and disparities. J Natl Med Assoc 112(2):229–232

Calleja MA, Albanell J, Aranda E, García-Foncillas J, Feliu A, Rivera F et al (2023) Budget impact analysis of bevacizumab biosimilars for cancer treatment in adult patients in Spain. Eur J Hosp Pharm 30(e1):e40–e47

Candelaria M, Gonzalez D, Fernández Gómez FJ, Paravisini A, Del Campo GA, Pérez L et al (2018) Comparative assessment of pharmacokinetics, and pharmacodynamics between RTXM83™, a rituximab biosimilar, and rituximab in diffuse large B-cell lymphoma patients: a population PK model approach. Cancer Chemother Pharmacol 81(3):515–527

Celltrion USA (28 Sep 2022) Celltrion USA receives U.S. FDA approval for its oncology biosimilar Vegzelma® (bevacizumab-adcd) for the treatment of six types of cancer. News release. https://bwnews.pr/3BQP6om. Accessed 28 Sep 2022

Cornes P, Gascon P, Vulto AG, Aapro M (2020) Biosimilar Pegfilgrastim: improving access and Optimising practice to supportive care that enables cure. BioDrugs 34(3):255–263

Danylesko I, Sareli R, Varda-Bloom N, Yerushalmi R, Shem-Tov N, Magen H et al (2021) Long-acting granulocyte colony-stimulating factor pegfilgrastim (lipegfilgrastim) for stem cell mobilization in multiple myeloma patients undergoing autologous stem cell transplantation. Int J Hematol 114(3):363–372

Deechongkit S, Aoki KH, Park SS, Kerwin BA (2006) Biophysical comparability of the same protein from different manufacturers: a Case study using Epoetin alfa from Epogen® and Eprex®. J Pharm Sci 95(9):1931–1943

Du S, Zhao Y, Lv C, Wei M, Gao Z, Meng X (2020) Applying serum proteins and MicroRNA as novel biomarkers for early-stage cervical cancer detection. Sci Rep 10(1):9033

European Medicine Agency (n.d.) Biosimilar. https://www.ema.europa.eu/en/human-regulatory/overview/biosimilar-medicines-overview

European Medicines Agency (n.d.-a) vegzelma-epar-medicine-overview_en.pdf. Vegzelma | European Medicines Agency (europa.eu)

European Medicines Agency (n.d.-b) ontruzant-epar-summary-public_en.pdf. Ontruzant | European Medicines Agency (europa.eu)

European Medicines Agency (n.d.-c) kanjinti-epar-medicine-overview_en.pdf. Kanjinti | European Medicines Agency (europa.eu)

European Medicines Agency (n.d.-d) trazimera-epar-medicine-overview_en.pdf. Trazimera | European Medicines Agency (europa.eu)

European Medicines Agency (n.d.-e) herzuma-epar-summary-public_en.pdf. Herzuma | European Medicines Agency (europa.eu)

Farhat F, Torres A, Park W, De Lima LG, Mudad R, Ikpeazu C et al (2018) The concept of Biosimilars: from characterization to evolution—a narrative review. Oncologist 23(3):346–352

FDA (n.d.-a) Biosimilar Basics for Patients_FDA.pdf. Biosimilar Basics for Patients | FDA

FDA (n.d.-b) Biosimilar. https://www.fda.gov/drugs/therapeutic-biologics-applications-bla/biosimilars

FDA (n.d.-c) FDA approves first biosimilar for the treatment of cancer. Available from: https://www.fda.gov/news-events/press-announcements/fda-approves-first-biosimilar-treatment-cancer

FDA (n.d.-d) FDA approves Herceptin biosimilar as US patent expires. https://pharmaphorum.com/news/fda-approves-herceptin-biosimilar-as-us-patent-expires

FDA (n.d.-e) FDA approves Herzuma as a biosimilar to Herceptin. https://www.fda.gov/drugs/fda-approves-herzuma-biosimilar-herceptin

Franceschetti A, Caldeira R (2018) Treatment approach for non-Hodgkin lymphoma patients since first biosimilars of rituximab approved in EU5. J Clin Oncol 36:112

Goldsmith D, Dellanna F, Schiestl M, Krendyukov A, Combe C (2018) Epoetin Biosimilars in the treatment of renal anemia: what have we learned from a decade of European experience? Clin Drug Investig 38(6):481–490

Gravel P, Naik A, Le Cotonnec JY (2012) Biosimilar rhG-CSFs: how similar are they? Targ Oncol 7(S1):3–16

Halim LA, Márquez M, Maas-Bakker RF, Castañeda-Hernández G, Jiskoot W, Schellekens H (2018) Quality comparison of biosimilar and copy Filgrastim products with the innovator product. Pharm Res 35(11):226

Herzuma (n.d.). https://www.herzuma.com/about-herzuma
Hull R, Mbele M, Makhafola T, Hicks C, Wang S, Reis R et al (2020) Cervical cancer in low and middle-income countries (review). Oncol Lett 20(3):2058–2074
Jacobs I, Ewesuedo R, Lula S, Zacharchuk C (2017) Biosimilars for the treatment of cancer: a systematic review of published evidence. BioDrugs 31(1):1–36
Jeon Y, Lee N, Baek S, Choi J, Jhee S, Lee H (2022) A randomized, double-blind, placebo- and active-controlled, escalating single-dose study to evaluate the safety, tolerability, pharmacokinetic, and Pharmacodynamic profiles of subcutaneous Eflapegrastim in healthy Japanese and Caucasian subjects. Drugs R D 22(1):71–87
Jurczak W, Moreira I, Kanakasetty GB, Munhoz E, Echeveste MA, Giri P et al (2017) Rituximab biosimilar and reference rituximab in patients with previously untreated advanced follicular lymphoma (ASSIST-FL): primary results from a confirmatory phase 3, double-blind, randomised, controlled study. Lancet Haematol 4(8):e350–e361
Jurczak W, Długosz Danecka M, Buske C (2019) Rituximab biosimilars for lymphoma in Europe. Expert Opin Biol Ther 19(10):1045–1056
Keeratichamroen S, Subhasitanont P, Chokchaichamnankit D, Weeraphan C, Saharat K, Sritana N, et al. (2020) Identification of potential cervical cancer serum biomarkers in Thai patients. Oncol Lett [Internet]. [cited 2023 Dec 1]. http://www.spandidos-publications.com/10.3892/ol.2020.11519
Kirchhoff CF, Wang XM, Conlon HD, Anderson S, Ryan AM, Bose A (2017) Biosimilars: key regulatory considerations and similarity assessment tools. Biotech Bioeng 114(12):2696–2705
Kirsch-Stefan N, Guillen E, Ekman N, Barry S, Knippel V, Killalea S et al (2023) Do the outcomes of clinical efficacy trials matter in regulatory decision-making for Biosimilars? BioDrugs 37(6):855–871
Koh WJ, Abu-Rustum NR, Bean S, Bradley K, Campos SM, Cho KR et al (2019) Cervical cancer, version 3.2019, NCCN clinical practice guidelines in oncology. J Natl Compr Cancer Netw 17(1):64–84
Konstantinidou S, Papaspiliou A, Kokkotou E (2019) Current and future roles of biosimilars in oncology practice (Review). Oncol Lett [Internet]. [cited 2023 Dec 3]. http://www.spandidos-publications.com/10.3892/ol.2019.11105
Kurki P, Kang HN, Ekman N, Knezevic I, Weise M, Wolff-Holz E (2022) Regulatory evaluation of Biosimilars: refinement of principles based on the scientific evidence and clinical experience. BioDrugs 36(3):359–371
Lamanna WC, Heller K, Schneider D, Guerrasio R, Hampl V, Fritsch C et al (2019) The in-use stability of the rituximab biosimilar Rixathon®/Riximyo® upon preparation for intravenous infusion. J Oncol Pharm Pract 25(2):269–278
Li X, Hu SY, He Y, Hernandez Donoso L, Qu KQ, Van Kriekinge G et al (2018) Systematic literature review of risk factors for cervical cancer in the Chinese population. Women's Health (Lond Engl) 14:174550651881659
Lupin (n.d.) Lupin announces the launch of cetuximab. https://www.lupin.com/lupin-announces-the-launch-of-cetuximab/
Mehrotra R, Yadav K (2022) Cervical cancer: formulation and implementation of Govt of India guidelines for screening and management. Indian J Gynecol Oncolog 20(1):4
Minghetti P, Rocco P, Cilurzo F, Del Vecchio L, Locatelli F (2012) The regulatory framework of biosimilars in the European Union. Drug Discov Today 17(1–2):63–70
Moorkens E, Vulto AG, Huys I, Dylst P, Godman B, Keuerleber S et al (2017) Policies for biosimilar uptake in Europe: an overview. PLoS One 12(12):e0190147. Bochenek T, editor
Mulcahy A, Hlavka J, Case S (2017) Biosimilar cost Savings in the United States: initial experience and future potential [internet]. RAND Corporation; [cited 2023 Dec 2]. https://www.rand.org/pubs/perspectives/PE264.html
Nakov P, Zesch T (2016) Computational semantic analysis of language: SemEval-2014 and beyond. Lang Resour Eval 50(1):1–4
Paplomata E, Nahta R (2018) ABP 980: promising trastuzumab biosimilar for HER2-positive breast cancer. Expert Opin Biol Ther 18(3):335–341

Pierpont TM, Limper CB, Richards KL (2018) Past, present, and future of rituximab—the World's first oncology monoclonal antibody therapy. Front Oncol 8:163

Piezzo M, D'Aniello R, Avallone I, Barba B, Cianniello D, Cocco S et al (2021) Uptake of Trastuzumab Biosimilars for the treatment of HER2-positive breast cancer: a real-world experience from a cancer center. Pharmaceutics 13(5):684

Pivot X, Bondarenko I, Nowecki Z, Dvorkin M, Trishkina E, Ahn JH et al (2018) A phase III study comparing SB3 (a proposed trastuzumab biosimilar) and trastuzumab reference product in HER2-positive early breast cancer treated with neoadjuvant-adjuvant treatment: final safety, immunogenicity and survival results. Eur J Cancer 93:19–27

Rahi V, Ram P, Kumar P (2022) Filgrastim, a recombinant form of granulocyte Colony-stimulating factor, ameliorates 3-nitropropionic acid and haloperidol-induced striatal neurotoxicity in rats. Neurotox Res 40(6):2089–2102

Robblee J, Collins BE, Kaundinya G, Bosques CJ (2015a) Inventors; Momenta Pharmaceuticals Inc, assignee. Methods related to bevacizumab. United States patent application US 14/403,835. 2015 May 28

Robblee J, Collins BE, Kaundinya G, Bosques CJ (2015b) Inventors; Momenta Pharmaceuticals Inc, assignee. Methods related to trastuzumab. United States patent application US 14/403,811. 2015 Apr 16

Robblee J, Collins BE, Kaundinya G, Bosques CJ (2015c) Inventors; Momenta Pharmaceuticals Inc, assignee. Methods related to rituximab. United States patent application US 14/404,229. 2015 May 21

Šarenac T, Mikov M (2019) Cervical cancer, different treatments and importance of bile acids as therapeutic agents in this disease. Front Pharmacol 10:484

Schellekens H (2009) Assessing the bioequivalence of biosimilars. Drug Discov Today 14(9–10):495–499

Sefuthi T, Nkonki L (2022) A systematic review of economic evaluations of cervical cancer screening methods. Syst Rev 11(1):162

Singh RS, Bansal BK, Gupta DK (2017) Relationship between teat morphological traits and subclinical mastitis in Frieswal dairy cows. Trop Anim Health Prod 49(8):1623–1629

Sörgel F, Schwebig A, Holzmann J, Prasch S, Singh P, Kinzig M (2015) Comparability of biosimilar filgrastim with originator filgrastim: protein characterization, pharmacodynamics, and pharmacokinetics. BioDrugs 29(2):123–131

Stubbs MJ, Low R, McGuckin S, Newton R, Thomas M, Westwood JP et al (2019) Comparison of rituximab originator (MabThera) to biosimilar (Truxima) in patients with immune-mediated thrombotic thrombocytopenic purpura. Br J Haematol 185(5):912–917

Taddei M, Cheung J, Gutka H (2019) Inventors; Outlook Therapeutics Inc, assignee. In: Buffered formulations of bevacizumab. United States Patent Application US 16/066,464. 2019 Jan 17

Tan AR et al (2021) Fixed-dose combination of pertuzumab and trastuzumab for subcutaneous injection plus chemotherapy in HER2-positive early breast cancer (FeDeriCa): a randomised, open-label, multicentre, noninferiority, phase 3 study. Lancet Oncol 22(1):85–97. https://doi.org/10.1016/S1470-2045(20)30536-2

Tekalign T, Teshome M (2022) Prevalence and determinants of late-stage presentation among cervical cancer patients, a systematic review and meta-analysis. PLoS One 17(4):e0267571. Ekwunife OI, editor

TOI (n.d.) Alkem launches world's 1st biosimilar for Cetuximab. https://timesofindia.indiatimes.com/business/india-business/alkem-launches-worlds-1st-biosimilar-for-cetuximab/articleshow/100659457.cms?from=mdr

Truxima (n.d.). https://www.truxima.com/

Waller CF, Vutikullird A, Lawrence TE, Shaw A, Liu MS, Baczkowski M et al (2018) A pharmacokinetics phase 1 bioequivalence study of the trastuzumab biosimilar MYL-1401O *vs.* EU-trastuzumab and US-trastuzumab. Br J Clin Pharm 84(10):2336–2343

World Health Organization (n.d.) Cervical cancer. https://www.who.int/news-room/fact-sheets/detail/cervical-cancer

Xu X, Zhang S, Xu T, Zhan M, Chen C, Zhang C (2022) Efficacy and safety of bevacizumab Biosimilars compared with reference biologics in advanced non-small cell lung cancer or metastatic colorectal cancer patients: a network meta-analysis. Front Pharmacol 13:880090

Yang BB, Kido A (2011) Pharmacokinetics and Pharmacodynamics of Pegfilgrastim. Clin Pharmacokinet 50(5):295–306

Yang J, Chaudhry BI, Yue AT, Roth JA, Kelton JM, Shelbaya A, et al. (2023) The impact of biosimilar use on total cost of care and provider financial performance in the medicare oncology care model: a population-based simulation study. Adv Ther [Internet]. [cited 2023 Dec 2]. https://link.springer.com/10.1007/s12325-023-02703-x

Biosimilars in Prostate Cancer

6

Priyal Patel, Yash Patel, Samir Gunvantbhai Patel,
Archita Samir Patel, Arvind Ganpule,
and Sachchida Nand Pandey

Abstract

Prostate cancer, a significant global health challenge, affects millions of men worldwide, with its incidence and mortality rates on the rise. This chapter provides a comprehensive overview of prostate cancer, covering its etiology, pathogenesis, and current treatment approaches, with a focus on emerging biological therapies. The text delves into the diverse factors contributing to prostate cancer development, including age, race, genetics, diet, obesity, and smoking. Additionally, it explores the intricate molecular pathways involved in prostate cancer progression, emphasizing the role of androgen receptors and apoptosis regulation. Furthermore, the chapter discusses the evolving landscape of prostate cancer treatment, encompassing surgical interventions, radiation therapy, androgen deprivation therapy (ADT), and emerging therapies such as biologics and immunotherapy. Specifically, it highlights the promising potential of biologics in prostate cancer management, including monoclonal antibodies, cancer vaccines, and immunomodulatory agents. The emergence of biosimilars as a cost-effective alternative for prostate cancer treatment is also discussed, underscoring the future prospects in improving accessibility and affordability of care. Overall, this chapter offers valuable insights into the multifaceted nature of prostate cancer and the evolving strategies like biosimilars that could be aimed at improving patient outcomes and reducing mortality rates.

P. Patel · Y. Patel · S. Gunvantbhai Patel · A. Samir Patel
Ramanbhai Patel College of Pharmacy, Charusat, Changa, Gujarat, India

A. Ganpule
Department of Urology, Muljibhai Patel Urological Hospital, Nadiad, Gujarat, India

S. N. Pandey (✉)
Department of Pathology, Muljibhai Patel Urological Hospital, Nadiad, Gujarat, India

S. Bhatt et al. (eds.), *Biosimilars for Cancer Treatment*,
https://doi.org/10.1007/978-981-97-3001-8_6

Keywords

ADT · Androgen · Biologics prostate · PTEN

6.1 Prostate Cancer

Prostate cancer (PCa) originates in the prostate gland, a tiny, walnut-shaped organ situated beneath the bladder in men. It commonly progresses slowly, initially staying localized within the prostate gland. However, in certain instances, it can become aggressive, spreading to distant areas like bones and lymph nodes. PCa represents a substantial global health challenge, with approximately 1.3 million fresh cases reported each year globally. Presently, there are roughly 10 million men living with a PCa diagnosis, with around 700,000 grappling with metastatic disease. The yearly mortality rate stemming from metastatic PCa surpasses 400,000, and forecasts suggest it will more than double by 2040. Moreover, a similar number of men are anticipated to endure treatment-related complications for more than 10 years following their initial diagnosis (Foreman et al. 2018; James et al. 2018; Sandhu et al. 2021). The worldwide occurrence and fatality rates of cancer are swiftly escalating. This complex trend arises from various factors, including population aging, overall population expansion, and changes in the prevalence and distribution of significant cancer risk elements. Numerous of these risk elements are intertwined with socio-economic advancement.

6.1.1 Incidence

Based on data from the GLOBOCAN (Global Cancer Observatory), around 1.27 million cases of PCa were estimated globally in 2018, with an occurrence rate of 29 per 100,000 men. PCa accounted for 7.1% of all cancer diagnoses during that period. This places PCa as one of the most prevalent cancers in men globally, making up 13% of all diagnosed cancers (Barsouk et al. 2020; Bray et al. 2018).

PCa is one of the most occurrent cancers diagnosed among American men in the Americas, significant portions of Africa, Oceania, and Europe. Regions with the highest incidence rates include Australia/New Zealand (86.4 per 100,000) and North-Western Europe. In contrast, South Central and Southeast Asia had the lowest rates (5.0 per 100,000) (Bray et al. 2018). In 2020, around 192,000 cases of PCa were recorded in the United States, making up 10.6% of all new cancer diagnoses. The lifetime risk of PCa occurrence in the US is roughly 12.1%, nearly four times higher than the global average (Barsouk et al. 2020).

In 2020, the global incidence of new PCa cases amounted to 1.41 million, constituting 7.3% of all cancer cases (Sung et al. 2021). In 2021, reports indicated 2.48 lakh fresh cases of PCa, along with a total of 34,130 deaths (https://www.cancer.org/cancer/types/prostate-cancer/about/key-statistics.html). Moreover, projections suggest that by 2040, these trends are anticipated to rise, reaching 2.43 million fresh cases along with 7.4 lakh mortality globally (Culp et al. 2020).

The implementation of prostate-specific antigen (PSA) screening in the 1990s highlighted a notable surge in PCa incidence rates in the United States, Australia, and Canada (Pernar et al. 2018). Research indicates that 1 out of 8 men face the risk of PCa development, with the likelihood increasing with age. Chances of PCa development are reported to be lower for those below 50 years in comparison to those aged 50–59 and 60–69 years (Pernar et al. 2018).

6.1.2 Etiology

Key factors contributing to the incidence of PCa include aging, race, family history, and genetic factors. In addition to these, diet, obesity, and smoking have also been identified as key contributors to PCa.

Age: PCa is frequently detected in men aged 60 and above. The rise in PSA testing and extended life expectancy has led to the diagnosis of PCa in elderly men. There is an increase in the risk of PCa after the age of 40 in individuals with a family history and after the age of 50 in men without a family history (Bray et al. 2018; Rawla 2019).

Race and ethnicity: Disparities based on race have been noted in PCa. African men exhibit a greater prevalence compared to white men (Chowdhury-Paulino et al. 2021). A U.S. study revealed that African Americans experience a 58% higher incidence and 144% higher mortality rate than white European ancestry men. Conversely, Hispanics documented a 14% lower incidence and 17% lower mortality rate than white men. The underlying factors contributing to these racial disparities in PCa include socioeconomic and biological factors. African American men possess the chr.8q24 mutation that is associated with an increased risk of PCa (Okobia et al. 2011; Robbins et al. 2011).

Family history and genetic factors: The likelihood of PCa development rises for individuals with a history of multiple family members affected at a younger age. Additionally, the risk increases by two to three times in individuals who have first-degree relatives with PCa (Kiciński et al. 2011). Genetic factors contribute to approximately 40% risk of PCa development. Breast cancer susceptibility gene-1 (*BRCA1*) and *BRCA2* gene mutations increase the susceptibility to PCa. Specifically, *BRCA2* gene mutation is linked to an increased risk of PCa by 8.6 times in males under the age of 65 (Castro et al. 2013; Junejo and AlKhateeb 2020). Other infrequent mutations associated with PCa include HOXB13, NBS1, and *CHEK2* (Goh and Eeles 2014).

Diet and Obesity: Certain food products carry an elevated risk for PCa, including saturated animal fat, red meat, and dairy proteins. Conversely, some diets, including soy, green tea, tomatoes, and lycopene, are reported to reduce the risk of cancer (Discacciati and Wolk 2014). Several meta-analyses indicate a significant correlation between obesity and an increased prevalence of PCa, regardless of changes in body mass index (BMI) (Allott et al. 2013). The correlation between obesity and PCa is reinforced by diverse biological mechanisms, including the onset of insulin resistance resulting from a sedentary lifestyle, fluctuations in sex steroid hormones, and metabolic changes (Cao and Giovannucci 2016).

Smoking: Various reports indicate a correlation between smoking and an increased number of fatalities attributed to PCa, with the extent of risk depending on the daily quantity of cigarettes smoked (De Nunzio et al. 2015). According to a report from California, the rate of smoking is decreasing at a rate of 3.5% annually, leading to a 2.5% reduction in PCa deaths. This suggests that reducing smoking habits can be advantageous in lowering cancer mortality rates (Jones et al. 2016).

6.1.3 Pathogenesis of Prostate Cancer

The PCa prognosis relies primarily on the androgen receptors (AR) stimulation by testosterone and dihydrotestosterone (DHT), resulting in translocating into the nucleus, and binds with androgen response elements (AREs) and initiating gene transcription responsible for regulating cellular differentiation, proliferation, and apoptosis. Elevated levels of androgens are associated with a decreased likelihood of a better disease prognosis. In contrast, individuals with lower serum androgen levels are more prone to relapse of PCa with more advanced pathology (Yuan et al. 2006).

The decrease in AR activation results in the activation and overexpression of alternative pathways. For example, the increase in insulin-like growth factor-1 (IGF-1), epidermal growth factor (EGF), and other growth factors subsequently trigger activation of ErbB2 and other tyrosine receptor kinases. Consequently, phosphatidylinositol 3-kinase (PI3K) is activated, initiating the subsequent PI3K-AKT-mTOR pathway. PI3K activation changes phosphatidylinositol 4,5-bisphosphate (PIP2) to phosphatidylinositol 3–5-triphosphate (PIP3), that sends protein kinase B (AKT) proteins to the cytoplasm of luminal cells. AKT signal disrupts the function of tuberous sclerosis 1/2 (TSC1/2), a GTP-binding protein RHEB inhibitor that activates the mechanistic target of rapamycin complex 1 (mTORC1), a kinase crucial for cell cycle regulation. mTORC1 inhibits autophagy and promotes cellular proliferation in PCa (He et al. 2021) (Fig. 6.1.).

The regulation of apoptosis is pivotal in PCa development, acting as a primary catalyst for the rapid proliferation of tumor cells. Essential to maintaining tissue balance, apoptosis operates through two primary pathways: intrinsic and extrinsic (Ali and Kulik 2021). The intrinsic pathway, centered on mitochondria, is triggered by intracellular harm, including DNA damage prompted by chemoradiotherapy. In response to intracellular insults, pro-apoptotic B-cell lymphoma 2 (Bcl-2) clan members like Bax, Bid, and Bad are activated, releasing cytochrome c (Cyc-c) from the mitochondria. Cyc-c, together with pro-caspase 9 and apoptotic protease activating factor-1 (Apaf-1), forms an apoptosome, that activates caspase-3, ultimately inducing apoptosis (Kang and Reynolds 2009).

On the other hand, the extrinsic pathway involves the binding of apoptosis-activating ligands, (like tumor necrosis factor (TNF)) binding with cell-surface

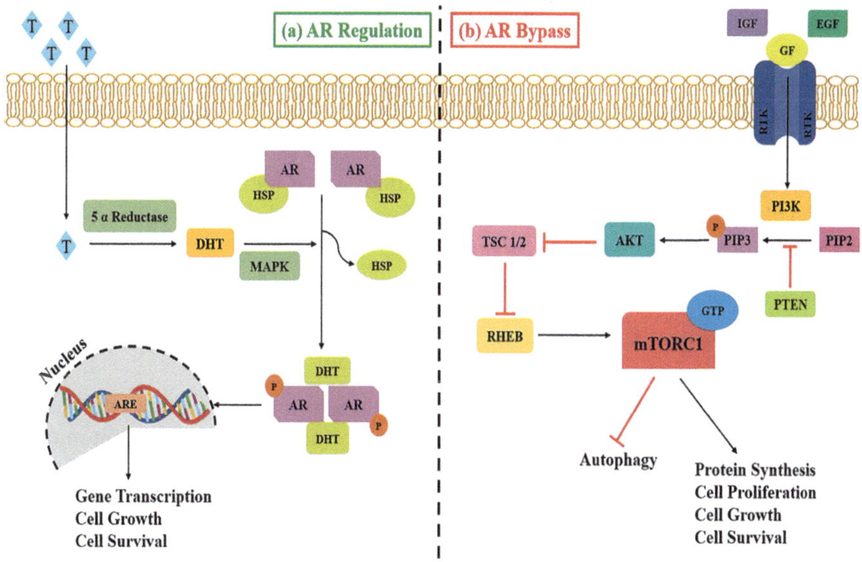

Fig. 6.1 The role of AR in the progression of PCa(**a**) Androgenic regulation: In the cytoplasm, the function of ARs is controlled through the interaction with ligands and heat shock proteins (HSP). Testosterone (T) enters the cytoplasm of cells sensitive to androgens and undergoes conversion to DHT by 5-α reductase. The binding of the DHT ligand triggers detachment from HSP, followed by receptor phosphorylation by MAPK. Subsequently, dimerization occurs. The AR dimer then moves into the nucleus, which attaches to androgen response elements (AREs) in the DNA, initiating the transcription of genes crucial for cell growth and survival. (**b**) AR bypass pathway: Diminished AR inhibition results in the elevation of Tyrosine Receptor Kinases (RTK), such as ErbB2, due to factors like IGF, GF, and EGF. Activation of RTK triggers the activation of PI3K, which phosphorylates PIP2 to PIP3. This process is counteracted by the tumor suppressor PTEN. PIP3 activates AKT, which indirectly hampers the activity of the cell survival protein mTORC1 by inhibiting TSC1/2, consequently suppressing the GTP-binding protein RHEB. mTORC1 plays a crucial role in protein translation and synthesis. Hence, inhibition of AR and suppression of PTEN leads to excessive activation of the PIP3-AKT-mTOR pathway, offering an alternative pathway for cell survival in CRPCa

death receptors like CD95 (Apo-1/Fas) or TNF-related apoptosis-inducing ligand (TRAIL) receptors. These receptors lead to the formation of death-inducing signaling complexes (DISCs) with intracellular caspases, leading to the activation of effector caspases like caspase-3, which initiate proteolysis and cleavage of intracellular substrates, ultimately causing cell death (Diaz Arguello and Haisma 2021).

Dysregulation of these apoptosis regulation pathways can lead to increased cell survival, resistance to cell death, and promotion of tumor growth and progression in PCa. Mutations in tumor suppressor genes like p53, overexpression of anti-apoptotic factors like Bcl-2, and apoptotic signaling alterations can disrupt the equilibrium between cell survival and death, contributing toward the pathogenicity of PCa and castration-resistant prostate cancer (CRPCa) (Plati et al. 2008).

6.1.4 Prostate Cancer and Molecular Pathways

PCa is a complex disease with a multifaceted etiology and progression. Its onset entails the disruption of diverse signaling pathways pivotal for instigating, advancing, and metastasizing prostate tumors. The pathogenesis of PCa is associated with various significant signaling pathways, including the AR pathway, the Wnt pathway, the MAPK pathway, the TGF-*β*/SMAD pathway, the JAK/STAT pathway, and the PI3K/AKT/mTOR pathway (da Silva et al. 2013).

6.1.5 Androgen Receptor Signaling Pathway

It facilitates the differentiation of epithelial cells into male urogenital architecture and codes for essential proteins for prostate function and the onset and continuity of spermatogenesis. AR also functions as a nuclear receptor serving as a transcription factor, persistently stimulating gene expression and governing cell growth and viability. Upon binding with androgens like testosterone and dihydrotestosterone, AR triggers downstream signaling pathways, orchestrating the expression of genes essential for cell proliferation, survival, and differentiation (Lonergan and Tindall 2011).

AR gene amplification is a key genetic alteration in CRPCa, occurring in approximately 80% of cases. AR overexpression is responsible for amplification/transcriptional changes in genes/translational upregulation of genes and is a significant cause of CRPCa. AR orchestrates the expression of numerous genes by binding the AR-ligand complex to androgen receptor binding sites (ARBSs) or androgen-responsive elements (AREs) in DNA, consequently upregulating androgen-regulated genes implicated in disease progression (Castoria et al. 2011; Green et al. 2012).

Distinct AR mutations have been identified, each possessing unique characteristics that influence the advancement of CRPCa. For example, mutations in the region between residues 701 and 730, such as L701H, V715M, and V730M, enable adrenal androgens, glucocorticoids, and progesterone resistance (Green et al. 2012; Waltering et al. 2012). Another prevalent mutation, T877A, situated between residues 874 and 910, alters AR ligands particularly by modifying the stereo-chemical environment of the binding pocket, thereby broadening the range of ligands capable of binding to AR (Waltering et al. 2012). This enables other hormones, such as Dehydroepiandrosterone, estrogen, progesterone, cortisone, and cortisol, to activate AR. Additionally, mutations within the region spanning residues 670–678 augment transactivation AR activity in response to DHT (Green et al. 2012; Waltering et al. 2012). These mutations broaden the specificity of AR, allowing it to respond to nonandrogenic molecules and bypass the requirement for a ligand, ultimately contributing to the androgen-refractory growth of diseased cells.

Targeting the AR pathway has proven to be a successful therapeutic approach for treating PCa. Androgen deprivation therapy (ADT) is designed to lower androgen levels or obstruct AR signaling. However, resistance to ADT frequently emerges, resulting in the development of CRPCa. To address this challenge, novel agents directed at the AR pathway have been developed to combat drug resistance in CRPCa. These include next-generation antiandrogens like enzalutamide and apalutamide, which hinder AR signaling by impeding androgen binding and the nuclear translocation of AR. Moreover, inhibitors of androgen synthesis, such as abiraterone acetate, have been formulated to diminish androgen levels and inhibit AR activation in CRPCa. These targeted therapies have effectively overcome drug resistance and impeded cancer progression in CRPCa by suppressing AR signaling and downstream pathways associated with cell proliferation and survival (da Silva et al. 2013).

6.1.6 PI3K/AKT/mTOR Signaling Pathway

The PI3K/AKT pathway is commonly deregulated during PCa, with up-regulation of PI3K/AKT/mTOR signaling observed in 30–50% of cases, often due to phosphatase and tensin homolog (PTEN) function loss (Cantley 2002). PTEN causes dephosphorylation of phosphatidylinositol (3,4,5)-trisphosphate (PIP3) to Phosphatidylinositol 4,5-bisphosphate (PIP2), negatively controlling PI3K/AKT signaling. Loss of PTEN function, either through gene deletion or other mechanisms such as microRNA action or nuclear exclusion, contributes to increased cell growth and resistance to apoptosis in PCa (Di Cristofano et al. 1998; Morgan et al. 2009).

Genetic mutations associated with the PI3K/AKT signaling pathway that accelerate PCa progression include:

- **Loss of PTEN function:** PTEN often exhibits haploinsufficiency in PC, where its genetic dosage correlates with disease advancement. Complete loss of PTEN function is associated with heightened PC progression, evident in engineered mouse models. Notably, complete PTEN inactivation in the prostate of these models results in a noninvasive PC phenotype, indicating that additional mutations may facilitate the development of invasive tumors. In-vivo studies have demonstrated that mutations in p53 or the cyclin-dependent kinase inhibitor p27KIP1, alongside PTEN loss, contribute to the emergence of more aggressive PC (Di Cristofano et al. 2001).
- **MicroRNA action on PTEN:** The action of microRNAs (miRNAs) on PTEN inactivation has been described, with the characterization of miR-22 and miR-106b ~ 25 as PTEN-targeting miRNAs aberrantly expressed in PCa. These miRNAs contribute to PTEN function loss, leading to increased cell growth and resistance to apoptosis (Poliseno et al. 2010).
- **Nuclear exclusion of PTEN:** Nuclear exclusion of PTEN is essential for tumor development, including PCa. Studies have elucidated the involvement of nuclear PTEN in PCa development, highlighting its impact on modifying sensitivity to mTOR inhibitors, thus exacerbating disease advancement (Song et al. 2008).

These genetic mutations and disruptions in the PI3K/AKT signaling pathway contribute to increased cell growth and resistance to apoptosis in PCa. PTEN function loss results in hyperactivation of AKT, promoting cell survival and proliferation.

6.1.7 JAK/STAT Pathway

The JAK/STAT pathway is commonly disrupted in PCa, typically as a result of genetic mutations and modifications in its components. Among the primary genetic mutations linked to the JAK/STAT, signaling pathway is the continuous activation of STAT3. Persistent activation of STAT3 has been associated with the expression of the *BRCA1* in specific tumor cell lines, and mutations in *BRCA* genes demonstrated increased susceptibility to PCa. Both *BRCA1* and *BRCA2* play roles in biological processes such as DNA repair, control of cell-cycle checkpoints, and transcriptional regulation (Tutt and Ashworth 2002; Yu and Jove 2004). BRCA1 interacts with JAK1/2 within PCa cells, triggering STAT3 phosphorylation, increasing cell proliferation, and suppressing cell death.

Constitutive activation of STAT3, a vital component of the JAK/STAT pathway, is observed in PCa cell lines and primary tumors, and it has been associated with cancer progression and resistance to apoptosis. STAT3 activation induces the expression of antiapoptotic genes such as *Bcl-2*, *Bcl-XL*, *Survivin*, and *Mcl-1*, promoting cell survival and inhibiting apoptotic cell death in PCa. Additionally, STAT3 regulates vascular endothelial growth factor (VEGF) expression, proangiogenic factors involved in tumor invasion and metastasis (Frank 2007). Activation of STAT3 prompts the upregulation of antiapoptotic genes like Bcl-2, Bcl-XL, Survivin, and Mcl-1, promoting cell survival while impeding apoptotic cell death in PCa. Overall, targeting the JAK/STAT pathway for PCa holds promise as a therapeutic strategy for inhibiting cancer progression and overcoming drug resistance. By inhibiting the aberrant signaling of this pathway, it may be possible to suppress tumor growth, induce apoptosis, and potentially overcome drug resistance in PCa (Frank 2007).

6.1.8 MAPK/ERK Signaling Pathway

The MAPK/ERK signaling pathway is significantly involved in PCa progression, and genetic mutations associated with this pathway can accelerate the progression of the disease. The MAPK/ERK pathway activates and is often deregulated with AKT signaling at the advanced stage of PCa (Gioeli et al. n.d.; Kinkade et al. 2008). The upstream events that lead to the MAPK signaling activation are possibly related to aberrant growth factor signaling, such as KGF, EGF, FGFs, and IGF-1, that are generally overexpressed in PCa (Carey et al. 2007). Members of the Ras family rarely report a mutation in PCa; these growth factors stimulate Ras and the MEK/ERK pathway, leading to increased cell growth and proliferation. Additionally, the expression of activated forms of Ras or Raf in the mouse prostate epithelium has been shown to result in PCa formation, indicating the role of MAPK pathway activation in PCa progression (Jeong et al. 2008; Palanisamy et al. 2010). Concurrent activation of the ERK and AKT signaling pathways promotes PCa and CRPCa both in-vitro and in-vivo while inhibiting these pathways stops cell proliferation and leads to Bcl-2 and Bim upregulation (Hour et al. 2012; Kinkade et al. 2008). Therefore, the MAPK pathway presents a promising target for PCa treatment if its modulation could be aligned with others, such as PI3K/AKT signaling.

6.1.9 TGF-β/SMAD Signaling Pathway

The TGF-β/SMAD signaling pathway is a critical pathway that regulates cellular functions, like embryonic development, cell growth, differentiation, migration, adhesion, and apoptosis (Blobe et al. 2000). The pathway is initiated when activation ligands, such as TGF-β, bind to their specific receptors, leading to the recruitment and activation of SMAD proteins, the main effector proteins of this pathway. SMADs transmit signals from the TGF-β superfamily of ligands to the nucleus, where they activate or suppress target gene transcription (Massagué et al. 2000; Schmierer and Hill 2007). Alterations in the TGF-β/SMAD signaling pathway have been implicated in PCa progression. Enhanced TGF-β levels are associated with disease progression, and TGF-β-mediated growth inhibition and increased motility in CRPCa cells. Metastatic CRPCa cells exhibit increased sensitivity to TGF-β1-mediated growth inhibition and downregulation of cyclin D, as well as robust phosphorylation and nuclear translocation of Smad2 and Smad3 (Miles et al. 2012).

Gene mutations associated with the TGF-β/SMAD pathway have been implicated in hastening PCa advancement. For instance, activation of TGF-β/SMAD signaling has been correlated with restraining proliferation and fostering migration in PCa cells. Moreover, Nodal, another TGF-β ligand, is excessively expressed in certain PCa cells and can impede proliferation and migration in these cells (Vo and Khan 2011). Additionally, activin A, recognized for impeding PCa cell growth and stimulating cell death, and identified as a facilitator of bone metastasis in PCa, potentially through SMAD signaling and concurrent AR gene transcription elevation (H. Y. Kang et al. 2009).

Focusing on the TGF-β pathway has emerged as a promising therapeutic approach to combat drug resistance and hinder cancer advancement in PCa. Notably, elevated levels of TGF-β1 have been observed in the blood of PCa patients and are linked to bone metastasis, which aligns with an unfavorable clinical prognosis (Adler et al. 1999; Wikström et al. 1998). TGF-β type I receptor inhibition has been shown to provide anti-tumoral effects against PCa cells after growth in bone tissue, suggesting the potential of targeting this pathway for therapeutic intervention (Wan et al. 2012). Furthermore, TGF-β1, produced by activated CD4+ T cells, has been implicated in tumor evasion from immune surveillance, indicating the complex role of the TGF-β pathway in the tumor microenvironment and its potential as a therapeutic target (Donkor et al. 2012; Wan et al. 2012). Additionally, evidence suggests that SMAD2, a key mediator of TGF-β-induced apoptosis, and SMAD-independent signaling through BMP-10 may also be potential targets for therapeutic intervention in PCa (Ye et al. 2009). Thus, the TGF-β pathway plays a key role in PCa progression, and inhibiting it can potentially overcome drug resistance and inhibit cancer progression.

6.1.10 Wnt Signaling Pathway

The Wnt signaling pathway is crucial in various cellular processes, like embryonic development, cell proliferation, differentiation, and tissue homeostasis (Whyte et al. 2012). In PCa, the Wnt signaling pathway has been found to play a significant role in disease progression. Overexpression of Wnt ligands and β-catenin gene overexpression is associated with advanced PCa (Yardy and Brewster 2005). Mutant forms of β-catenin affecting GSK3β-dependent phosphorylation sites have been detected in radical prostatectomy specimens, and loss of PTEN, common in advanced PCa, can lead to the stabilization and increased levels of β-catenin. Another mechanism contributing to increased β-catenin expression in PCa could be the loss of PTEN, a general occurrence in advanced PCa. PTEN loss results in the activation of the PI3K and downstream AKT signaling pathways (Willert et al. 2003). Additionally, certain members of the Wnt pathway, such as Frizzled-4 (FZD4, a Wnt receptor), are co-expressed in human PCa samples alongside the ETS-related gene (ERG). ETS transcription factor gene fusions, primarily involving ERG, are detected in approximately 50% of all PCas. The upregulation of FZD4 in ERG-positive PCa induces an epithelial-to-mesenchymal transition, a critical step toward metastasis (Gupta et al. 2010).

The Wnt signaling pathway holds a crucial role in PCa advancement, and genetic mutations associated with this pathway, including mutant forms of β-catenin and loss of PTEN, can accelerate the progression of PCa. Understanding these genetic mutations and their impact on the Wnt signaling pathway is essential for developing targeted therapeutic strategies for PCa.

6.2 Current Treatment for Prostate Cancer

PCa, the most common cancer among males, manifests with diverse progression patterns. Advances in medicine have yielded a broad spectrum of treatment choices, enabling tailored plans according to each patient's prognosis. While these options facilitate effective disease management, they typically do not provide a cure, and recurrence is common. This chapter delves into the contemporary realm of PCa treatment, encompassing established practices, emerging therapies, and critical considerations for selecting the optimal approach.

Treatment approaches employed for various stages of PCa encompass surgical interventions, radiation therapy, androgen deprivation therapy (ADT), taxanes, and sipuleucel-T (Fig. 6.2.).

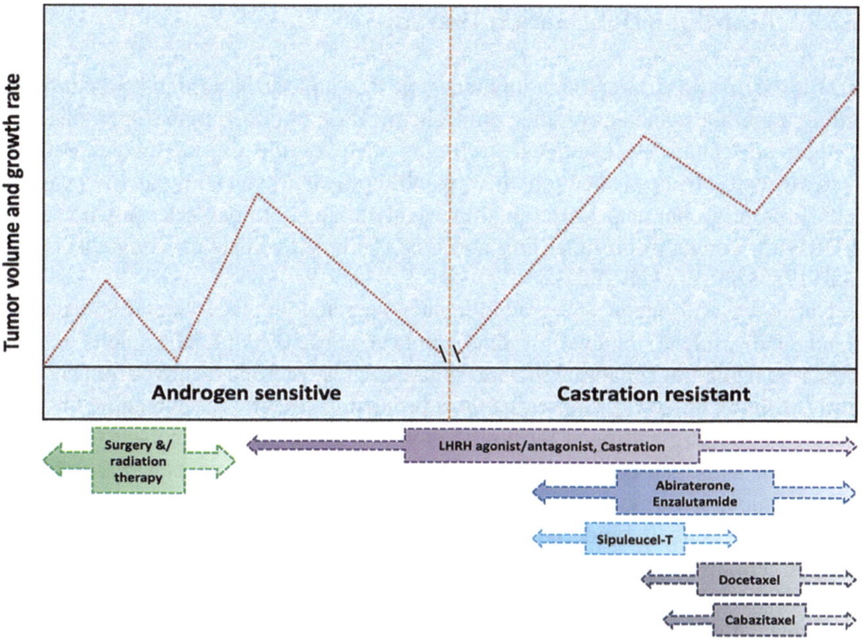

Fig. 6.2 PCa progression and treatment strategyThe majority of PCa cases are diagnosed before they spread to other areas of the body. Localized tumors can be treated with surgery or radiotherapy. For metastatic PCa, the standard treatment is ADT. However, many cases of metastatic PCa will eventually relapse, resulting in the development of castration-resistant (CR) PCa. Initially, CR PCa can be managed with antiandrogens like enzalutamide and abiraterone, or with immunotherapy agents such as Sipuleucel-T. As the disease progresses further, docetaxel and cabazitaxel may be added to the treatment regimen, along with abiraterone and enzalutamide, if the patient hasn't received them before

6.2.1 Surgery and Radiation Therapy

Currently, surgery and radiotherapy stand as the frontline treatments for localized PCa. While active surveillance is preferred for certain low-risk cases, other options are available for localized disease (Filson et al. 2015). Surgery, external beam radiation therapy (EBRT), and brachytherapy are standard courses of treatment. EBRT is often used for high-risk diseases, while brachytherapy is suitable for low-risk cases (Goy et al. 2020; Smith et al. 2015). Primary surgery provides a viable option for histopathological tumor analysis (Bach et al. 2014). Research has shown that surgery outperforms watchful waiting concerning mortality, disease progression, and metastasis (Bill-Axelson et al. 2011). However, mortality rates don't notably vary among active surveillance, surgery, or radiotherapy for low-risk patients (Hamdy et al. 2016). However, these treatments are not fail-safe, as there is still a risk of metastatic disease relapse and progression to CRPCa.

6.2.2 Androgen Deprivation Therapy

ADT is the first line of treatment for metastatic PCa and can be administered through different methods, such as orchiectomy, chemical castration, antiandrogen therapy, or their combinations. Chemical castration involves the use of medications to decrease testosterone production, thereby inhibiting androgen stimulation of cancer cells. Presently, chemical castration utilizes luteinizing hormone-releasing hormone (LHRH) or gonadotropin-releasing hormone (GnRH) agonists or antagonists (Liu et al. 1985; Siegel et al. 2022).

Normally, androgens are predominantly secreted by the testes and adrenal glands. However, it's noteworthy that approximately 50% of CRPCa cells exhibit intracrine regulation, enabling them to synthesize testosterone de novo from cholesterol (Miller et al. 2021). Consequently, a promising strategy for managing CRPCa involves inhibiting androgen biosynthesis within these cells. To bolster the effectiveness of androgen deprivation therapy (ADT), antiandrogens can be combined with chemical or surgical castration. This combination suppresses androgen stimulation of PCa cells by hindering the synthesis or utilization of androgen signaling. Notable antiandrogens include androgen biosynthesis inhibitors, like abiraterone acetate. These medications disrupt the function of CYP17A, an enzyme critical in key stages of androgen production.

Consequently, they serve as potent treatment modalities for PCa, mainly targeting tumors capable of intra-tumoral androgen synthesis. Another category of antiandrogens comprises AR blockers, which act as antagonists to the androgen receptor (AR), hindering its translocation to the nucleus and binding to DNA. The AR blocker bicalutamide (Casodex) and the second-generation agent enzalutamide (Xtandi) received FDA approval in 1995 and 2012, respectively, for managing PCa (Dannah R. Miller et al. 2021).

6.2.3 Taxanes

Taxanes, a class of anticancer medications, work by stabilizing microtubules, which inhibit cell division and ultimately result in the demise of rapidly proliferating tumor cells. Paclitaxel (Taxol), the inaugural and most extensively utilized anti-microtubule agent, received FDA approval in 1998 for treating advanced cancer. Docetaxel (Taxotere), among the limited drugs endorsed by the FDA for treating CRPCa, has exhibited a 2.4-month survival advantage when combined with prednisone compared to mitoxantrone. Cabazitaxel (Jevtana), another member of the taxane family, is employed to combat docetaxel resistance in various cancers. Notably, resistance to taxanes may emerge due to the heightened expression of the ABC1 transporter P-glycoprotein. It's imperative to recognize that taxanes can induce severe adverse effects in patients due to their considerable toxicity (Debes and Tindall 2004; Dillard et al. 2008).

6.2.4 Biological Therapy

Various biological therapies have been developed, including immune checkpoint inhibitors, monoclonal antibodies, adoptive cell therapy, cancer vaccines, monoclonal antibodies, and cytokines, for enhanced specificity and better outcomes. Immunotherapy represents a promising area of biological treatment, leveraging the immune system to slow down tumor growth by targeting tumor-associated antigens (TAAs) or interfering with molecular pathways that facilitate tumor expansion. Ipilimumab, Pembrolizumab, and Nivolumab exemplify immunotherapies that bolster immune cell activity, rendering them more targeted and effective (De Bono et al. 2010; Lombard et al. 2017). Sipuleucel-T, an immunological product derived from peripheral blood mononuclear cells through leukapheresis, received FDA approval in 2010 for use as a first- or second-line therapy in treating asymptomatic or minimally symptomatic metastatic CRPCa (Miller et al. 2021). Moreover, the PROSTVAC immunotherapy protocol involves an initial immunotherapy employing a recombinant vaccinia vector, followed by several booster doses utilizing a recombinant fowlpox vector. This approach employs a heterologous prime-boost strategy. Poxviruses, which are double-stranded DNA viruses belonging to the poxvirus family, replicate within the cytoplasm of infected cells without integrating into the host cell genome (Kantoff et al. 2010a). Upon subcutaneous injection and subsequent uptake by dendritic cells, PROSTVAC's inherent immunogenicity triggers a potent immune response against the viral protein, which includes the tumor antigen PSA encoded within it (Singh et al. 2015). Olaparib and Rucaparib, poly-ADP ribose polymerase inhibitors (PARPi), are emerging as promising therapeutic choices for PCa treatment (Taylor et al. 2023).

6.2.5 Combination Treatments

PCa frequently exhibits resistance to and progression on the diverse therapies mentioned earlier. Combining current treatment strategies could potentially enhance tumor suppression and prolong patient lifespan. Many of these combinations incorporate ADT, while combinations with radiation, chemotherapy, and immunotherapy hold considerable potential. ADT, established as the standard treatment for metastatic PCa, offers a wider array of effective treatment combinations than the other FDA-approved drugs for CRPCa. It's increasingly recommended to employ ADT before, during, and after radiation therapy with intermediate- or high-risk disease. Combining ADT with radiotherapy results in improved overall survival, biochemical-free survival, distant metastasis-free survival, and disease-free survival rates compared to radiotherapy alone (Fessas et al. 2017; Vieweg 2007).

Combining ADT with chemotherapy presents another potential treatment alternative to effectively target both androgen-sensitive (AS) and androgen-independent (AI) PCa cells. Combining ADT with docetaxel has been shown to extend survival by 5.7 months. Indeed, it also raises the risk of severe side effects and deaths attributed to docetaxel toxicity (Bolla et al. 2013). On the contrary, research indicates that combining ADT with docetaxel yields a survival benefit of approximately 10 months compared to ADT alone (Shipley et al. 2017). Immunotherapy, particularly cancer vaccines, presents an additional hopeful avenue for combination therapy with ADT. An example is the PSA-targeted poxviral vaccine, PROSTVAC-IF, that initially demonstrated a 44% reduction in death rates and an 8.5-month increase in survival for PCa patients when used alone (Gravis et al. 2013). In conclusion, combining these approaches will help achieve our primary objective of decreasing the mortality associated with PCa. While improving existing combinations is crucial, as no single agent or combination has proven to cure patients with CRPCa, substantial advancements are ongoing, leading to prolonged survival for those with advanced PCa.

6.3 Biologics and Prostate Cancer

Biologics are medications derived from living organisms or their components, including proteins, antibodies, nucleic acids, and cellular therapies. These drugs target specific molecular pathways involved in disease processes, offering precision and efficacy in treatment. Biologics have revolutionized the management of various diseases, including cancer, autoimmune disorders, infectious diseases, and genetic disorders (Makurvet 2021).

Biologics are complex molecules produced through living organisms or biological systems. They often consist of large, structurally complex proteins or nucleic acids, which are challenging to synthesize using traditional chemical methods. Biologics typically target specific molecular targets or pathways involved in disease pathogenesis, leading to more targeted and precise therapeutic effects than traditional small-molecule drugs. Due to their biological origin, biologics may elicit immune responses in some patients, producing anti-drug antibodies (Kling 2012). Managing immunogenicity is a critical consideration in the development and clinical use of biological therapies. Biologics are produced through sophisticated biotechnological processes, often involving recombinant DNA technology, cell culture systems, or transgenic organisms (Burnett and Rossi 2012). Manufacturing biologics requires stringent quality control measures to ensure product consistency and safety (Škalko-Basnet 2014).

Biologics can be engineered to target molecules or pathways involved in disease pathogenesis, specifically minimizing off-target effects and improving therapeutic efficacy. By targeting specific cellular receptors, enzymes, or signaling molecules, biologics can precisely disrupt disease processes, leading to more effective treatment outcomes. For example, monoclonal antibodies such as trastuzumab target the HER2 receptor in breast cancer cells, inhibiting tumor growth and improving patient outcomes (Mann and Kullberg 2016). Biologics often exhibit reduced toxicity compared to traditional chemotherapeutic agents due to their targeted mechanism of action. By selectively targeting diseased cells or pathways, biologics can minimize adverse effects and improve tolerability, enhancing patients' quality of life. This advantage is particularly evident in treating autoimmune diseases, where biologics can modulate aberrant immune responses without causing widespread immunosuppression (Gibbons and Hyrich 2009). Biologics offer opportunities for personalized medicine approaches by targeting specific molecular abnormalities or biomarkers associated with individual patients' diseases. Using companion diagnostics or biomarker testing, clinicians can identify patients most likely to benefit from biologic therapy, optimizing treatment selection and improving patient outcomes. This personalized approach has been particularly successful in oncology, where targeted biological therapies have revolutionized the treatment of cancers with specific molecular alterations, such as EGFR mutations in non-small cell lung cancer (Lynch et al. 2004). Some biological therapies exert their effects by modulating the immune system by enhancing anti-tumor immune responses or suppressing aberrant immune activation in autoimmune diseases. Immunomodulatory biologics, such as checkpoint inhibitors or cytokine therapies, have shown significant efficacy in various cancers by unleashing the body's immune system to recognize and eliminate tumor cells. Additionally, biologics targeting pro-inflammatory cytokines or immune cells have demonstrated efficacy in diseases characterized by dysregulated immune responses, such as rheumatoid arthritis or inflammatory bowel disease (Pardoll 2012). It also offers access to novel mechanisms that may not be achievable with traditional small-molecule drugs, expanding the therapeutic options available to patients. Biologics can address previously unexplored medical needs and provide innovative solutions for complex diseases by using diverse biological pathways and targets in living organisms. Examples include gene therapies, oncolytic viruses, and cell-based therapies, which harness the power of genetic engineering and cellular biology to deliver targeted and durable treatment effects (Rosenberg et al. 2011).

Immunotherapy is a promising investigational approach to biological therapy, leveraging the immune system to slow down malignant growth by inhibiting TAAs or targeting molecular pathways which facilitate tumor growth. Immunotherapy has shown promising results in treating various types of cancer, including melanoma, lung cancer, and bladder cancer (Fay and Graff 2020). It is a rapidly evolving field of cancer treatment that continues to offer new and innovative ways to harness the power of the immune system in fighting cancer. Different types of immunotherapy approaches include immune checkpoint inhibitors, monoclonal antibodies, cancer vaccines, adoptive cell therapy, and cytokines. Each approach uniquely enhances the immune response against cancer cells (Vieweg 2007).

Monoclonal antibody-based therapies were among the first immunotherapies to gain FDA approval for the treatment of solid tumors and blood cancers. These antibodies can be modified by attaching radioisotopes, cytotoxic agents, or immunotoxins, serving as vehicles to deliver these compounds directly to cancer cells in a targeted manner. In PCa research, therapeutic antibodies are being explored to target various TAAs such as PSA, mucin, prostate-specific membrane antigen (PSMA), as well as cell surface receptors like EGFR, and human epidermal growth factor receptor 2 (HER2) (Vieweg and Jackson 2004). For instance, ipilimumab is a monoclonal antibody that targets cytotoxic T-lymphocyte-associated protein 4 (CTLA-4), a protein expressed on activated T cells that acts as a negative regulator of T-cell function.

By blocking CTLA-4, ipilimumab enhances T-cell activation and promotes an anti-tumor immune response. Moreover, Pembrolizumab and nivolumab are immune checkpoint inhibitors that target the programmed cell death protein 1 (PD-1) receptor on T-cells (Kapadia and Fong 2005; Weber 2005). By blocking the interaction between PD-1 and its ligands (PD-L1 and PD-L2) on cancer cells, these antibodies prevent the inhibition of T-cell activity, allowing the immune system to recognize and attack cancer cells (Fessas et al. 2017). Additionally, CP-675,206 and MDX-010 antibodies block human CTLA-4 and enhance vaccine-induced antitumor responses by potentiating the immune response against cancer cells (Hurwitz et al. 2000).

Several vaccine approaches have developed for the management of PCa, including Sipuleucel-T, an FDA-approved therapeutic cancer vaccine for metastatic-CRPCa (mCRPCa), PROSTVAC vaccine (PSA-TRICOM) that uses a recombinant vaccinia virus and a fowlpox virus to deliver prostate-specific antigen (PSA) to the immune system and aims to enhance anti-PSA immunity in patients with advanced PCa, leading to stabilized PSA levels and reduced clinical progression (Kantoff et al. 2010b; Lubaroff 2012; Mulders et al. 2015), DCVax®-Prostate is a dendritic cell-based vaccine that has shown safety and induced robust immune responses in patients with hormone-refractory, nonmetastatic PCa (Fishman 2009), and GVAX vaccine employs irradiated PC cells that are genetically altered to generate granulocyte-macrophage colony-stimulating factor (GM-CSF) and stimulate immune cells to target and destroy cancer cells (Simons and Sacks 2006).

6.4 Future of Biosimilars in Prostate Cancer

Biologics have reshaped the treatment of multiple acute and chronic conditions in recent decades, including hormonal deficiencies, autoimmune and inflammatory diseases, as well as solid tumors and hematological malignancies. Nonetheless, the development of biological drugs is both costly and time-intensive. Consequently, the entire biopharmaceutical industry is redirecting its attention toward the development of "Biosimilars" (Mestre-Ferrandiz et al. 2016). Biosimilars are biopharmaceuticals closely resembling existing reference products in terms of safety and efficacy. While sharing similar amino acid sequences with their reference counterparts, biosimilars possess distinct attributes like isoform profiles, glycosylation sites, 3D structures, and protein aggregation (Janjigian et al. 2018; Kaida-Yip et al. 2018).

As discussed before, multiple biologics are employed in treating PCa. Yet, as far as we are aware, no biosimilars are currently available on the market or undergoing clinical trials.

In the future, the development and availability of biosimilars for the treatment of PCa hold promising potential. While they are not yet widely available in the market, the landscape is evolving rapidly. Many biological drugs currently in use for various conditions have the potential to be developed as biosimilars for PCa treatment. Moreover, the availability of anti-cancer biosimilars targeting molecular pathways relevant to PCa could further expand treatment options. By leveraging existing knowledge and infrastructure, researchers and pharmaceutical companies can expedite the development of these biosimilars, potentially bringing them to market more quickly.

In summary, while biosimilars for PCa treatment are not currently widely available, the future holds promise for their development and utilization. With ongoing research, technological advancements, and regulatory support, biosimilars have the potential to improve the accessibility and affordability of treatment for PCa patients.

References

Adler HL, McCurdy MA, Kattan MW, Timme TL, Scardino PT, Thompson TC (1999) Elevated levels of circulating interleukin-6 and transforming growth factor-beta 1 in patients with metastatic prostatic carcinoma. J Urol 161(1):182–187. https://doi.org/10.1016/S0022-5347(01)62092-5

Ali A, Kulik G (2021) Signaling pathways that control apoptosis in prostate cancer. Cancers 13(5):1–35. https://doi.org/10.3390/CANCERS13050937

Allott EH, Masko EM, Freedland SJ (2013) Obesity and prostate cancer: weighing the evidence. Eur Urol 63(5):800–809. https://doi.org/10.1016/J.EURURO.2012.11.013

Bach C, Pisipati S, Daneshwar D, Wright M, Rowe E, Gillatt D, Persad R, Koupparis A (2014) The status of surgery in the management of high-risk prostate cancer. Nat Rev Urol 11(6):342–351. https://doi.org/10.1038/nrurol.2014.100

Barsouk A, Padala SA, Vakiti A, Mohammed A, Saginala K, Thandra KC, Rawla P, Barsouk A (2020) Epidemiology, staging and management of prostate cancer. Med Sci 8:28. https://doi.org/10.3390/MEDSCI8030028

Bill-Axelson A, Holmberg L, Ruutu M, Garmo H, Stark JR, Busch CD, Nordling S, Häggman M, Andersson S-O, Bratell S, Spångberg A, Palmgren J, Steineck G, Adami H-O, Johansson J-E (2011) Radical prostatectomy versus watchful waiting in early prostate cancer. N Engl J Med 364(18):1708–1717. https://doi.org/10.1056/NEJMOA1011967

Blobe GC, Schiemann WP, Lodish HF (2000) Role of transforming growth factor beta in human disease. N Engl J Med 342(18):1350–1358. https://doi.org/10.1056/NEJM200005043421807

Bolla M, Verry C, Long JA (2013) High-risk prostate cancer: Combination of high-dose, high-precision radiotherapy and androgen deprivation therapy. Current Opinion in Urology 23(4):349–354. https://doi.org/10.1097/MOU.0B013E328361EBFD

Bray F, Ferlay J, Soerjomataram I, Siegel RL, Torre LA, Jemal A (2018) Global cancer statistics 2018: GLOBOCAN estimates of incidence and mortality worldwide for 36 cancers in 185 countries. CA Cancer J Clin 68(6):394–424. https://doi.org/10.3322/CAAC.21492

Burnett JC, Rossi JJ (2012) RNA-based therapeutics: current progress and future prospects. Chem Biol 19(1):60–71. https://doi.org/10.1016/j.chembiol.2011.12.008

Cantley LC (2002) The phosphoinositide 3-kinase pathway. Science 296(5573):1655–1657. https://doi.org/10.1126/SCIENCE.296.5573.1655

Cao Y, Giovannucci E (2016) Obesity and prostate cancer. Recent Results Cancer Res 208:137–153. https://doi.org/10.1007/978-3-319-42542-9_8/COVER

Carey AM, Pramanik R, Nicholson LJ, Dew TK, Martin FL, Muir GH, Morris JDH (2007) Ras-MEK-ERK signaling cascade regulates androgen receptor element-inducible gene transcription and DNA synthesis in prostate cancer cells. Int J Cancer 121(3):520–527. https://doi.org/10.1002/IJC.22715

Castoria G, D'Amato L, Ciociola A, Giovannelli P, Giraldi T, Sepe L, Paolella G, Barone MV, Migliaccio A, Auricchio F (2011) Androgen-induced cell migration: role of androgen receptor/Filamin A association. PLoS One 6(2):e17218. https://doi.org/10.1371/JOURNAL.PONE.0017218

Castro E, Goh C, Olmos D, Saunders E, Leongamornlert D, Tymrakiewicz M, Mahmud N, Dadaev T, Govindasami K, Guy M, Sawyer E, Wilkinson R, Ardern-Jones A, Ellis S, Frost D, Peock S, Evans DG, Tischkowitz M, Cole T et al (2013) Germline BRCA mutations are associated with higher risk of nodal involvement, distant metastasis, and poor survival outcomes in prostate cancer. J Clin Oncol 31(14):1748. https://doi.org/10.1200/JCO.2012.43.1882

Chowdhury-Paulino IM, Ericsson C, Vince R, Spratt DE, George DJ, Mucci LA (2021) Racial disparities in prostate cancer among black men: epidemiology and outcomes. Prostate Cancer Prostatic Dis 25(3):397–402. https://doi.org/10.1038/s41391-021-00451-z

Culp MBB, Soerjomataram I, Efstathiou JA, Bray F, Jemal A (2020) Recent Global Patterns in Prostate Cancer Incidence and Mortality Rates. European Urology 77(1):38–52. https://doi.org/10.1016/J.EURURO.2019.08.005

De Bono JS, Oudard S, Ozguroglu M, Hansen S, MacHiels JP, Kocak I, Gravis G, Bodrogi I, MacKenzie MJ, Shen L, Roessner M, Gupta S, Sartor AO (2010) Prednisone plus cabazitaxel or mitoxantrone for metastatic castration-resistant prostate cancer progressing after docetaxel treatment: A randomised open-label trial. Lancet 376(9747):1147–1154. https://doi.org/10.1016/S0140-6736(10)61389-X

De Nunzio C, Andriole GL, Thompson IM, Freedland SJ (2015) Smoking and prostate cancer: A systematic review. Eur Urol Focus 1(1):28–38. https://doi.org/10.1016/J.EUF.2014.10.002

Debes JD, Tindall DJ (2004) Mechanisms of androgen-refractory. Prostate Cancer 351(15):1488–1490. https://doi.org/10.1056/NEJMP048178

Di Cristofano A, Pesce B, Cordon-Cardo C, Pandolfi PP (1998) Pten is essential for embryonic development and tumour suppression. Nat Genet 19(4):348–355. https://doi.org/10.1038/1235

Di Cristofano A, De Acetis M, Koff A, Cordon-Cardo C, Pandolfi P (2001) Pten and p27KIP1 cooperate in prostate cancer tumor suppression in the mouse. Nat Genet 27(2):222–224. https://doi.org/10.1038/84879

Diaz Arguello OA, Haisma HJ (2021) Apoptosis-inducing TNF superfamily ligands for cancer therapy. Cancers 13(7). https://doi.org/10.3390/CANCERS13071543

Dillard PR, Lin MF, Khan SA (2008) Androgen-independent prostate cancer cells acquire the complete steroidogenic potential of synthesizing testosterone from cholesterol. Mol Cell Endocrinol 295(1–2):115–120. https://doi.org/10.1016/J.MCE.2008.08.013

Discacciati A, Wolk A (2014) Lifestyle and dietary factors in prostate cancer prevention. Recent Results Cancer Res 202:27–37. https://doi.org/10.1007/978-3-642-45195-9_3/COVER

Donkor MK, Sarkar A, Li MO (2012) Tgf-β1 produced by activated CD4+ T cells antagonizes T cell surveillance of tumor development. Onco Targets Ther 1(2):162–171. https://doi.org/10.4161/ONCI.1.2.18481

Fay EK, Graff JN (2020) Immunotherapy in prostate cancer. Cancers 12(7):1752. https://doi.org/10.3390/CANCERS12071752

Fessas P, Lee H, Ikemizu S, Janowitz T (2017) A molecular and preclinical comparison of the PD-1–targeted T-cell checkpoint inhibitors nivolumab and pembrolizumab. Semin Oncol 44(2):136–140. https://doi.org/10.1053/J.SEMINONCOL.2017.06.002

Filson CP, Marks LS, Litwin MS (2015) Expectant management for men with early stage prostate cancer. CA Cancer J Clin 65(4):264–282. https://doi.org/10.3322/CAAC.21278

Fishman M (2009) A changing world for DCvax: a PSMA loaded autologous dendritic cell vaccine for prostate cancer. Expert Opin Biol Ther 9(12):1565–1575. https://doi.org/10.1517/14712590903446921

Foreman KJ, Marquez N, Dolgert A, Fukutaki K, Fullman N, McGaughey M, Pletcher MA, Smith AE, Tang K, Yuan CW, Brown JC, Friedman J, He J, Heuton KR, Holmberg M, Patel DJ, Reidy P, Carter A, Cercy K et al (2018) Forecasting life expectancy, years of life lost, and all-cause and cause-specific mortality for 250 causes of death: reference and alternative scenarios for 2016-40 for 195 countries and territories. Lancet (London, England) 392(10159):2052–2090. https://doi.org/10.1016/S0140-6736(18)31694-5

Frank DA (2007) STAT3 as a central mediator of neoplastic cellular transformation. Cancer Lett 251(2):199–210. https://doi.org/10.1016/J.CANLET.2006.10.017

Gibbons LJ, Hyrich KL (2009) Biologic therapy for rheumatoid arthritis: clinical efficacy and predictors of response. BioDrugs 23(2):111–124. https://doi.org/10.2165/00063030-200923020-00004/METRICS

Gioeli D, Mandell JW, Petroni GR, Frierson HF, Weber MJ (n.d.) Activation of mitogen-activated protein kinase associated with prostate cancer progression. http://aacrjournals.org/cancerres/article-pdf/59/2/279/3238693/ch029900279o.pdf

Goh CL, Eeles RA (2014) Germline genetic variants associated with prostate cancer and potential relevance to clinical practice. Recent Results Cancer Res 202:9–26. https://doi.org/10.1007/978-3-642-45195-9_2/COVER

Goy BW, Burchette R, Soper MS, Chang T, Cosmatos HA (2020) Ten-year treatment outcomes of radical prostatectomy vs external beam radiation therapy vs brachytherapy for 1503 patients with intermediate-risk prostate cancer. Urology 136:180–189. https://doi.org/10.1016/J.UROLOGY.2019.09.040

Gravis G, Fizazi K, Joly F, Oudard S, Priou F, Esterni B, Latorzeff I, Delva R, Krakowski I, Laguerre B, Rolland F, Théodore C, Deplanque G, Ferrero JM, Pouessel D, Mourey L, Beuzeboc P, Zanetta S, Habibian M et al (2013) Androgen-deprivation therapy alone or with docetaxel in non-castrate metastatic prostate cancer (GETUG-AFU 15): A randomised, open-label, phase 3 trial. Lancet Oncol 14(2):149–158. https://doi.org/10.1016/S1470-2045(12)70560-0

Green SM, Mostaghel EA, Nelson PS (2012) Androgen action and metabolism in prostate cancer. Mol Cell Endocrinol 360(1–2):3–13. https://doi.org/10.1016/J.MCE.2011.09.046

Gupta S, Iljin K, Sara H, Mpindi JP, Mirtti T, Vainio P, Rantala J, Alanen K, Nees M, Kallioniemi O (2010) FZD4 as a mediator of ERG oncogene-induced WNT signaling and epithelial-to-mesenchymal transition in human prostate cancer cells. Cancer Res 70(17):6735–6745. https://doi.org/10.1158/0008-5472.CAN-10-0244/656273/P/FZD4-AS-A-MEDIATOR-OF-ERG-ONCOGENE-INDUCED-WNT

Hamdy FC, Donovan JL, Lane JA, Mason M, Metcalfe C, Holding P, Davis M, Peters TJ, Turner EL, Martin RM, Oxley J, Robinson M, Staffurth J, Walsh E, Bollina P, Catto J, Doble A, Doherty A, Gillatt D et al (2016) 10-year outcomes after monitoring, surgery, or radiotherapy for localized prostate cancer. N Engl J Med 375(15):1415–1424. https://doi.org/10.1056/NEJMOA1606220/SUPPL_FILE/NEJMOA1606220_DISCLOSURES.PDF

He Y, Sun MM, Zhang GG, Yang J, Chen KS, Xu WW, Li B (2021) Targeting PI3K/Akt signal transduction for cancer therapy. Signal Transduc Target Ther 6(1). https://doi.org/10.1038/S41392-021-00828-5

Hour MJ, Tsai SC, Wu HC, Lin MW, Chung JG, Wu JB, Chiang JH, Tsuzuki M, Yang JS (2012) Antitumor effects of the novel quinazolinone MJ-33: inhibition of metastasis through the MAPK, AKT, NF-κB and AP-1 signaling pathways in DU145 human prostate cancer cells. Int J Oncol 41(4):1513–1519. https://doi.org/10.3892/ijo.2012.1560

Hurwitz AA, Foster BA, Kwon ED, Truong T, Choi EM, Greenberg NM, Burg MB, Allison JP (2000) Combination immunotherapy of primary prostate cancer in a transgenic mouse model using CTLA-4 blockade 1. Cancer Res 60:2444. http://aacrjournals.org/cancerres/article-pdf/60/9/2444/3240675/ch090002444p.pdf

James SL, Abate D, Abate KH, Abay SM, Abbafati C, Abbasi N, Abbastabar H, Abd-Allah F, Abdela J, Abdelalim A, Abdollahpour I, Abdulkader RS, Abebe Z, Abera SF, Abil OZ, Abraha HN, Abu-Raddad LJ, Abu-Rmeileh NME, Accrombessi MMK et al (2018) Global, regional, and national incidence, prevalence, and years lived with disability for 354 diseases and injuries for 195 countries and territories, 1990-2017: a systematic analysis for the Global Burden of Disease Study 2017. Lancet 392(10159):1789–1858. https://doi.org/10.1016/S0140-6736(18)32279-7

Janjigian YY, Bissig M, Curigliano G, Coppola J, Latymer M (2018) Talking to patients about biosimilars. Future Oncol 14(23):2403–2414. https://doi.org/10.2217/FON-2018-0044/ASSET/IMAGES/LARGE/FIGURE1.JPEG

Jeong JH, Wang Z, Guimaraes AS, Ouyang X, Figueiredo JL, Ding Z, Jiang S, Guney I, Kang GH, Shin E, Hahn WC, Loda MF, Abate-Shen C, Weissleder R, Chin L (2008) BRAF activation initiates but does not maintain invasive prostate adenocarcinoma. PLoS One 3(12):e3949. https://doi.org/10.1371/JOURNAL.PONE.0003949

Jones MR, Joshu CE, Kanarek N, Navas-Acien A, Richardson KA, Platz EA (2016) Peer reviewed: cigarette smoking and prostate cancer mortality in four US states, 1999–2010. Prev Chronic Dis 13(4). https://doi.org/10.5888/PCD13.150454

Junejo NN, AlKhateeb SS (2020) BRCA2 gene mutation and prostate cancer risk: comprehensive review and update. Saudi Med J 41(1):9. https://doi.org/10.15537/SMJ.2020.1.24759

Kaida-Yip F, Deshpande K, Saran T, Vyas D (2018) Biosimilars: review of current applications, obstacles, and their future in medicine. World J Clin Cases 6(8):161. https://doi.org/10.12998/WJCC.V6.I8.161

Kang MH, Reynolds CP (2009) Bcl-2 inhibitors: targeting mitochondrial apoptotic pathways in cancer therapy. Clin Cancer Res 15(4):1126. https://doi.org/10.1158/1078-0432.CCR-08-0144

Kang HY, Huang HY, Hsieh CY, Li CF, Shyr CR, Tsai MY, Chang C, Chuang YC, Huang KE (2009) Activin A enhances prostate cancer cell migration through activation of androgen receptor and is overexpressed in metastatic prostate cancer. J Bone Miner Res 24(7):1180–1193. https://doi.org/10.1359/JBMR.090219

Kantoff PW, Higano CS, Shore ND, Berger ER, Small EJ, Penson DF, Redfern CH, Ferrari AC, Dreicer R, Sims RB, Xu Y, Frohlich MW, Schellhammer PF (2010a) Sipuleucel-T immunotherapy for castration-resistant prostate cancer. N Engl J Med 363(5):411–422. https://doi.org/10.1056/NEJMOA1001294/SUPPL_FILE/NEJMOA1001294_DISCLOSURES.PDF

Kantoff PW, Schuetz TJ, Blumenstein BA, Michael Glode L, Bilhartz DL, Wyand M, Manson K, Panicali DL, Laus R, Schlom J, Dahut WL, Arlen PM, Gulley JL, Godfrey WR (2010b) Overall survival analysis of a phase II randomized controlled trial of a poxviral-based PSA-targeted immunotherapy in metastatic castration-resistant prostate cancer. J Clin Oncol 28(7):1099. https://doi.org/10.1200/JCO.2009.25.0597

Kapadia D, Fong L (2005) CTLA-4 blockade: autoimmunity as treatment. J Clin Oncol 23(35):8926–8928. https://doi.org/10.1200/JCO.2005.07.012

Kiciński M, Vangronsveld J, Nawrot TS (2011) An epidemiological reappraisal of the familial aggregation of prostate cancer: a meta-analysis. PLoS One 6(10):e27130. https://doi.org/10.1371/JOURNAL.PONE.0027130

Kinkade CW, Castillo-Martin M, Puzio-Kuter A, Yan J, Foster TH, Gao H, Sun Y, Ouyang X, Gerald WL, Cordon-Cardo C, Abate-Shen C (2008) Targeting AKT/mTOR and ERK MAPK signaling inhibits hormone-refractory prostate cancer in a preclinical mouse model. J Clin Invest 118(9):3051–3064. https://doi.org/10.1172/JCI34764

Kling J (2012) Fresh from the biotech pipeline--2011: drug approvals were up in 2011, reversing the trend of the last decade. Jim Kling reports. Nat Biotechnol 30(2):128–132. https://go.gale.com/ps/i.do?p=AONE&sw=w&issn=10870156&v=2.1&it=r&id=GALE%7CA281520386&sid=googleScholar&linkaccess=fulltext

Liu J, Geller J, Albert J, Kirshner M (1985) Acute effects of testicular and adrenal cortical blockade on protein synthesis and dihydrotestosterone content of human prostate tissue. J Clin Endocrinol Metabol 61(1):129–133. https://doi.org/10.1210/JCEM-61-1-129

Lombard AP, Liu C, Armstrong CM, Cucchiara V, Gu X, Lou W, Evans CP, Gao AC (2017) ABCB1 mediates cabazitaxel–docetaxel cross-resistance in advanced prostate cancer. Mol Cancer Ther 16(10):2257–2266. https://doi.org/10.1158/1535-7163.MCT-17-0179/87190/AM/ABCB1-MEDIATES-CABAZITAXEL-DOCETAXEL-CROSS

Lonergan P, Tindall D (2011) Androgen receptor signaling in prostate cancer development and progression. J Carcinogen 10. https://doi.org/10.4103/1477-3163.83937

Lubaroff DM (2012) Prostate cancer vaccines in clinical trials. Expert Rev Vaccines 11(7):857–868. https://doi.org/10.1586/ERV.12.54

Lynch TJ, Bell DW, Sordella R, Gurubhagavatula S, Okimoto RA, Brannigan BW, Harris PL, Haserlat SM, Supko JG, Haluska FG, Louis DN, Christiani DC, Settleman J, Haber DA (2004) Activating mutations in the epidermal growth factor receptor underlying responsiveness of non–Small-cell lung cancer to Gefitinib. N Engl J Med 350(21):2129–2139. https://doi.org/10.1056/NEJMOA040938/SUPPL_FILE/NEJMOA040938_APPENDIX.PDF

Makurvet FD (2021) Biologics vs. small molecules: drug costs and patient access. Med Drug Discov 9:100075. https://doi.org/10.1016/J.MEDIDD.2020.100075

Mann K, Kullberg M (2016) Trastuzumab-targeted gene delivery to Her2-overexpressing breast cancer cells. Cancer Gene Ther 23(7):221–228. https://doi.org/10.1038/cgt.2016.21

Massagué J, Blain SW, Lo RS (2000) TGFβ signaling in growth control, cancer, and heritable disorders. Cell 103(2):295–309. https://doi.org/10.1016/S0092-8674(00)00121-5

Mestre-Ferrandiz J, Towse A, Berdud M (2016) Biosimilars: how can payers get long-term savings? PharmacoEconomics 34(6):609–616. https://doi.org/10.1007/S40273-015-0380-X/METRICS

Miles FL, Tung NS, Aguiar AA, Kurtoglu S, Sikes RA (2012) Increased TGF-β1-mediated suppression of growth and motility in castrate-resistant prostate cancer cells is consistent with Smad2/3 signaling. Prostate 72(12):1339–1350. https://doi.org/10.1002/PROS.22482

Miller DR, Ingersoll MA, Teply BA, Lin M-F (2021) Combination treatment options for castration-resistant prostate cancer. Exon Publications, pp 59–79. https://doi.org/10.36255/EXONPUBLICATIONS.PROSTATECANCER.COMBINATIONTREATMENT.2021

Morgan TM, Koreckij TD, Corey E (2009) Targeted therapy for advanced prostate cancer: inhibition of the PI3K/Akt/mTOR pathway. Curr Cancer Drug Targets 9(2):237–249. https://doi.org/10.2174/156800909787580999

Mulders PF, De Santis M, Powles T, Fizazi K (2015) Targeted treatment of metastatic castration-resistant prostate cancer with sipuleucel-T immunotherapy. Cancer Immunol Immunother 64(6):655–663. https://doi.org/10.1007/S00262-015-1707-3/TABLES/1

Okobia MN, Zmuda JM, Ferrell RE, Patrick AL, Bunker CH (2011) Chromosome 8q24 variants are associated with prostate cancer risk in a high risk population of African ancestry. Prostate 71(10):1054–1063. https://doi.org/10.1002/PROS.21320

Palanisamy N, Ateeq B, Kalyana-Sundaram S, Pflueger D, Ramnarayanan K, Shankar S, Han B, Cao Q, Cao X, Suleman K, Kumar-Sinha C, Dhanasekaran SM, Chen YB, Esgueva R, Banerjee S, Lafargue CJ, Siddiqui J, Demichelis F, Moeller P et al (2010) Rearrangements of the RAF kinase pathway in prostate cancer, gastric cancer and melanoma. Nat Med 16(7):793–798. https://doi.org/10.1038/nm.2166

Pardoll DM (2012) The blockade of immune checkpoints in cancer immunotherapy. Nat Rev Cancer 12(4):252–264. https://doi.org/10.1038/nrc3239

Pernar CH, Ebot EM, Wilson KM, Mucci LA (2018) The epidemiology of prostate cancer. Cold Spring Harb Perspect Med 8(12):a030361. https://doi.org/10.1101/cshperspect.a030361

Plati J, Bucur O, Khosravi-Far R (2008) Dysregulation of apoptotic signaling in cancer: molecular mechanisms and therapeutic opportunities. J Cell Biochem 104(4):1124. https://doi.org/10.1002/JCB.21707

Poliseno L, Salmena L, Riccardi L, Fornari A, Song MS, Hobbs RM, Sportoletti P, Varmeh S, Egia A, Fedele G, Rameh L, Loda M, Pandolfi PP (2010) Identification of the miR-106b~25 microRNA cluster as a proto-oncogenic PTEN-targeting intron that cooperates with its host gene MCM7 in transformation. Science Signaling 3(117). https://doi.org/10.1126/SCISIGNAL.2000594/SUPPL_FILE/3_RA29_SM.PDF

Rawla P (2019) Epidemiology of prostate cancer. World J Oncol 10(2):63. https://doi.org/10.14740/WJON1191

Robbins CM, Hooker S, Kittles RA, Carpten JD (2011) EphB2 SNPs and sporadic prostate cancer risk in African American men. PLoS One 6(5):e19494. https://doi.org/10.1371/JOURNAL.PONE.0019494

Rosenberg SA, Yang JC, Sherry RM, Kammula US, Hughes MS, Phan GQ, Citrin DE, Restifo NP, Robbins PF, Wunderlich JR, Morton KE, Laurencot CM, Steinberg SM, White DE, Dudley ME (2011) Durable complete responses in heavily pretreated patients with metastatic melanoma using T-cell transfer immunotherapy. Clin Cancer Res 17(13):4550–4557. https://doi.org/10.1158/1078-0432.CCR-11-0116/84227/AM/DURABLE-COMPLETE-RESPONSES-IN-HEAVILY-PRETREATED

Sandhu S, Moore CM, Chiong E, Beltran H, Bristow RG, Williams SG (2021) Prostate cancer. Lancet (London, England) 398(10305):1075–1090. https://doi.org/10.1016/S0140-6736(21)00950-8
Schmierer B, Hill CS (2007) TGFβ–SMAD signal transduction: molecular specificity and functional flexibility. Nat Rev Mol Cell Biol 8(12):970–982. https://doi.org/10.1038/nrm2297
Shipley WU, Seiferheld W, Lukka HR, Major PP, Heney NM, Grignon DJ, Sartor O, Patel MP, Bahary J-P, Zietman AL, Pisansky TM, Zeitzer KL, Lawton CAF, Feng FY, Lovett RD, Balogh AG, Souhami L, Rosenthal SA, Kerlin KJ, Sandler HM (2017) Radiation with or without Antiandrogen Therapy in Recurrent Prostate Cancer. New England Journal of Medicine 376(5):417–428. https://doi.org/10.1056/NEJMOA1607529/SUPPL_FILE/NEJMOA1607529_DISCLOSURES.PDF
Siegel RL, Miller KD, Fuchs HE, Jemal A (2022) Cancer statistics, 2022. CA Cancer J Clin 72(1):7–33. https://doi.org/10.3322/CAAC.21708
da Silva HB, Amaral EP, Nolasco EL, de Victo NC, Atique R, Jank CC, Anschau V, Zerbini LF, Correa RG (2013) Dissecting major signaling pathways throughout the development of prostate cancer. Prostate Cancer 2013:1–23. https://doi.org/10.1155/2013/920612
Simons JW, Sacks N (2006) Granulocyte-macrophage colony-stimulating factor–transduced allogeneic cancer cellular immunotherapy: the GVAX® vaccine for prostate cancer. Urol Oncol 24(5):419–424. https://doi.org/10.1016/J.UROLONC.2005.08.021
Singh P, Pal SK, Alex A, Agarwal N (2015) Development of PROSTVAC immunotherapy in prostate cancer. Http://Dx.Doi.Org/10.2217/Fon.15.120, 11(15):2137–2148. https://doi.org/10.2217/FON.15.120
Škalko-Basnet N (2014) Biologics: the role of delivery systems in improved therapy. Biologics 8:107–114. https://doi.org/10.2147/BTT.S38387
Smith GD, Pickles T, Crook J, Martin AG, Vigneault E, Cury FL, Morris J, Catton C, Lukka H, Warner A, Yang Y, Rodrigues G (2015) Brachytherapy improves biochemical failure–free survival in low- and intermediate-risk prostate cancer compared with conventionally fractionated external beam radiation therapy: a propensity score matched analysis. Int J Radiat Oncol Biol Phys 91(3):505–516. https://doi.org/10.1016/J.IJROBP.2014.11.018
Song MS, Salmena L, Carracedo A, Egia A, Lo-Coco F, Teruya-Feldstein J, Pandolfi PP (2008) The deubiquitinylation and localization of PTEN are regulated by a HAUSP–PML network. Nature 455(7214):813–817. https://doi.org/10.1038/nature07290
Sung H, Ferlay J, Siegel RL, Laversanne M, Soerjomataram I, Jemal A, Bray F (2021) Global cancer statistics 2020: GLOBOCAN estimates of incidence and mortality worldwide for 36 cancers in 185 countries. CA Cancer J Clin 71(3):209–249. https://doi.org/10.3322/CAAC.21660
Taylor AK, Kosoff D, Emamekhoo H, Lang JM, Kyriakopoulos CE (2023) PARP inhibitors in metastatic prostate cancer. Front Oncol 13:1159557. https://doi.org/10.3389/fonc.2023.1159557
Tutt A, Ashworth A (2002) The relationship between the roles of BRCA genes in DNA repair and cancer predisposition. Trends Mol Med 8(12):571–576. https://doi.org/10.1016/S1471-4914(02)02434-6
Vieweg J (2007) Immunotherapy for advanced prostate cancer. Rev Urol 9(Suppl 1):S29
Vieweg J, Jackson A (2004) Antigenic targets for renal cell carcinoma immunotherapy. Expert Opin Biol Ther 4(11):1791–1801. https://doi.org/10.1517/14712598.4.11.1791
Vo BT, Khan SA (2011) Expression of nodal and nodal receptors in prostate stem cells and prostate cancer cells: autocrine effects on cell proliferation and migration. Prostate 71(10):1084–1096. https://doi.org/10.1002/PROS.21326
Waltering KK, Urbanucci A, Visakorpi T (2012) Androgen receptor (AR) aberrations in castration-resistant prostate cancer. Mol Cell Endocrinol 360(1–2):38–43. https://doi.org/10.1016/J.MCE.2011.12.019
Wan X, Li ZG, Yingling JM, Yang J, Starbuck MW, Ravoori MK, Kundra V, Vazquez E, Navone NM (2012) Effect of transforming growth factor beta (TGF-β) receptor I kinase inhibitor on prostate cancer bone growth. Bone 50(3):695–703. https://doi.org/10.1016/J.BONE.2011.11.022
Weber JS (2005) Tumor regression and autoimmunity in cytotoxic T lymphocyte-associated antigen 4 blockade-treated patients. Ann Surg Oncol 12(12):957–959. https://doi.org/10.1245/ASO.2005.08.917/METRICS

Whyte JL, Smith AA, Helms JA (2012) Wnt signaling and injury repair. Cold Spring Harbor Perspect Biol 4(8). https://doi.org/10.1101/cshperspect.a008078

Wikström P, Stattin P, Franck-Lissbrant I, Damber J-E, Bergh A (1998) Transforming growth factor 1 is associated with angiogenesis, metastasis, and poor clinical outcome in prostate cancer. Prostate 37, 19:–29

Willert K, Brown JD, Danenberg E, Duncan AW, Weissman IL, Reya T, Yates JR, Nusse R (2003) Wnt proteins are lipid-modified and can act as stem cell growth factors. Nature 423(6938):448–452. https://doi.org/10.1038/nature01611

Yardy GW, Brewster SF (2005) Wnt signalling and prostate cancer. Prostate Cancer Prostatic Dis 8(2):119–126. https://doi.org/10.1038/sj.pcan.4500794

Ye L, Kynaston H, Jiang WG (2009) Bone morphogenetic protein-10 suppresses the growth and aggressiveness of prostate cancer cells through a Smad independent pathway. J Urol 181(6):2749–2759. https://doi.org/10.1016/J.JURO.2009.01.098

Yu H, Jove R (2004) The STATs of cancer—new molecular targets come of age. Nat Rev Cancer 4(2):97–105. https://doi.org/10.1038/nrc1275

Yuan X, Li T, Wang H, Zhang T, Barua M, Borgesi RA, Bubley GJ, Lu ML, Balk SP (2006) Androgen receptor remains critical for cell-cycle progression in androgen-independent CWR22 prostate cancer cells. Am J Pathol 169(2):682. https://doi.org/10.2353/AJPATH.2006.051047

Biosimilars in Gastric Cancer

7

Arghya Kusum Dhar and Sidhartha S. Kar

Abstract

Gastric cancer, often known as GC, continues to be one of the most common forms of cancer in the world. Its toll in 2020 exceeded one million new cases and positioned it at the fourth spot in terms of worldwide fatalities. Since its inception, substantial study has been carried out to enhance the prognosis of GC therapy. The advancement of GC therapy in the past few years has rendered biosimilars a viable therapeutic option. Comparability testing is a crucial step in the manufacturing of biosimilars to ensure that the finished product meets the safety, efficacy, and quality standards of the reference product. The biosimilar versions of trastuzumab such as Ontruzant (SB3), Trazimera (PF-05280014), Herzuma (CT-P6), Kanjinti (ABP980), etc.. have demonstrated effectiveness in the treatment of GC, making them a competitive substitute for the reference medicine. By including biosimilars into GC therapy regimens, it is possible to enhance patient access to quality care while taking healthcare expenses into account.

Keywords

Biosimilar · Gastric cancer · Trastuzumab · Safety · Cost-effectiveness

A. K. Dhar
School of Pharmacy, The Neotia University, Sarisa, West Bengal, India

S. S. Kar (✉)
Faculty of Pharmacy, C. V. Raman Global University, Bhubaneswar, Odisha, India

S. Bhatt et al. (eds.), *Biosimilars for Cancer Treatment*,
https://doi.org/10.1007/978-981-97-3001-8_7

7.1 Introduction and Background

With an anticipated 769,000 deaths and over 1 million new cases in 2020, gastric cancer (GC) continues to be a major global health concern, ranking fifth in incidence and fourth in mortality rates, respectively (Sung et al. 2021). The link between H. pylori infection and antral/body GCGC has just been recognized in the last few years. Furthermore, the likelihood of acquiring stomach cancer is substantially raised by an infection with H. pylori. In addition to a persistent H. pylori infection, other risk factors for GC include smoking, consuming too much salt, and being an alcoholic (Guggenheim and Shah 2013; Sano 2017). Patients with localized advanced GC (AGC) have witnessed an improvement in survival rates when they have undergone various perioperative therapy regimens, including as neo-adjuvant, adjuvant, or a combination of the two. As a first-line treatment option for stomach cancer that cannot be surgically removed, a combination of platinum compounds and fluoropyrimidine-based chemotherapy has been shown to improve symptoms, prolong survival, and overall quality of life (Obermannova and Lordick 2016). Increasing toxicity while improving patient survival is possible by adding a third drug, like docetaxel, to the platinum–fluoropyrimidine combination (Van Cutsem et al. 2006).

The human epidermal growth factor receptor 2 (HER2) is an important player in the development of a number of human malignancies (Neve et al. 2001). Twenty-five percent to 24% of stomach and gastroesophageal junction cancers also have HER2 overexpression or amplification (Van Cutsem et al. 2015). Response rates, progression-free survival (PFS), and overall survival (OS) were all enhanced in the Trastuzumab for GC (ToGA) trial when the HER2-targeting monoclonal antibody trastuzumab was added to platinum-fluoropyrimidine treatment in HER2-positive AGC. This was in comparison to chemotherapy alone (Bang et al. 2010a). Compared to the treatment group, individuals allocated to trastuzumab with chemotherapy had a median OS of 13.8 months (95% CI: 0.60–0.91; $P = 0.0046$). For HER2-positive AGC, the gold standard for first-line treatment is trastuzumab with platinum-based chemotherapy. The anti-HER2 target has become the gold standard in cancer treatment thanks to the proliferation of innovative immunotherapies and molecular-targeted therapies, which greatly improve chances of survival of patients.

More patients may be able to affordably use these GC treatments thanks to the recent discovery of biosimilar drugs for trastuzumab and other comparable biological medications. Data from biosimilar immunomodulatory drugs is the primary basis for the estimation that, with increasing usage of biosimilar medications, patients might save about $54 billion over a decade (Mulcahy et al. 2018). Savings predictions, however, are highly dependent on regulatory and usage variables whose precise nature is still uncertain. The term "biosimilar medication" refers to synthetic pharmaceuticals that mimic the pharmacokinetic (PK) and pharmacodynamic (PD) characteristics of their biologically inspired counterparts. When it comes to safety, purity, and potency, there are no discernible discrepancies between a reference product and its biosimilar; furthermore, there are only small variations in components that do not have any clinically significant impact on the efficacy of the drug

(Rugo et al. 2019). The total amount of evidence that supports the suggested medicine as a biosimilar product usually determines how quickly these drugs are approved.

The prospective biosimilar needs a thorough analytical evaluation of its structural and functional properties. The amino acid sequence of the molecule, structural conformation, and target binding factors are all part of the evaluation process. Other aspects include clinical investigation of PK and PD properties, evidence of clinical efficacy and safety that is comparable to the original product, and nonclinical evaluation of activity (Rugo et al. 2019; Santos et al. 2019). Biosimilar drugs, including those for the reference trastuzumab, have been authorized for use in stomach malignancies based on the combined data that supports structural and functional similarity and identical pharmacologic characteristics and efficacy.

7.1.1 Involvement of Human Epidermal Growth Factor Receptor 2 (HER2) in GC

A member of the family of receptors known as epidermal growth factor receptors (EGFRs), HER2, sometimes called Neu or ErbB2, is encoded by the gene ERBB2 on chromosome 17. Extracellular domains (ECDs), transmembrane domains, and intracellular kinase domains are shared features among the four members of this family (HER1, EGFR, HER2, HER3, and HER4) (Tebbutt et al. 2013; Olayioye 2001). Crucial for tumour cell proliferation, apoptosis, adhesion, migration, and differentiation, a number of signal transduction pathways are initiated when various ligands bind to the ECD (Gravalos and Jimeno 2008). It was in 1984 that the 185 kD transmembrane glycoprotein known as the HER2 receptor was initially identified (Schechter et al. 1984).

HER2 cannot bind ligands on its own and must form heterodimers with either HER1 or HER3 in order to be stimulated (Moasser 2007). Major signalling pathways include Ras/MAPK and PI3K/Akt, and HER-2 activates downstream pathways through heterodimerization and tyrosine kinase autophosphorylation facilitated signal transduction. The most active HER signalling dimer is the HER2–HER3 heterodimer, which is essential for HER2-driven tumour oncogenic transformation (Zhu et al. 2021). There is a strong relationship between the pathophysiology of different tumours and these pathways, which regulate cell proliferation, differentiation, and survival (Rezatabar et al. 2019; Yuan et al. 2020; Noorolyai et al. 2019). One main way that HER2-specific antibodies disrupt HER2 signalling is by preventing the formation of heterodimers between growth factor receptors and ErbB-2 (Klapper et al. 1997).

Although HER2 protein overexpression and/or gene amplification are present in various quantities in breast and GCs, ovarian, colon, and pancreatic cancers, the overexpression of HER2 protein is more noticeable in breast and gastric malignancies (Yan et al. 2015).

The death rate and likelihood of distant metastases are higher for HER2-positive breast cancer, which has also been the subject of more research. One of the first molecularly targeted medications, trastuzumab, has a better prognosis and is now prescribed to all patients with early stage HER2-positive cancer (Choong et al. 2020; Dawood et al. 2010; Figueroa-Magalhaes et al. 2014). It was approved in 1998. A recent worldwide study found that 8.8% of Chinese GC patients tested positive for HER2; the overall HER2-positive rate for GC is 17.9% (Abrahao-Machado and Scapulatempo-Neto 2016; Dan et al. 2018). There was a correlation between HER2 overexpression and tumour location, tumour differentiation, lymph node status, venous invasion, lymphatic invasion, Bormann classification, and Lauren's classification in patients. Gastric corpus cancer is less prevalent than gastroesophageal junction cancer (GEJ) among gastric tumours with strong HER2 protein expression. There is a higher prevalence of HER2 positive in the intestinal type compared to the diffuse or mixed types, and the tumours are mostly tubular in shape, with a bigger proportion of carcinomas that are well or moderately differentiated and a lower proportion that are poorly differentiated. Having said that, neither gender nor age nor clinical stages (Liang et al. 2014; Takehana et al. 2002; Van Cutsem et al. 2015; Lauren 1965) show any association. Despite the poor prognosis linked with HER2 positive in breast cancer, the significance of HER2 in predicting GC prognosis remains controversial. Research has shown that HER2 expression does not affect prognosis.

7.2 Monoclonal Antibodies for GC therapy

7.2.1 Trastuzumab

After undergoing anti-HER2 therapy, patients with HER2-positive GC have experienced improved clinical outcomes. The humanized recombinant monoclonal antibody trastuzumab inhibits angiogenesis, DNA repair, and apoptosis by binding preferentially to HER2 ECD IV and decreasing HER2 receptor expression. Contrarily, trastuzumab involves an IgG1 Fc structure that can attack target cells by antibody-dependent cellular cytotoxicity (ADCC) (Spector and Blackwell 2009). The median OS was 16 months for patients who received cisplatin, fluorouracil, and trastuzumab in combination, compared to 11 months for patients who received chemotherapy alone. Patients with high HER2 protein expression tend to have a better chance of survival, which is the fundamental reason behind this (Smyth et al. 2020; Bang et al. 2010b).

As tumour immunotherapy develops further, combining immune checkpoint inhibitors will be considered a potentially effective treatment option. In HER2-positive gastroesophageal junction cancer (GEJ) patients, a phase II study (NCT03409848) found that first-line chemotherapy in conjunction with PD-1 inhibitors nivolumab and trastuzumab had encouraging results (Stein et al. 2022; Janjigian et al. 2021). It is also safe to combine trastuzumab with pembrolizumab and chemotherapy (Janjigian et al. 2020; Price-Schiavi et al. 2002). As a result of the addition of pembrolizumab to trastuzumab and chemotherapy in a phase III trial

(NCT03615326), tumour size was significantly reduced and objective response rates were improved (Janjigian et al. 2021).

As trastuzumab resistance develops in the majority of patients, the issue has come to the forefront. Tumour heterogeneity seems to have a major role in trastuzumab resistance. Alterations in the receptor tyrosine kinase-RAS-PI3K signalling pathway are also linked to treatment failure with anti-HER2 therapy. Mucins, which are proteins found on the surface of cells, also lessen the interaction between trastuzumab and the HER2 receptor. Therefore, the inhibitory action of the medication is prevented (Price-Schiavi et al. 2002). Numerous novel medications and treatments are being developed to address this issue.

7.2.2 Pertuzumab

In contrast to trastuzumab, the recombinant humanized anti-HER2 antibody pertuzumab has anticancer effects that are unique to it. Pertuzumab attaches to the HER2 ECD II receptor. Thus, pertuzumab can limit ligand-dependent HER2 signalling by inhibiting dimerization of HER2 with other HER family members, particularly by efficiently blocking HER2–HER3 heterodimerization (Metzger-Filho et al. 2013; Yamashita-Kashima et al. 2011; Oh and Bang 2016). Patients with HER2-positive metastatic GC who take pertuzumab in addition to trastuzumab and chemotherapy do not have a better chance of surviving, even though pertuzumab greatly increases survival rates for patients with metastatic breast cancer.

7.2.3 Margetuximab

The HER2-targeting antibody margetuximab has a modified FCg domain. An alteration to the Fc region renders it more capable of attaching to the activating Fc receptor, FcgRIIIA (CD16A), while simultaneously decreasing its affinity for the inhibitory Fc receptor, FcgRIIB (CD32B). There was an uptick in responses (Markham 2021).

Compared to trastuzumab with the wild-type Fc domain, margetuximab is more powerful and cytotoxic because CD16A is an Fc receptor crucial in mediating ADCC effects. The anticancer effects of margetuximab and anti-PD-1 medications, such as pembrolizumab and retifanlimab, are enhanced in a phase Ib/II study (NCT02689284) (Catenacci et al. 2020). Those whose tumours produce low levels of HER2 or who have CD16A low binding alleles are also more affected (Nordstrom et al. 2011). Another advantage of margetuximab over trastuzumab is that it can increase the ADCC action, leading to a more potent death of tumour cells (Bang et al. 2017). Margetuximab and retifanlimab with or without chemotherapy and margetuximab and tebotelimab alongside chemotherapy are being evaluated in the current Phase II/III randomized study (MAHOGANY) in patients with HER2-positive gastric or GEJ cancer (Catenacci et al. 2021). Various biosimilars of trastuzumab approved for GC therapy are shown in Table 7.1.

Table 7.1 Approved trastuzumab gastric cancer (GC) biosimilars (Joshi et al. 2022)

Biosimilar agent	Brand name	Year of approval	Innovator
EG12014		2023	EirGenix Inc.
SB3	Ontruzant	2019	Samsung Bioepis
PF- 05280014	Trazimera	2019	Pfizer
ABP980	Kanjinti	2019	Amgen and Allergan
Q5114	Ogivri	2018	Mylan S.A.S.
CT-P6	Herzuma	2018	Celltrion Healthcare

7.3 Biosimilars of Trastuzumab for GC Therapy

Despite approval of three trastuzumab biosimilars for GC by the U.S. FDA, market launch and availability has been halted due to remaining patents on Herceptin, the trastuzumab originator drug marketed by Genentech, South San Francisco, CA, USA. The trastuzumab patent is set to expire mid-2019 making it very likely that trastuzumab biosimilars will be launched shortly after this in the United States. Sales of trastuzumab biosimilars are expected to lead to cost savings overall (Miller and Schwartzberg 2019).

SB3 (Ontruzan) is the initial biosimilar of the standard anti-HER2 antibody trastuzumab to receive approval in the European Union. It is approved for use in all the same indications as reference trastuzumab, which are HER2-positive early breast cancer, metastatic breast cancer, and metastatic GC. SB3 exhibits comparable physicochemical and PD characteristics to the reference trastuzumab. The PK comparability of the two drugs has been demonstrated in healthy individuals and women with HER2-positive early stage or locally advanced breast cancer. SB3 showed clinical effectiveness comparable to reference trastuzumab in women with HER2-positive early stage or locally advanced breast cancer (Lamb 2018). The tolerability, immunogenicity, and safety profiles of SB3 closely resembled those of the reference trastuzumab. Reference trastuzumab plays a well-established function in treating HER2-positive malignancies, and SB3 offers an effective biosimilar option for individuals needing trastuzumab treatment. SB3 attaches to subdomain IV of the HER2 receptor, preventing ligand-induced HER2 signalling and the breakdown of its extracellular domain. Approved for use in conjunction with capecitabine or 5-fluorouracil and cisplatin in adult patients with HER2+ metastatic adenocarcinoma of the stomach or gastro-oesophageal junction who have not undergone previous anti-cancer treatment for their metastatic disease (Lamb 2018).

CT-P6, a trastuzumab biosimilar developed by Celltrion Inc. in Incheon, South Korea, displays comparable PKs to the original trastuzumab in healthy individuals (Park et al. 2022). CT-P6 received approval for treating HER2-positive early stage breast cancer patients following the phase 3 trial outcomes (Park et al. 2022). CT-P6 has also been applied to treat metastatic breast cancer and metastatic GC. Approving trastuzumab biosimilars could increase utilization of trastuzumab by patients in clinical settings (Park et al. 2022).

ABP 980, developed by Amgen in the USA and Europe, is a biosimilar to trastuzumab, an anti-HER2 monoclonal antibody marketed by Genentech in the USA and Roche in Germany. ABP 980 has been approved in the United States, the European Union, and Japan for all the authorized uses of the trastuzumab reference product (RP), such as treating HER2-overexpressing breast cancer, metastatic GC, and gastroesophageal junction adenocarcinoma (Kolberg et al. 2021).

An extensive analytical assessment of ABP 980 was carried out using advanced methods to detect any possible variances in structure and function compared to trastuzumab RP (Kolberg et al. 2021). The study revealed a high degree of analytical resemblance between the two in terms of physicochemical properties, including primary structure, amino acid sequence, post-translational changes, more advanced structure, the amount of protein, and product-related impurities and substances (Kolberg et al. 2021). ABP 980 shares identical amino acid sequence, molecular mass (matching theoretical mass), and N-linked glycosylation at N300 in the identical glycopeptide with trastuzumab RP. Functional similarity assessments of ABP 980 and trastuzumab RP included examinations of Fab-mediated, Fc-mediated, and combined Fab- and Fc-mediated activities. Biological activity of trastuzumab is mainly achieved by binding to the extracellular domain of HER2, leading to internalization and activation of downstream signalling events. Evaluation of ABP 980 showed comparable binding to HER2's extracellular domain, potency, and antibody-mediated cytotoxicity activities in cell-based experiments (Kolberg et al. 2021).

TRAZIMERA, another biosimilar of trastuzumab, is approved for use in conjunction with cisplatin and capecitabine or 5-fluorouracil to treat individuals with HER2-overexpressing metastatic gastric or gastroesophageal junction adenocarcinoma who have not been treated for metastatic disease before (Joshi et al. 2022; Barbier et al. 2019).

OGIVRI is authorized for use in conjunction with chemotherapy (cisplatin and either capecitabine or 5-fluorouracil) to treat HER2-positive metastatic cancer of the stomach or gastroesophageal junction in patients who have not been treated previously for their metastatic condition (Joshi et al. 2022; Barbier et al. 2019).

EG12014, a biosimilar of trastuzumab, has been authorized by the European Commission for treating HER2-positive breast and metastatic GC in the European Union, the same as trastuzumab's approved uses in the EU (Joshi et al. 2022; Barbier et al. 2019).

7.4 Mechanisms of Action of Trastuzumab and Its Biosimilars

Herceptin® (Trastuzumab) is a monoclonal antibody that specifically attaches to the outer part of the HER2. Figure 7.1 illustrates the impact of trastuzumab and its biosimilars on HER2 receptor binding.

Trastuzumab has a strong and specific binding to sub-domain IV, which is a location near the cell membrane of the extracellular domain of HER2. Binding of trastuzumab and its biosimilars to HER2 blocks HER2 signalling without the need for a

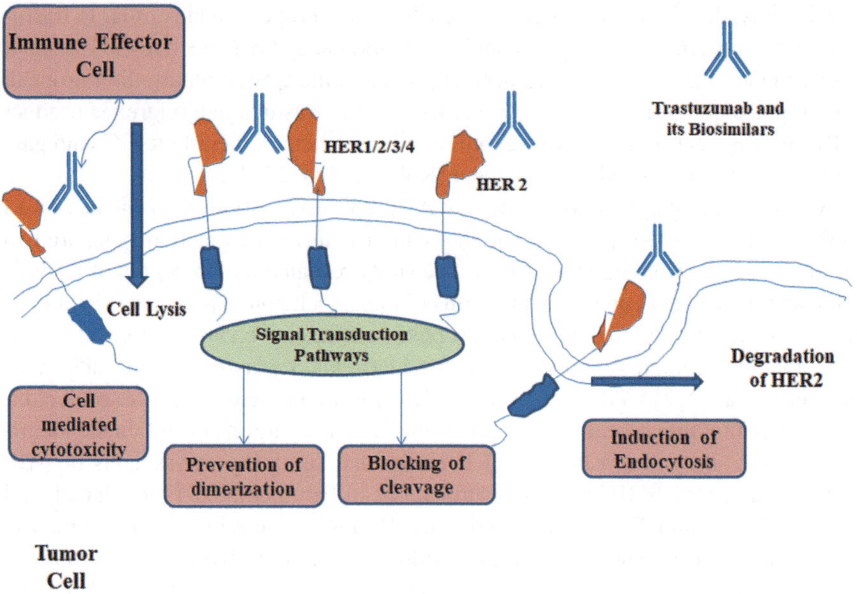

Fig. 7.1 Impact of trastuzumab and its biosimilars on HER2 receptor binding (Bachu et al. 2022). (**a**) Binding of Trastuzumab and its biosimilars with HER2 receptors leads to endocytosis and subsequent destruction of the receptors. The antobody and its biosimilars (**b**) inhibit receptor cleavage and (**c**) prevent dimerization with other HER receptors. (**d**) Cell-mediated cytotoxicity is triggered by interaction to both HER2 receptors and immunological effector cells (Bachu et al. 2022)

ligand and stops the proteolytic breakdown of its extracellular domain, which is a way HER2 is activated. Trastuzumab and its biosimilars have been demonstrated to hinder the growth of human cancer cells that have an excessive amount of HER2, both in laboratory tests and in animal studies. Trastuzumab is a powerful inducer of antibody-dependent cell-mediated cytotoxicity (ADCC) (Bachu et al. 2022). In vitro studies have demonstrated that trastuzumab-induced ADCC is more effective on cancer cells that overexpress HER2 compared to those that do not (Bachu et al. 2022).

7.5 Manufacturing and Development Process for Biosimilars Targeting GC

Developing and producing generic-equivalent products for small-molecule medicines is quite simple. Simply put, the generic medication needs to prove it has the same chemical makeup as the innovator product and show through a bioavailability study that their PK qualities are equivalent (Schellekens and Ryff 2002).

Biosimilars are not identical to the original products they are based on. The European Medicines Agency (EMEA) has accepted this perspective as the foundation for its biosimilar approval standards (European Medicines Agency 2005). Biosimilars contain active components that are not identical to those in the original product. Biopharmaceuticals are more intricate than small-molecule medicines due to their complex protein structure and the numerous manufacturing methods involved. Biopharmaceuticals' manufacturing techniques have a significant impact on the end product, unlike traditional medications (Schellekens 2002). The characteristics of biopharmaceuticals rely on the manufacturing procedures (Schellekens 2002).

Variations in production procedures, protein source, and extraction/purification methods lead to diversity in the produced biopharmaceuticals (Schellekens and Ryff 2002; Schellekens 2002; Deicher and Horl 2004; Hesse and Wagner 2000; Schellekens 2004). The different cellular expression methods utilized in protein production naturally generate a range of isoforms (European Medicines Agency 2014). Minor adjustments in production techniques can lead to modifications in the 3D structure of protein, the number of acid-base variations, and the glycosylation profile (Schellekens and Ryff 2002; European Medicines Agency 2005).

Biosimilar producers will not be able to utilize the manufacturing procedures of original products as they are considered confidential information. Biosimilar firms will be unable to accurately copy any protein product (Schellekens and Ryff 2002). Furthermore, due to the limitations of analytical tools in identifying or predicting all biological and clinical features of proteins, discrepancies between biopharmaceutical products may go unnoticed (Schellekens and Ryff 2002).

Regulatory guidelines should consider the distinctions between biosimilars and their reference products and require strict pharmacovigilance. The significance of the manufacturing procedure on the characteristics of biopharmaceuticals is shown by the variation in content and biological activity of products manufactured in various countries.

The development of biosimilars begins with a thorough analysis of physical and chemical properties as well as biological characteristics, which involves comparing quality features, and then conducting comparative preclinical investigations (Declerck et al. 2017; Vulto and Jaquez 2017). Clinical comparison testing is necessary to verify similar PKs and to validate similar efficacy and safety compared to the reference product (Sano 2017). The biosimilarity exercise focuses more on data from physicochemical and biological characterization of the candidate and comparative analytical testing with the reference product, rather than on clinical trials, in comparison to the approval pathway for a new biological (Vulto and Jaquez 2017; Declerck and Farouk Rezk 2017). The progression and scope of each phase in clinical development are determined by the quality of data gathered in the preceding stages of the comparability assessment. The clinical package typically includes a phase 1 trial and at least one phase 3 study for one of the approved indications of the reference product.3 Confirmatory PK and PD investigations may be adequate to show clinical biosimilarity in some situations (Declerck and Farouk Rezk 2017). After the process is completed, the biosimilar is assessed based on the whole body of evidence for biosimilarity (Vulto and Jaquez 2017).

7.6 Biosimilars Targeting GC in Clinical Studies

Almost half of the 262 biosimilar agents approved by the end of 2019 in 40 countries were in the field of oncology, consisting of 66 cancer-supportive drugs and 63 anticancer biosimilars (Declerck and Farouk Rezk 2017). About 19 clinical trials spanning 13 distinct candidate biosimilars and non-originator biologicals were found to be pertinent to the development of trastuzumab biosimilars. Comparative nonclinical research is conducted after a thorough physicochemical and biological screening, which includes a comparison of quality parameters, to begin the biosimilar development process (Declerck et al. 2017). Moreover, clinical comparative testing is necessary to verify equivalent efficacy and safety as well as similar PKs to the generic product. Each clinical development step's characteristics and extent are determined by the type and quantity of data collected in the antecedent comparability exercise steps (Park et al. 2020). For an approved indication of the reference product, the clinical trial typically includes a phase 1 trial and a minimum of one phase 3 study. Confirmatory PD and PK investigations may be adequate in certain situations to prove clinical biosimilarity (Agostini et al. 2015). In the end, the evaluation of the biosimilar hinges on every aspect of the evidence that backs up its biosimilarity. At least seven of the several trastuzumab biosimilar candidates that target stomach cancer are now in clinical development (Uhlig and Goll 2017). The PKs of generic trastuzumab and the biosimilar trastuzumab, CT-P6 (Herzuma), are identical in healthy individuals. After a phase 3 trial, CT-P6 was given permission to be used to treat early stage breast cancer patients who tested positive for HER2 (Barbier et al. 2019). Treatments for metastatic GC and breast cancer have also been extended to include CT-P6 (Park et al. 2022). More people may have access to trastuzumab in clinical settings with the approval of trastuzumab biosimilars. Recent findings from clinical trials have substantiated the efficacy of Truxima (rituximab), a treatment for B-cell lymphoma, and Herzuma (trastuzumab), which is prescribed for GC. When combined with chemotherapy and the antibody medication pembrolizumab (Keytruda), Herzuma has exhibited effectiveness in the treatment of HER2-positive GC patients. In a study, conducted at Seoul, South Korea, 43 patients were given pembrolizumab (200 mg IV, day 1), biosimilar trastuzumab (6 mg/kg, day 1), capecitabine (100 mg/m^2 bid, day 1 to day 14), and cisplatin (80 mg/m^2, day 1 every 3 weeks). Patients had an overall response rate of 76.7%, and no additional side effects were reported (Esteva et al. 2017). The findings, which demonstrated a noteworthy decrease in tumour size in more than 95% of patients, were presented at the American Society of Clinical Oncology Annual Meeting in 2020.

7.7 Pharmacovigilance and Safety

Because there isn't a lot of clinical data on biosimilars that target stomach cancer when they are first approved, it is very important to collect safety data after the drugs have been approved. There is a chance that differences in biosimilars' safety and effectiveness might not be seen in the pre-approval process when the product is only given to a small group of patients for a set amount of time (European Medicines Agency 2005). It is important to be able quickly identify the difference between biosimilar products that target stomach cancer and the reference generics. Given that biopharmaceuticals may pose risks, like immunogenicity and the chance for clinically significant variations between products, stringent pharmacovigilance programmes are needed to keep an eye on all of them, including innovator products and biosimilars, for safety and efficacy issues after they've been granted approval. To address difficulties unique to each biopharmaceutical category, such post-marketing programmes will need to be customized. Pharmacovigilance programmes are well-established by both the Food and Drug Administration. in the United States and the European Medicines Agency in Europe for tracking adverse occurrences related to pharmaceuticals (Stebbing et al. 2017; GaBI Online—Generics and Biosimilars Initiative 2020).

Eudra Vigilance is the moniker of the EMEA network that helps people report and evaluate possible adverse incidents that might happen during and after a product is approved for sale. Healthcare workers' spontaneous reporting is one aspect of pharmacovigilance. It's best for these reports to have as much information as possible about the drug, such as the type of adverse event, dose given, lot number, and proprietary name. Adverse events, however, are often not documented or just partially reported. Additionally, reporting regulations differ between nations (European Medicines Agency 2004).

The post-approval obligations made by manufacturers to regulatory authorities for providing follow-up monitors over safety typically include pharmacovigilance programmes that they have created and implemented (Stebbing et al. 2017; GaBI Online—Generics and Biosimilars Initiative 2020). Through the use of these programmes, adverse event frequency can be measured and major or previously unreported adverse events can be reported more quickly. Retrospective or prospective observational studies, patient registries, and pharmacoepidemiologic research may be a part of these programmes. The EMEA mandates that pharmacovigilance programmes for biosimilars include a continuous approach to the observation and assessment of safety problems in order to facilitate prompt and precise actions (European Medicines Agency 2004). Programmes for pharmacovigilance may be helpful in identifying new safety concerns with biosimilars meant to treat stomach cancer in advance. Programmes for pharmacovigilance are crucial for evaluating a product's safety in particular patient populations with stomach cancer. It is very important to follow these steps for the safe use of biosimilars in treatment cases in which the product may not have been officially tested (US Food and Drug Administration 2005). For biopharmaceutical goods, automated substitution is inappropriate for a number of reasons. As previously said, biosimilars are not generic replica of innovator drugs, and at the time of approval, biosimilars will have little clinical experience. Clinical results may be impacted by subtle variations between innovator drugs and biosimilars. Furthermore, patients may receive several biopharmaceutical drugs throughout therapy if automatic substitution is permitted. Such actions would complicate the process of gathering data for pharmacovigilance. Even if someone switched from one biopharmaceutical to another without keeping track of it, if an adverse effect happened, the pharmacovigilance review would not be able to link it to a specific product. Instead, it could be related to the wrong product. Clinicians must know precisely which biopharmaceutical product their patients are getting for safety monitoring purposes. The EMEA's policy statement stresses that the specific drug given to the patient should be clearly identified as to aid alongside pharmacovigilance monitoring (Hedenmalm and Spigset 1996). INN allocation is managed by a World Health Organization committee, acting on the recommendation of a worldwide expert advisory council, rather than the EMEA. It seems sensible that this committee give biopharmaceuticals unique INNs. Accurate biopharmaceutical prescription and dispensing would be made easier by this. In order to enable doctors and chemists to make knowledgeable judgements, the EMEA ought to think about mandating full labelling for biosimilars. Labelling should incorporate unique safety and efficacy data for biosimilars, as they are not interchangeable with reference products. Additionally, indications that are moreover based on data extrapolation should be noted in the labelling.

7.8 Current Regulatory Landscape

Before the Food and Drug Administration in the United States, biosimilars were overseen by the European Medicines Agency. Subsequently, they were scrutinized by the World Health Organization. These rules spell out the requirements that biosimilars must meet in order to be licensed. These include important preliminary studies that compare the biosimilar to the approved biologic originator (European Medicines Agency 2006a). More than 40 biosimilars have been given a positive judgement by the European Medicines Agency and subsequently authorized by the European Commission since the first biosimilar was authorized in the Europe in 2006 (Barbier et al. 2019). The FDA has given approval for more than 10 biosimilars since 2015 (Francescon et al. 2016). Future years will see a significant increase in the number of authorized biosimilars together with a gradual loss of exclusivity for biological reference products, particularly in cancer (Declerck et al. 2017). The European Medicines Agency has created regulatory norms and the European Union has designed a general legal approach for the licensing of biosimilars (U.S. Food and Drug Administration n.d.). The approval process differs depending on the product due to the significant variations amongst biopharmaceutical goods. In particular, the amount of clinical data that is available may be affected by how variable efficacy measures are and how many validated surrogate markers are available (IMS Institute for Healthcare Informatics 2016). Biosimilars will often be approved once it can be shown that they are as safe and effective as an innovator reference product in a relevant patient population—a process known as "comparability". According to EMEA standards, data extrapolation from one therapeutic indication to another will be allowed with appropriate justification, enabling the use of biosimilars in applications for which they have not been well researched. The efficacy and safety of biosimilar drugs must be monitored after approval by pharmacovigilance programmes and immunogenicity testing, as mandated by EMEA standards due to the infrequent but occasionally significant adverse event associations associated with biopharmaceuticals. The U.S. Food and Drug Administration has not yet created regulations pertaining to biosimilars because there is presently no legal route in the country for their approval. Healthcare workers should understand the distinctions between "comparability" and therapeutic equivalency, sometimes known as substitutability. When a generic product and its reference product have the same chemical makeup and are bioequivalent—that is, they have similar PK profiles—it is referred to as therapeutic equivalency. Comparability in the context of biosimilars refers to the idea that while the biosimilar and reference products are not the same, their efficacy and safety are comparable. Therapeutic equivalency is not ensured by studies demonstrating therapeutic similarity (European Medicines Agency 2006b). Without undergoing clinical testing for every indication, a biosimilar candidate may be evaluated for approval for one or more that the reference product is approved for. The term "extrapolation of indications" refers to a regulatory concept (Mellstedt et al. 2008). The primary justification for extending data to additional indications is to prevent needless clinical research (Weise et al. 2014). Extrapolation is a well-established regulatory concept that is used, for instance, when a new formulation of a product with a licence is created, in addition to the context of biosimilars (Tinsley et al. 2018). The EMA has already approved extrapolation for trastuzumab biosimilars from early breast cancer to metastatic breast cancer and metastatic GC (SB3, ABP 980, and CT-P6) and from metastatic breast cancer to early breast cancer and metastatic GC (MYL-1401O). This is based on all the evidence that supports biosimilarity (European Medicines Agency 2013).

7.9 Pharmacoeconomic Evaluation

In 2020, there were almost 1.8 million cases of stomach cancer each five years, as reported by Globocan. The market is divided into two segments based on illness indication: gastrointestinal stromal tumours and GC/gastroesophageal junction cancer (GC/GEJC). The size of the worldwide market for stomach cancer treatments was projected to be USD 3.83 billion in 2021 and is projected to grow to USD 4.25 billion in 2022. From 2022 to 2030, the stomach cancer treatment market is projected to expand at a compound yearly growth rate of 12.86%, with a market value of USD 11.2 billion (Fig. 7.2). In 2021, Asia Pacific retained a dominant market share of over 55.0% in the GC therapy market. This is mostly because stomach cancer is more common in this part of the world. For example, in 2020, there were approximately 138,000 new cases in Japan and 479,000 in China. Due to the high expense of GC treatments, healthcare systems worldwide have developed a variety of strategies aimed at trying and keep costs under control; these strategies frequently lead to what appear to be arbitrary access limits for patients. Even at the regional level, there has been evidence of substantial variation in patient access to cancer medications. Compared to small-molecule medications, biopharmaceuticals are more costly, but their use is booming. The fact that biosimilars are probably going to result in cost savings is a significant advantage. The impact of biosimilar cost

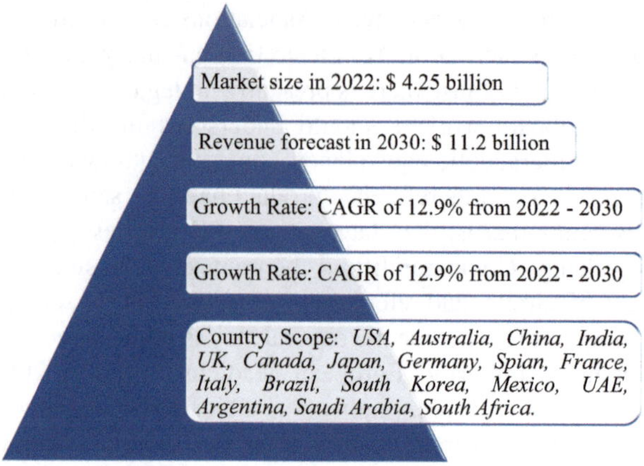

Fig. 7.2 Market report on scope of gastric cancer treatment (Konstantinidou et al. 2020)

pressures on innovative product prices is unknown, though, since manufacturers may change their pricing policies before patents expire in order to offset their investment expenses. Furthermore, due to the high costs associated with both product launch and production, the cost savings associated with biosimilars are probably not going to be as large as those associated with small-molecule generics. It is also important to remember that medications only cover a small portion of the entire expense of treating people with stomach cancer (Pivot et al. 2018). One apparent benefit of biosimilars is their capacity to lower the purchase price of medications for stomach cancer, making them more accessible and affordable for patients. The utilization of these medications in clinical practice will therefore be influenced by economic factors. Biosimilars are expected to save 20–40% in the USA, and savings of 10–35% have been documented in the European Union (Bordeleau 2006). This stands in contrast to the usual 70–80% reduction linked to generic small-molecule medications. The goal behind the development of biosimilars was to reduce healthcare expenses. The starting price of a biosimilar is usually 20–30% less than that of the reference product (Camacho et al. 2014). They also offer further private privileges and competitiveness to the formerly exclusive area (Jacobs et al. 2017). According to a study, the CT-P6 treatments in the neoadjuvant scenario cost about €1500 less than the reference trastuzumab (RTZ). Despite the fact that this could seem like a tiny savings, it should be noted that fewer neoadjuvant cycles are required here than in mother situations. A follow-up period of eight years on average was included in the 2020 update of the CLEOPATRA investigation (Migliavacca Zucchetti et al. 2019). Additionally, biosimilars promote rivalry among businesses, which lowers costs (Barman et al. 2024). By moving to new biosimilars like rituximab, etanercept, and infliximab, the U.K. National Health System saved £210 million from 2017 to 2018 (Zelenetz 2016). The patents on nearly 20 biologics, including trastuzumab and bevacizumab, will expire by 2023. Herceptin's patents expired in July 2014 in Europe and June 2019 in the USA. Furthermore, the United States patent for avastin expired in July 2019 (Robinson 2018). The total budget savings estimates varied depending on the scenario in evaluations conducted using the base case scenario for the three years after CT-P6 adoption. These estimates ranged from €290 million to €562 million. From 27,145 to 44,810 additional patients were treated as a result of savings. Budget savings in the first year alone were expected to be between €53 million and €272 million, with 4,982–21,730 additional patients treated as a result of the savings across a range of scenarios and substitution rates. The introduction of Herceptin biosimilars, however, is anticipated to limit market expansion. In the United States, biosimilars including Biocon/Mylan's Ogivri and Celltrion/Teva's Herzuma were approved in December 2017 and December 2018, respectively. Furthermore, Ontruzant, a trastuzumab biosimilar made by Samsung Bioepis, received approval in January 2019, while Trazimera received approval in March 2019 in the United States. Sales of Herceptin fell by 14% in 2019 and 49% in 2020 across all indications as a result of the introduction of biosimilars (Konstantinidou et al. 2020).

7.10 Challenges to Develop GC Targeting Biosimilars

When biosimilars for GC hit the market, clinicians will have to deal with a lot of challenges. Biosimilars are new, non-innovative drugs that will have limited clinical data sets to support them upon licensure; they are not generic biopharmaceuticals. Generic comparable product development and production is comparatively simple for small-molecule medications. All that is needed is a bioavailability study that demonstrates the PK features of the standard and generic products are identical, along with proof that the generic product is made up of the same chemicals as the original product. Conversely, biosimilars aren't generic versions of the original products (Schellekens and Ryff 2002). The European Medicines Agency adopted this viewpoint, and its biosimilar approval guidelines are based on it. This is due to the fact that a biosimilar's active components differ from those of the originator product. Biopharmaceuticals are significantly more complex than small-molecule medicines since they are complex proteins that need extensive manufacturing techniques. In contrast to traditional pharmaceuticals, there is a direct correlation between the features of the finished product and the techniques used in the production of biopharmaceuticals (Schellekens 2002). It is true that the manufacturing procedures have an impact on the qualities of biopharmaceuticals. The resulting biopharmaceuticals are heterogeneous because to variations in the manufacturing processes, protein sources, and extraction/purification procedures (Deicher and Horl 2004). For instance, a range of isoforms are naturally produced by the distinct cellular-expression systems used in the production of proteins. Furthermore, minute modifications in the process of production might result in changes to the protein's three-dimensional structure, number of acid–base variations, and glycosylation profile. Since this is proprietary information, producers of biosimilars will not have access to the innovator drugs' manufacturing procedures. As a result, producers of biosimilars will be unable to accurately duplicate any protein product (Schellekens and Ryff 2002). Changes in biopharmaceutical products can also be overlooked because there aren't yet analytical methods that are capable of identifying or predict all of the biological and therapeutic properties of proteins. Regulations need to require strict pharmacovigilance and take into consideration the variations between biosimilars and their reference medicines (Hollmann et al. 2018). Of 400 healthcare workers surveyed in 2013, including chemists, nurses, and oncologists, most said they didn't know much about biosimilars in the current medication pipeline and didn't know how they were regulated (Mellstedt et al. 2008). Similar findings were found in a survey performed by the National Comprehensive Cancer Network of 277 healthcare professionals, including doctors, nurses, and chemists (Krejcik 2013). Two out of three doctors, three out of four nurses, and two out of six chemists said they needed more information before they would prescribe, distribute, or give biosimilars in their practices. The multidisciplinary Biosimilars Working Group's NCCN recommendations included educating policymakers and healthcare professionals about biosimilars (Zelenetz et al. 2011). Respondents were especially intrigued in learning more about comparative clinical objectives, including safety and efficacy, PK similarity, chemical/physical similarity, and guideline and compendia inclusion. Various variables besides the need for thorough instructional programmes on biosimilar therapy can impact the incorporation of biosimilars into oncology practices (Rugo et al. 2016). These include selecting the patient group for efficacious evaluation, selecting the efficacious endpoints, extending the indication, ensuring immunogenicity safety post-approval, and ensuring interchangeability and automatic substitution. The acceptance of biosimilars in clinical practice will be influenced by payers, physicians, and other healthcare stakeholders. The NCCN study says that oncologists must trust the evidence that supports biosimilarity and look for the data that supports FDA approval of biosimilars (Yang et al. 2019). Therefore, it is the responsibility of manufacturers to offer the greatest level of transparency in their development programmes.

7.11 Conclusion and Future Prospectives

Biosimilar agents will undoubtedly become more prevalent in the treatment of GC therapy since many of them are either in development or have already received approval. Even though biosimilars offer a lot of advantages, it's critical that they be introduced properly. When selecting a biopharmaceutical product, it is crucial to take into account any significant problems about the use of biosimilars in GC patients. To make prudent choices, clinicians need to have a solid awareness of the problems surrounding biosimilars. Clinicians should be most concerned about understanding that biosimilars are not generic versions of originators. The European Medicines Agency will approve biosimilars as safe and effective medicines; nevertheless, they will not be the same as innovator medications. Consequently, it is appropriate to consider the substitution or switching between innovative drugs and biosimilars as a change in clinical treatment of GC treatment. Due to the lack of clinical experience with biosimilars at the time of their approval toward GC treatment, these potential therapeutic differences might not come to light until after approval. Therefore, in-depth pharmacovigilance programmes are required to gather this information and create a database outlining each product's clinical use. Automatic substitution should be banned in order to guarantee that these pharmacovigilance programmes create an accurate database. This is because clinicians must closely monitor GC patients who are prescribed biopharmaceuticals, and obtaining such authorizations would complicate proper pharmacovigilance. This is particularly relevant when biopharmaceuticals are used as medicines that aren't part of a study in both clinical trials and everyday life. Also, there should be caution about applying clinical data for biosimilars from one type of cancer to another. A risk/benefit study is what data extrapolation comes down to in the end. The ramifications for the patient should a biosimilar's efficacy or safety profile diverge from that of the innovator medication must be weighed against the hypothesis that it will be effective for indications outside those for which it has undergone formal testing. Case-by-case extrapolation is needed because biopharmaceutical product may have a higher risk-to-benefit ratio for some patient groups, like stem cell donors, than for others. Furthermore, to the clinical data that support extrapolation, the labelling on the package should make it clear in the event data extrapolation is permissible or not. In conclusion, minimizing any possible worries about the use of biosimilars requires knowledge. Physicians and chemists in particular need to understand the distinctions between innovator biopharmaceuticals and biosimilars. If you move from one complex biopharmaceutical drug to another (from innovator to biosimilar or vice versa), this should be seen as a change in how you manage GC clinically.

References

Abrahao-Machado LF, Scapulatempo-Neto C (2016) HER2 testing in gastric cancer: an update. World J Gastroenterol 22(19):4619–4625. https://doi.org/10.3748/wjg.v22.i19.4619

Agostini C, Canonica G-W, Maggi E (2015) European medicines agency guideline for biological medicinal products: a further step for a safe use of biosimilars. Clin Mol Allergy 13(1):3. https://doi.org/10.1186/s12948-015-0010-3

Bachu R-D, Abou-Dahech M, Balaji S et al (2022) Oncology biosimilars: new developments and future directions. Cancer Rep 5(11):e1720. https://doi.org/10.1002/cnr2.1720

Bang Y-J, Van Cutsem E, Feyereislova A et al (2010a) Trastuzumab in combination with chemotherapy versus chemotherapy alone for treatment of HER2-positive advanced gastric or gastro-oesophageal junction cancer (ToGA): a phase 3, open-label, randomised controlled trial. Lancet 376:687–697

Bang Y-J, Van Cutsem E, Feyereislova A et al (2010b) Trastuzumab in combination with chemotherapy versus chemotherapy alone for treatment of HER2-positive advanced gastric or gastro-oesophageal junction cancer (ToGA): a phase 3, open-label, randomised controlled trial. Lancet 376(9742):687–697. https://doi.org/10.1016/s0140-6736(10)61121-x

Bang Y-J, Giaccone G, Im S-A et al (2017) First-in- human phase 1 study of margetuximab (MGAH22), an Fc-modified chimeric monoclonal antibody, in patients with HER2-positive advanced solid tumors. Ann Oncol 28(4):855–861. https://doi.org/10.1093/annonc/mdx002

Barbier L, Declerck P, Simoens S et al (2019) The arrival of biosimilar monoclonal antibodies in oncology: clinical studies for trastuzumab biosimilars. Br J Cancer 121(3):199–210. https://doi.org/10.1038/s41416-019-0480-z

Barman D, Bandyopadhyay T, Talukdar R (2024) Biosimilar in breast cancer: a narrative review. Cureus 16:e52828. https://doi.org/10.7759/cureus.528

Bordeleau L (2006) Pharmacoeconomics of systemic therapies for lung cancer. Treat Respir Med 5:129–141. https://doi.org/10.2165/00151829-200605020-00006

Camacho L-H, Frost C-P, Abella E et al (2014) Biosimilars 101: considerations for U.S. oncologists in clinical practice. Cancer Med 3:889–899. https://doi.org/10.1002/cam4.258

Catenacci DV-T, Kang Y-K, Park H et al (2020) Margetuximab plus pembrolizumab in patients with previously treated, HER2-positive gastro-oesophageal adenocarcinoma (CP-MGAH22–05): a single-arm, phase 1b–2 trial. Lancet Oncol 21(8):1066–1076. https://doi.org/10.1016/s1470-2045(20)30326-0

Catenacci D-V, Rosales M, Chung H-C et al (2021) Mahogany: margetuximab combination in HER2+ unresectable/metastatic gastric/gastroesophageal junction adenocarcinoma. Future Oncol 17(10):1155–1164. https://doi.org/10.2217/fon-2020-1007

Choong G-M, Cullen G-D, O'Sullivan C-C (2020) Evolving standards of care and new challenges in the management of HER2-positive breast cancer. CA Cancer J Clin 70(5):355–374. https://doi.org/10.3322/caac.21634

Dan H, Li Z, Xiangshan F et al (2018) HER2 status in gastric adenocarcinoma of Chinese: a multicenter study of 40,842 patients. Chin J Pathol 47(11):822–826. https://doi.org/10.3760/cma.j.issn.0529-5807.2018.11.002

Dawood S, Broglio K, Buzdar A-U et al (2010) Prognosis of women with metastatic breast cancer by HER2 status and trastuzumab treatment: an institutional-based review. J Clin Oncol 8(1):92–98. https://doi.org/10.1200/JCO.2008.19.9844

Declerck P, Farouk Rezk M (2017) The road from development to approval: evaluating the body of evidence to confirm biosimilarity. Rheumatology 56:iv4–iv13

Declerck P, Danesi R, Petersel D et al (2017) The language of biosimilars: clarification, definitions, and regulatory aspects. Drugs 77:671–677

Deicher R, Horl W-H (2004) Differentiating factors between erythropoiesis-stimulating agents: a guide to selection for anaemia of chronic kidney disease. Drugs 64:499–509

Esteva F-J, Lee S-J, Stebbing J et al (2017) Phase I clinical trial comparing PK and safety of trastuzumab and its biosimilar candidate CT-P6. Breast 32:S68–S69. https://doi.org/10.1016/S0960-9776(17)30232-1

European Medicines Agency (2004) Pharmacovigilance—Medicinal products for human use and veterinary medicinal products. http://eudravigilance.emea.eu.int/human/docs/Vol9en.pdf. Accessed 12 Feb 2024

European Medicines Agency (2005) Guideline on similar biological medicinal products. http://www.emea.eu.int/pdfs/human/biosimilar/043704en.pdf. Accessed 1 May 2006

European Medicines Agency (2006a) Guideline on similar biological medicinal products containing biotechnology-derived proteins as active substance: non-clinical and clinical issues. http://www.emea.eu.int/pdfs/human/biosimilar/4283205en.pdf. Accessed 16 Jan 2024

European Medicines Agency (2006b) Guideline on similar biological medicinal products containing biotechnology-derived proteins as active substance: quality issues. http://www.emea.eu.int/pdfs/human/biosimilar/4934805en.pdf. Accessed 25 Jan 2024

European Medicines Agency (2013) Concept paper on extrapolation of efficacy and safety in medicine development

European Medicines Agency (2014) Guideline on similar biological medicinal products

Figueroa-Magalhaes M-C, Jelovac D, Connolly R et al (2014) Treatment of HER2- positive breast cancer. Breast 23(2):128–136. https://doi.org/10.1016/j.breast.2013.11.011

Francescon S, Fornasier G, Baldo P (2016) Biosimilar oncology drugs in Europe: regulatory and pharmacovigilance considerations. Oncol Ther 4(2):173–182. https://doi.org/10.1007/s40487-016-0028-9

GaBI Online—Generics and Biosimilars Initiative. Trastuzumab biosimilar Herzuma approved in Canada [www.gabionline.net]. Pro Pharma Communications International, Mol; [cited 2020 Sep 11]. www.gabionline.net/Biosimilars/News/Trastuzumab-biosimilar-Herz-EuropeanMedicinesAgency-uma-approved-in-Canada

Gravalos C, Jimeno A (2008) HER2 in gastric cancer: a new prognostic factor and a novel therapeutic target. Ann Oncol 19(9):1523–1529. https://doi.org/10.1093/annonc/mdn169

Guggenheim D-E, Shah M-A (2013) Gastric cancer epidemiology and risk factors. J Surg Oncol 107(3):230–236. https://doi.org/10.1002/jso.23262

Hedenmalm K, Spigset O (1996) Peripheral sensory disturbances related to treatment with fluoroquinolones. J Antimicrob Chemother 37:831–837

Hesse F, Wagner R (2000) Developments and improvements in the manufacturing of human therapeutics with mammalian cell cultures. Trends Biotechnol 18:173–180

Hollmann S, Painter C, Hogan A et al (2018) PCN68—budget impact analysis of tisagenlecleucel for the treatment of paediatric and young adult patients with relapsed or refractory B-cell acute lymphoblastic leukaemia in England. Value Health 21:25. https://doi.org/10.1016/j.jval.2018.09.150

IMS Institute for Healthcare Informatics (2016) Delivering on the potential of biosimilar medicines: the role of functioning competitive markets introduction. GaBI Journal Editor. Patent expiry dates for best-selling biologicals. GaBI J 4: 178–179.

Jacobs I, Ewesuedo R, Lula S et al (2017) Biosimilars for the treatment of cancer: a systematic review of published evidence. BioDrugs 31:1–36. https://doi.org/10.1007/s40259-016-0207-0

Janjigian Y-Y, Maron S-B, Chatila W-K et al (2020) First-line pembrolizumab and trastuzumab in Her2-positive oesophageal, gastric, or gastro-oesophageal junction cancer: an open-label, single-arm, phase 2 trial. Lancet Oncol 21(6):821–831. https://doi.org/10.1016/s1470-2045(20)30169-8

Janjigian Y-Y, Kawazoe A, Yanez P et al (2021) The KEYNOTE- 811 trial of dual PD-1 and HER2 blockade in HER2-positive gastric cancer. Nature 600(7890):727–730. https://doi.org/10.1038/s41586-021-04161-3

Joshi D, Khursheed R, Gupta S et al (2022) Biosimilars in oncology: latest trends and regulatory status. Pharmaceutics 14(12):2721. https://doi.org/10.3390/pharmaceutics14122721. PMID: 36559215; PMCID: PMC9784530

Klapper L-N, Vaisman N, Hurwitz E et al (1997) A subclass of tumor-inhibitory monoclonal antibodies to ERBB-2/HER2 blocks crosstalk with growth factor receptors. Oncogene 14(17):2099–2109. https://doi.org/10.1038/sj.onc.1201029

Kolberg H-C, Demetriou G-S, Hanes V (2021) Totality of evidence supporting the use of ABP 980, a trastuzumab biosimilar: practical considerations. Oncol Ther 9(1):225–238. https://doi.org/10.1007/s40487-020-00129-x. Epub 2021 Jan 11. PMID: 33428085; PMCID: PMC8297514

Konstantinidou S, Papaspiliou A, Kokkotou E (2020) Current and future roles of biosimilars in oncology practice. Oncol Lett 19:45–51. https://doi.org/10.3892/ol.2019.11105

Krejcik C (2013) NACCME releases results of survey assessing medical education needs in biosimilars. http://www.prweb.com/releases/2013/5/prweb10769551.htm

Lamb Y-N (2018) SB3 (Ontruzant®): a trastuzumab biosimilar. BioDrugs 32(3):293–296. https://doi.org/10.1007/s40259-018-0282-5. PMID: 29796993

Lauren P (1965) The two histological main types of gastric carcinoma: diffuse and so- called intestinal-type carcinoma. an attempt at a histo-clinical classification. Acta Pathol Microbiol Scand 64:31–49. https://doi.org/10.1111/apm.1965.64.1.31

Liang J-W, Zhang J-J, Zhang T et al (2014) Clinicopathological and prognostic significance of HER2 overexpression in gastric cancer: a meta-analysis of the literature. Tumour Biol 35(5):4849–4858. https://doi.org/10.1007/s13277-014-1636-3

Markham A (2021) Margetuximab: first approval. Drugs 81(5):599–604. https://doi.org/10.1007/s40265-021-01485-2

Mellstedt H, Niederwieser D, Ludwig H (2008) The challenge of biosimilars. Ann Oncol 19:411–419. https://doi.org/10.1093/annonc/mdm345

Metzger-Filho O, Winer E-P, Krop I (2013) Pertuzumab: optimizing HER2 blockade. Clin Cancer Res 19(20):5552–5556. https://doi.org/10.1158/1078-0432.Ccr-13-0518

Migliavacca Zucchetti B, Nicolò E, Curigliano G (2019) Biosimilars for breast cancer. Expert Opin Biol Ther 19:1015–1021. https://doi.org/10.1080/14712598.2019.1638362

Miller E-M, Schwartzberg L-S (2019) Biosimilars for breast cancer: a review of HER2-targeted antibodies in the United States. Ther Adv Med Oncol 14(11):1758835919887044. https://doi.org/10.1177/1758835919887044. PMID: 31798693; PMCID: PMC6859678

Moasser MM (2007) The oncogene HER2: its signaling and transforming functions and its role in human cancer pathogenesis. Oncogene 26(45):6469–6487. https://doi.org/10.1038/sj.onc.1210477

Mulcahy A-W, Hlavka J-P, Case S-R (2018) Biosimilar cost savings in the United States: initial experience and future potential. Rand Health Q 7:3

Neve R-M, Lane H-A, Hynes N-E (2001) The role of overexpressed HER2 in transformation. Ann Oncol 12(suppl 1):S9–S13

Noorolyai S, Shajari N, Baghbani E et al (2019) The relation between PI3K/AKT signalling pathway and cancer. Gene 698:120–128. https://doi.org/10.1016/j.gene.2019.02.076

Nordstrom J-L, Gorlatov S, Zhang W et al (2011) Anti-tumor activity and toxicokinetics analysis of MGAH22, an anti-HER2 monoclonal antibody with enhanced fcg receptor binding properties. Breast Cancer Res 13(6):R123. https://doi.org/10.1186/bcr3069

Obermannova R, Lordick F (2016) Insights into next developments in advanced gastric cancer. Curr Opin Oncol 28(4):367–375. https://doi.org/10.1097/CCO.0000000000000289

Oh D-Y, Bang Y-J (2016) Pertuzumab in gastrointestinal cancer. Expert Opin Biol Ther 16(2):243–253. https://doi.org/10.1517/14712598.2016.1126578

Olayioye M-A (2001) Update on HER-2 as a target for cancer therapy: intracellular signaling pathways of ERBB2/HER-2 and family members. Breast Cancer Res 3(6):385–389. https://doi.org/10.1186/bcr327

Park H-Y, Wu D-W, Ma F et al (2020) Availability of anticancer biosimilars in 40 countries. Lancet Oncol 21:197–201. https://doi.org/10.1016/s1470-2045(19)30860-5

Park J-H, Yeo J-H, Kim Y-S et al (2022) Efficacy and safety of trastuzumab biosimilar (CT-P6) compared with reference trastuzumab in patients with HER2-positive advanced gastric cancer: a retrospective analysis. Am J Clin Oncol 45(2):61–65. https://doi.org/10.1097/COC.0000000000000887. PMID: 34991106; PMCID: PMC8781232

Pivot X, Bondarenko I, Nowecki Z et al (2018) Phase III, randomized, double-blind study comparing the efficacy, safety, and immunogenicity of SB3 (trastuzumab biosimilar) and reference trastuzumab in patients treated with neoadjuvant therapy for human epidermal growth factor receptor 2–positive early breast cancer. J Clin Oncol 36:968–974. https://doi.org/10.1200/JCO.2017.74.0126

Price-Schiavi S-A, Jepson S, Li P et al (2002) Rat Muc4 (sialomucin complex) reduces binding of anti-ERBB2 antibodies to tumor cell surfaces, a potential mechanism for herceptin resistance. Int J Cancer 99(6):783–791. https://doi.org/10.1002/ijc.10410

Rezatabar S, Karimian A, Rameshknia V, Parsian H, Majidinia M, Kopi TA et al (2019) RAS/MAPK signaling functions in oxidative stress, DNA damage response and cancer progression. J Cell Physiol 234(9):14951–14965. https://doi.org/10.1002/jcp.28334

Robinson J (2018) Ten things' pharmacists should know about biologics and biosimilars. Pharm J 8. Epub ahead of print. https://pharmaceutical-journal.com/article/feature/ten-things-pharmacists-should-know-about-biologics-and-biosimilars

Rugo H-S, Linton K-M, Cervi P et al (2016) A clinician's guide to biosimilars in oncology. Cancer Treat Rev 46:73–79. https://doi.org/10.1016/j.ctrv.2016.04.003

Rugo H-S, Rifkin R-M, Declerck P et al (2019) Demystifying biosimilars: development, regulation and clinical use. Future Oncol 15:777–790

Sano T (2017) Gastric cancer: Asia and the world. Gastric Cancer 20(Suppl 1):1–2. https://doi.org/10.1007/s10120-017-0694-9

Santos S-B, Sousa Lobo J-M, Silva A-C (2019) Biosimilar medicines used for cancer therapy in Europe: a review. Drug Discov Today 24:293–299

Schechter AL, Stern DF, Vaidyanathan L, Decker SJ, Drebin JA, Greene MI et al (1984) The neu oncogene: an erb-B-Related gene encoding a 185,000-Mr tumour antigen. Nature 312(5994):513–516. https://doi.org/10.1038/312513a0

Schellekens H (2002) Bioequivalence and the immunogenicity of biopharmaceuticals. Nat Rev Drug Discov 1:457–462

Schellekens H (2004) Biosimilar epoetins: how similar are they? Eur J Hosp Pharm 3:8–12

Schellekens H, Ryff J-C (2002) 'Biogenerics': the off-patent biotech products. Trends Pharmacol Sci 23:119–121

Smyth E-C, Nilsson M, Grabsch H-I et al (2020) Gastric cancer. Lancet 396(10251):635–648. https://doi.org/10.1016/s0140-6736(20)31288-5

Spector N-L, Blackwell K-L (2009) Understanding the mechanisms behind trastuzumab therapy for human epidermal growth factor receptor 2-positive breast cancer. J Clin Oncol 27(34):5838–5847. https://doi.org/10.1200/JCO.2009.22.1507

Stebbing J, Baranau Y, Baryash V et al (2017) CT-P6 compared with reference trastuzumab for HER2-positive breast cancer: a randomised, double-blind, active-controlled, phase 3 equivalence trial. Lancet Oncol 18:917–928

Stein A, Paschold L, Tintelnot J et al (2022) Efficacy of ipilimumab vs folfox in combination with nivolumab and trastuzumab in patients with previously untreated ERBB2-positive esophagogastric adenocarcinoma: the AIO intega randomized clinical trial. JAMA Oncol 8(8):1150–1158. https://doi.org/10.1001/jamaoncol.2022.2228

Sung H, Ferlay J, Siegel R-L, Laversanne M, Soerjomataram I, Jemal A et al (2021) Global cancer statistics 2020: Globocan estimates of incidence and mortality worldwide for 36 cancers in 185 countries. CA Cancer J Clin 71(3):209–249. https://doi.org/10.3322/caac.21660

Takehana T, Kunitomo K, Kono K et al (2002) Status of c-ERBB-2 in gastric adenocarcinoma: a comparative study of immunohistochemistry, fluorescence in situ hybridization and enzyme-linked immuno-sorbent assay. Int J Cancer 98(6):833–837. https://doi.org/10.1002/ijc.10257

Tebbutt N, Pedersen M-W, Johns T-G (2013) Targeting the ERBB family in cancer: couples therapy. Nat Rev Cancer 13(9):663–673. https://doi.org/10.1038/nrc3559

Tinsley S-M, Grande C, Olson K et al (2018) Potential of biosimilars to increase access to biologics: considerations for advanced practice providers in oncology. J Adv Pract Oncol 9(7):699–716. PMID: 31249718; PMCID: PMC6570521

U.S. Food and Drug Administration (n.d.) Biosimilar product information. https://www.fda.gov/drugs/biosimilars/biosimilar-product-information. Accessed 25 Jan 2024

Uhlig T, Goll G-L (2017) Reviewing the evidence for biosimilars: key insights, lessons learned and future horizons. Rheumatology 56(suppl_4):iv49–iv62. https://doi.org/10.1093/rheumatology/kex276

US Food and Drug Administration (2005) Guidance for industry—Good pharmacovigilance practices and pharmacoepidemiologic assessment. http://www.fda.gov/Cder/guidance/6359OCC.htm. Accessed 12 Feb 2024

Van Cutsem E, Moiseyenko V-M, Tjulandin S, Majlis A, Constenla M, Boni C et al (2006) Phase III study of docetaxel and cisplatin plus fluorouracil compared with cisplatin and fluorouracil as first-line therapy for advanced gastric cancer: a report of the V325 study group. J Clin Oncol 24(31):4991–4997. https://doi.org/10.1200/JCO.2006.06.8429

Van Cutsem E, Bang Y-J, Feng-Yi F et al (2015) HER2 screening data from ToGA: targeting HER2 in gastric and gastroesophageal junction cancer. Gastric Cancer 18:476–484

Vulto A-G, Jaquez O-A (2017) The process defines the product: what really matters in biosimilar design and production? Rheumatology 56:iv14–iv29

Weise M, Kurki P, Wolff-Holz E et al (2014) Biosimilars: the science of extrapolation. Blood 124:3191–3196

Yamashita-Kashima Y, Iijima S, Yorozu K et al (2011) Pertuzumab in combination with trastuzumab shows significantly enhanced antitumor activity in HER2-positive human gastric cancer xenograft models. Clin Cancer Res 17(15):5060–5070. https://doi.org/10.1158/1078-0432.Ccr-10-2927

Yan M, Schwaederle M, Arguello D et al (2015) HER2 expression status in diverse cancers: review of results from 37,992 patients. Cancer Metastasis Rev 34(1):157–164. https://doi.org/10.1007/s10555-015-9552-6

Yang J, Yu S, Yang Z et al (2019) Efficacy and safety of anti-cancer biosimilars compared to reference biologics in oncology: a systematic review and meta-analysis of randomized controlled trials. BioDrugs 33(4):357–371. https://doi.org/10.1007/s40259-019-00358-1

Yuan J, Dong X, Yap J, Hu J (2020) The MAPK and AMPK signalings: Interplay and implication in targeted cancer therapy. J Hematol Oncol 13(1):113. https://doi.org/10.1186/s13045-020-00949-4

Zelenetz A-D (2016) Biosimilars in oncology. Oncol Hematol Rev 12:22–28. https://doi.org/10.17925/OHR.2016.12.01.22

Zelenetz A-D, Ahmed I, Braud E-L et al (2011) NCCN biosimilars white paper: regulatory, scientific, and patient safety perspectives. J Natl Compr Cancer Netw 9 Suppl 4:S1–S22. https://doi.org/10.6004/jnccn.2011.0136

Zhu Y, Zhu X, Wei X, Tang C, Zhang W (2021) HER2-targeted therapies in gastric cancer. Biochim Biophys Acta Rev Cancer 1876(1):188549. https://doi.org/10.1016/j.bbcan.2021.188549

Biosimilar in Lung Cancer

8

Kuttiappan Anitha, Shvetank Bhatt, Santenna Chenchula, Girdhari Lal Gupta, and Ravindra Babu Pingili

Abstract

In recent years, the development and clinical integration of biosimilar therapies have garnered significant attention in the field of oncology. An updated summary of function of biosimilar in treatment of lung cancer (LC) is given in this thorough review. The availability, cost of efficient therapy alternatives are crucial because LC is primary reason for cancer-related deaths globally. Biosimilar has become a promising means of increasing patient access to life-saving therapies because they provide efficacy and safety profiles that are comparable to those of reference biologics. This study explores the current state of biosimilar in the treatment of LC, taking into account legislative frameworks, clinical studies, and empirical data. It also discusses important issues related to extrapolation, interchangeability, and pharmacovigilance. In particular, the financial effects and cost-effectiveness of biosimilar in the treatment of LC are explored, giving light on their potential to reduce financial burdens on healthcare systems and individuals alike.

The review also emphasizes certain biosimilar drugs and their therapeutic applications in LC, with a focus on important trials and results. The effect of

K. Anitha (✉) · G. L. Gupta · R. B. Pingili
Department of Pharmacology, School of Pharmacy and Technology Management (SPTM), SVKM's Narsee Monjee Institute of Management Studies (NMIMS) Deemed-to-University, Shirpur, India
e-mail: Kuttiappan.anitha@nmims.edu

S. Bhatt
School of Health Sciences and Technology, Dr. Vishwanath Karad MIT World Peace University, Pune, Maharashtra, India

S. Chenchula
Department of Pharmacology, AIIMS, Bhopal, Madhya Pradesh, India

S. Bhatt et al. (eds.), *Biosimilars for Cancer Treatment*,
https://doi.org/10.1007/978-981-97-3001-8_8

biosimilar on patient outcomes, quality of life, and the use of medical resources is given particular consideration.

Finally, the study provides information on current research projects and potential developments in the field of LC biosimilar medicines. Exciting prospects to transform therapy paradigms, improve patient access, and optimize healthcare budget allocation are presented by the emerging biosimilar landscape in oncology.

Keywords

Biosimilar · Lung cancer · oncology · Bevacizumab · VEGF · Clinical trials

8.1 Introduction to Biosimilars in Lung Cancer

Lung cancer (LC) accounts for the highest number of cancer-related deaths globally, with an approximate annual toll of 1.8 million lives. Around 57% of patients at time of first diagnosis have distant metastases, making surgical intervention impossible (Lyman et al. 2018a). Systemic therapy therefore acquires a crucial role in the all-encompassing treatment of LC. Vascular endothelial growth factor (VEGF) is a key player in controlling tumour angiogenesis and linked with augmented mortality, metastasis, and recurrence risks (Lyman et al. 2018b).

The field of oncology treatment has experienced a sea change due to the revolutionary medicinal class known as biosimilars. Biosimilars are highly alike versions of licensed biologic medications that are derived from living things and have similar efficacy and safety profiles (Kaida-Yip et al. 2018). With the advent of biosimilar, a promising opportunity to recover patient access to essential medicines and concurrently address the financial difficulties associated with cancer care has emerged (Shukla et al. 2020).

Biologic drugs have revolutionized the treatment of serious medical disorders over the past 20 years. The use of "targeted therapy" has outperformed the results of empirical palliative chemotherapy in terms of progression-free and overall survivals (OS), as well as more attractive toxicity profiles and improved quality of life (Ledón and Lage 2017). However, this class of treatments has significantly increased healthcare costs over time due to large cost related with biologic drugs and substantial rise in their consumption (He et al. 2016; Herndon et al. 2021).

The development of biosimilars is particularly significant in field of LC, a condition that is responsible for a sizable amount of cancer caused mortality worldwide. LC requires novel approaches to treatment as major cause of mortality (Casak et al. 2018). By providing a cheaper alternative to pricey reference biologics and so increasing the accessibility of efficient treatments, biosimilars offer a means of reaching this objective (Bloomfield et al. 2022).

This overview lays the groundwork for a thorough investigation of LC biosimilar medicines that covers regulatory frameworks, comparative efficacy and safety assessments, clinical studies, and factors for their incorporation into clinical practice (Leonard et al. 2019; Chopra and Lopes 2017). This study attempts to lay a basis for comprehending the changing landscape of oncology care by clarifying the core ideas behind biosimilar and their possible effects on the treatment of LC.

While having similar safety, purity, and potency to the reference product, a biosimilar is not an exact clone. While maintaining treatment efficacy, it presents a vital chance to lower patient expenses and lessen the financial burden on national healthcare systems (US Food and Drug Administration n.d.). The therapeutic application of a biosimilar does, however, require particular attention and focus due to its multifaceted character (Niederwieser and Schmitz 2011).

Biologics play a critical role in contemporary cancer treatment strategies and are strongly advised by both National Comprehensive Cancer Network (NCCN) and American Society for Clinical Oncology (ASCO) guidelines due to their demonstrated capacity to improve clinical outcomes, including OS (Chauhan and Malik n.d.). The introduction of biosimilar medications is a potential means of reducing the cost of cancer treatment on a global scale. These biosimilars, which are available for a limited number of biologic drugs, have encouraged price competition between the original product and its biosimilar counterparts, leading to a considerable decrease in treatment costs (Li et al. 2015; Markets and Markets 2020). According to projections, the release of 11 biosimilars, including bevacizumab-based Avastin®, trastuzumab-based Herceptin®, and rituximab-based Rituxan®, could result in potential savings of around $250 billion in the USA from 2014 to 2024 (Jacobs et al. 2016).

8.2 Biosimilar Concerns and Information Gaps

Until now, the introduction of therapeutic oncology biosimilar has faced significant obstacles, including disparities in approval timelines between Europe and the United States, persistent knowledge gaps among healthcare providers, and concerns related to safety, efficacy, and regulatory complexities (Mellstedt et al. 2008; Press and Lenz 2007). Despite a decade of clinical knowledge with oncology biosimilars, healthcare professionals in both Europe and the United States continue to grapple with limited familiarity and understanding of these alternative therapies (Coiffier 2017; Declerck et al. 2018). The results of a recent systematic review, incorporating surveys from 20 healthcare providers in United States and Europe, highlighted prevalent issues such as insufficient knowledge, concerns about safety and efficacy, and uncertainties regarding interchangeability and extrapolation (Azevedo et al. 2017). The study also revealed a lack of awareness among patients, with 70% reporting limited knowledge of biosimilars (Syrigos 2013).

Physicians, in particular, held misconceptions about biosimilars, mistakenly associating their lower cost and expedited FDA approval processes with increased risks of side effects and reduced effectiveness compared to reference drugs (Cohen et al. 2017). Ongoing studies and editorials further underscore healthcare professionals' reservations about the safety and efficacy of biosimilars compared to reference products. A comprehensive survey exploring barriers to biosimilar adoption in U.S. spanning healthcare leaders, clinician practices, and pharmaceutical industry, highlighted shared concerns, particularly regarding safety and efficacy, as well as rebate increases for reference products (Chopra and Lopes 2017; Greene et al. 2019). In summary, understanding with therapeutic oncology biosimilars has unveiled intricate challenges that extend beyond those encountered with supportive care oncology biosimilars. These challenges contribute to reluctance of healthcare providers to endorse, prescribe, and integrate biosimilars into clinical practice (Nabhan et al. 2018). Effectively addressing these challenges through continuous education on oncology biosimilars and targeted initiatives to bridge knowledge gaps is imperative, as emphasized in this summary by ASCO (Finn et al. 2020; Garcia et al. 2020).

8.3 Adverse Events Associated with Biosimilar in Lung Cancer

The utilization of biosimilars in LC treatment, generally acknowledged for their safety and efficacy, accompanied by potential adverse events that demand careful consideration. Infusion reactions, a common phenomenon with biologics, may present as fever, chills, nausea, and headache during or shortly after the administration of biosimilars (Xue et al. 2023). It is imperative to closely monitor these reactions to ensure patient safety. Immunogenicity, involving the development of antibodies against the biosimilar, could occur and influence treatment efficacy and safety (Vaddepally et al. 2022). Hematologic effects, such as anaemia, neutropenia, and thrombocytopenia, pose challenges that require regular monitoring of blood cell counts (Thatcher et al. 2019a). The immunosuppressive characteristics of certain biosimilars used in LC treatment elevate the risk of infections, necessitating vigilance in recognizing and managing infectious complications encounter gastrointestinal issues, encompassing nausea, vomiting, diarrhoea, and constipation. Reported skin reactions, like rash or dermatitis, underscore the need for dermatological monitoring (Yang et al. 2019). Fatigue and asthenia further contribute to considerations affecting patients' overall well-being. Biosimilars used in LC treatment may entail a risk of pulmonary toxicity, presenting symptoms such as cough, shortness of breath, and changes in lung function (Ngo and Chen 2021). Continuous vigilance, robust monitoring, and a comprehensive understanding of the individual biosimilar adverse event profile are essential for optimizing the safety and efficacy of these therapeutic agents in LC management (Santos et al. 2019). Healthcare professionals should refer to prescribing information and relevant clinical data for the latest updates on adverse events associated with specific biosimilars.

8.4 Biosimilar Agents in Lung Cancer Treatment

8.4.1 Bevacizumab

A monoclonal antibody is a class of medication that includes bevacizumab. Laboratory-produced monoclonal antibodies can replicate immune system's capability to battle hazardous infections like bacteria or viruses. They made to identify and cling to particular proteins on cell surfaces (Xu et al. 2022).

VEGF is a protein that the body produces naturally. It is essential to course of angiogenesis, which is growth of new blood vessels. Angiogenesis is essential for healthy bodily functions like wound healing and growth, but it's also essential for cancer cells to multiply and spread throughout the body (Monk et al. 2017). The purpose of bevacizumab is to target and obstruct VEGF's function. By attaching to VEGF, it prevents the protein from interacting with cell surface receptors (Jacobs et al. 2017). The inability of VEGF to attach to its receptors prevents it from telling cells to develop new blood vessels. Bevacizumab interferes with angiogenesis by inhibiting VEGF. This is crucial in the treatment of cancer since cancers require blood flow to develop and spread (Xiao et al. 2022).

8.4.2 Bevacizumab Mechanism of Action

Angiogenesis, crucial for tumours larger than 1 mm, is the process through which cancer develops new blood vessels for oxygen supply. Without neovascularization, tumour growth halts. Mechanisms include endothelial sprouting, co-option of existing vasculature, vasculogenic mimicry, and direct release of proangiogenic factors by cells. The VEGF family and its receptors are promising targets, especially VEGF-A, elevated in breast, lung, colon, uterus, and ovarian cancers, linked to poor prognosis. VEGF-A binding to receptors triggers angiogenesis and Bevacizumab inhibits this, neutralizing VEGF's activity, leading to regression of tumour vascularization and inhibition of new vasculature formation, ultimately suppressing tumour growth (Fig. 8.1).

The tumour's growth can be slowed down or even stopped in its tracks without a constant supply of blood and nutrients. The drug bevacizumab used to deprive cancers of their blood supply. Various biosimilars with their FDA indications and usage have been shown in Table 8.1. They deprived of the nourishment and oxygen they require to thrive and spread. Additionally, by enhancing drug transport to the tumour location, bevacizumab may help chemotherapy become more effective (Sandler et al. 2006). Bevacizumab has authorized for use in treating a number of cancers, including colorectal, lung, breast, and other cancers. It frequently works well with other cancer therapies like chemotherapy (Luo et al. 2022).

They prioritize improving patients' access to efficient treatments for LC. Biosimilars have the potential to provide access to essential therapeutics by offering more cost substitutes for the reference biologic medications (Luo et al. 2023). Additionally, to guarantee that patients receive the best care possible,

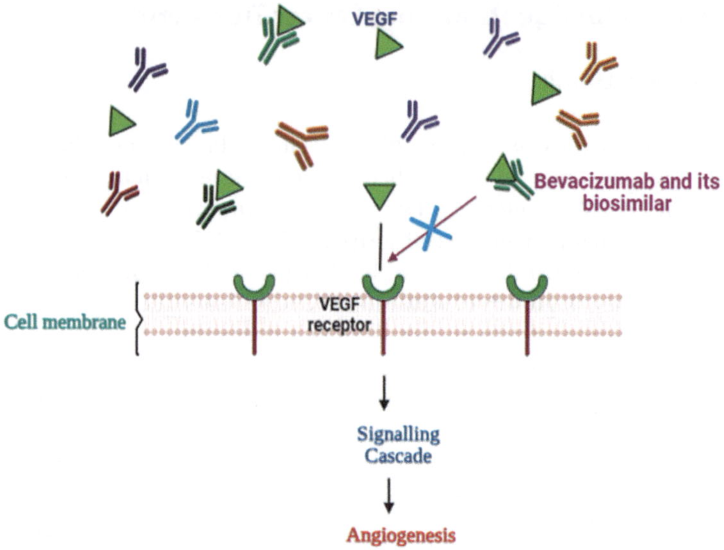

Fig. 8.1 Bevacizumab functions by specifically attaching to circulating VEGF, preventing its interaction with cell surface receptors. This interference results in a decrease in the development of microvascular structures within tumour blood vessels, consequently restricting the blood flow to the tumour tissues

Table 8.1 Various biosimilars with their FDA indications and usage

Reference product	Biosimilars	Manufacturer	Year of FDA approval	FDA indications and usage	Reference
Bevacizumab (Avastin)	Bevacizumab-awwb (Mvasi)	Amgen/ Allergan	2017	Colorectal, lung, glioblastoma, cervical, and ovarian cancers	Bevacizumab-awwb information (n.d.)
Bevacizumab (Avastin)	Bevacizumab-bvzr (Zirabev)	Pfizer	2019	Colorectal, lung, glioblastoma, cervical, and ovarian cancers	Bevacizumab-bvzr information (n.d.)
Bevacizumab (Avastin)	Bevacizumab-awwb (Zirabev)	Samsung Bioepis/ Merck	2020	Colorectal, lung, glioblastoma, cervical, and ovarian cancers	Jin et al. (2021)

biosimilars must have effectiveness and safety profiles that are equivalent to those of their reference counterparts. The financial burden of LC treatment is another important area of emphasis. Biosimilars provide the possibility of lowering the overall cost of therapy through the introduction of market competition, hence

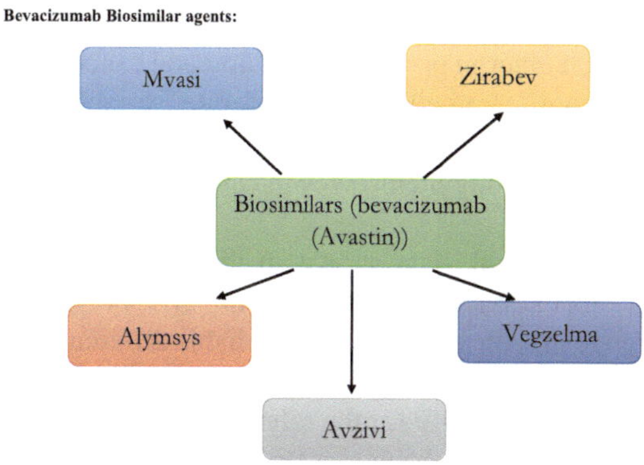

Fig. 8.2 Different biosimilars of avastin

improving financial sustainability within healthcare systems (Jin et al. 2021). Rigid clinical equivalence studies are carried out to prove the high similarity between biosimilars and their reference biologic medications in order to accomplish these goals (Ma et al. 2021; Thatcher et al. 2019b). Finally, before biosimilars are incorporated into clinical practice, regulatory approval procedures are crucial in guaranteeing their efficacy and quality. It is significant to note that there may have been advancements in the area of biosimilars since my previous update, thus it advised to seek credible medical sources or healthcare specialists for the most recent information (Huang et al. 2020). Different biosimilars of avastin have been shown in Fig. 8.2.

8.4.3 Bevacizumab Biosimilar Agents

8.5 Clinical Trials of Bevacizumab Biosimilars

Saleem et al. reported that while biosimilars have a bright future, gaining mass-market availability will be difficult due to the accompanying costs. Nonetheless, when important stakeholders develop a better knowledge and acceptance of biosimilars, rules that are more accommodating and legislation may emerge. This, in turn, has the potential to increase biosimilar utilization in United States (Saleem et al. 2020).

Xiao et al. states that study stands out as the first independent meta-analysis reporting on use of biosimilar bevacizumab in treatment of NSCLC patients, comparing efficacy and safety of biosimilar bevacizumab to reference bevacizumab (Avastin®). According to the data, biosimilar bevacizumab has equivalent efficacy and safety profiles to original bevacizumab.

Xu et al. reported that seven separate types of bevacizumab biosimilars were identified in treatment protocol for patients with NSCLC, with equivalent efficacy and safety characteristics. The degree of certainty in these findings ranged from low to moderate. Furthermore, three distinct biosimilars were regarded clinically equal to each other in persons with colorectal cancer (CRC), with evidence certainty ranging from extremely low to moderate. Notably, the effectiveness and safety of bevacizumab biosimilars were shown to be comparable not only to reference biologics but also among themselves in cases with advanced NSCLC or metastatic CRC (Xu et al. 2022).

Zhao et al. data shown in the biosimilar cohort, treatment-emergent adverse events (TEAEs) occurred at a rate of 76.41%, slightly surpassing the 71.65% incidence in reference group, though this difference did not reach statistical significance ($P = 0.098$). Occurrence of Grade 3 or higher TEAEs was 22.14% for the biosimilar group and 19.49% for the reference group, with no significant disparity ($P = 0.324$). The bevacizumab biosimilar established efficacy comparable to bevacizumab in patients facing locally advanced and advanced NSCLC. Moreover, it displayed a suitable toxicity profile, devoid of any newly identified adverse events. These findings suggest that individuals excluded from clinical trials can also experience beneficial outcomes from the utilization of bevacizumab biosimilar (Zhao et al. 2023; Reinmuth et al. 2019).

8.6 Ongoing Research and Future Prospects

Future prospects for biosimilars in the field of LC treatment hold great promise and have the potential to have a significant impact on clinical practice in a number of ways. The development of a greater variety of biosimilars that target different features of LC, such as biologics that target particular pathways involved in tumour growth and progression, is a goal of ongoing research (García et al. 2020). The availability of treatment for a larger population of LC patients may be improved by the growth of biosimilar choices. The cost of treating LC may be greatly decreased with the use of biosimilars (Bloomfield et al. 2022). The goal of ongoing research is to show that biosimilars are clinically equivalent to their reference biologics in order to promote their adoption and support the sustainability of healthcare systems. Research into the creation of LC biosimilars continues to centre on rigorous clinical studies (Cohen et al. 2007). These studies seek to demonstrate that biosimilars and their reference biologics are comparable in terms of efficacy, safety, and immunogenicity.

Future studies might look towards creating biosimilars that are especially suited to some subtypes or genetic mutations linked to LC. Individual patients may benefit from more effective and targeted treatments as a result of this individualized approach (Trukhin et al. 2021; Barlesi et al. 2013). An area of interest is examining the compatibility and efficacy of mixing biosimilars with other targeted treatments or immunotherapies. This method might result in beneficial synergistic effects and better LC therapy outcomes (Niho et al. 2012; Syrigos et al. 2021). To give a more thorough assessment of the effect of biosimilars on the wellbeing of LC patients, future research may concentrate a higher emphasis on evaluating patient-reported outcomes, including quality of life and treatment satisfaction (Reck et al. 2020).

8.7 Discussion

A significant advancement in oncology has been the incorporation of biosimilars into the treatment of LC. These remarkably identical variations of well-known biologic medications have enormous potential for enhancing patient access to effective treatments while addressing the financial difficulties related to cancer care. Clinical trials and real-world investigations have shown that biosimilars have efficacy and safety profiles that are equivalent to those of their reference biologics, giving patients and healthcare professionals alike peace of mind. Because of this equivalence, which is supported by strict regulatory procedures, biosimilar products are held to the same high standards as their original counterparts. Additionally, there is a significant possibility for cost reductions with biosimilars, providing a way to optimize healthcare resources and improve treatment accessibility for a larger patient group. However, thorough education campaigns to boost patient and healthcare professional confidence, open communication about the advantages and limitations of biosimilars, and the development of unambiguous clinical guidelines are necessary for successful integration. Biosimilars have the potential to change LC treatment by providing high-quality care that is both universally accessible and financially viable by negotiating these issues. This development not only highlights oncology's advancements but also signifies a more open-minded and patient-centred approach to cancer treatment.

Biosimilars have completely changed the way LC is treated. They are an important addition to the arsenal against this deadly illness due to their demonstrated efficacy, tolerability, and cost-effectiveness. Biosimilars can be effortlessly incorporated into clinical practice by adhering to established criteria, educating healthcare personnel and patients, and upholding strict regulatory control. They therefore have the potential to dramatically increase LC patients' access to high-quality care, which would eventually result in improved outcomes and a higher standard of living. The continuous development and use of biosimilars in the treatment of LC mark a crucial achievement in the struggle against this terrible disease.

8.8 Conclusion

In conclusion, development of biosimilars in treatment of LC has great potential for revolutionizing the medical landscape. Biosimilars provide a viable alternative to reference biologics, increasing access to essential medications thanks to their equivalent efficacy, safety profiles, and significant cost reductions. Adherence to strict regulatory standards, open communication with patients, and thorough education and training of healthcare professionals are all required for the incorporation of biosimilars into clinical practice. Informed consent procedures, pharmacovigilance programs, and clear guidelines are crucial for a successful adoption of biosimilars. Additionally, the financial effects of biosimilars offer a chance to better allocate healthcare funds for the benefit of patients. Our comprehension of how biosimilars affect LC treatment will be furthered by ongoing study and data collection from actual patients. Biosimilars are a crucial component of the comprehensive care given to people fighting LC through collaboration amongst healthcare stakeholders, ultimately enhancing results and furthering the study of oncology.

References

Azevedo V, Dörner T, Strohal R, Isaacs J, Castañeda-Hernández G, Gonçalves J, McInnes I (2017) Biosimilars: considerations for clinical practice. Considerations Med 1:13–18

Barlesi F, Scherpereel A, Rittmeyer A, Pazzola A, Ferrer Tur N, Kim JH (2013) Randomized phase III trial of maintenance bevacizumab with or without pemetrexed after first-line induction with bevacizumab, cisplatin, and pemetrexed in advanced nonsquamous non-small-cell lung cancer: AVAPERL (MO22089). J Clin Oncol 31(24):3004–3011

Bevacizumab-awwb information (n.d.). https://www.accessdata.fda.gov/scripts/cder/daf/index.cfm?event=overview.process&applno=761028

Bevacizumab-bvzr information (n.d.). https://www.accessdata.fda.gov/drugsatfda_docs/label/2019/761099s000lbl.pdf

Bloomfield D, D'Andrea E, Nagar S, Kesselheim A (2022) Characteristics of clinical trials evaluating biosimilars in the treatment of cancer: a systematic review and meta-analysis. JAMA Oncol 8(4):537–545. https://doi.org/10.1001/jamaoncol.2021.7230

Casak SJ, Lemery SJ, Chung J et al (2018) FDA's approval of the first biosimilar to bevacizumab. Clin Cancer Res 24:4365–4370

Chauhan MK, Malik S (n.d.) Regulatory guidelines for the approval of biosimilars in India, Europe, Brazil, and China: a comprehensive overview. http://innovareacademics.in/journals/index.php/ijpps/article/view/11753

Chopra R, Lopes G (2017) Improving access to cancer treatments: the role of biosimilars. J Glob Oncol 3:596–610

Cohen MH, Gootenberg J, Keegan P, Pazdur R (2007) FDA drug approval summary: bevacizumab (Avastin) plus carboplatin and paclitaxel as first-line treatment of advanced/metastatic recurrent nonsquamous non-small cell lung cancer. Oncologist 12(6):713–718. https://doi.org/10.1634/theoncologist.12-6-713

Cohen H, Beydoun D, Chien D et al (2017) Awareness, knowledge, and perceptions of biosimilars among specialty physicians. Adv Ther 33:2160–2172

Coiffier B (2017) Preparing for a new generation of biologic therapies: understanding the development and potential of biosimilar cancer therapeutics. Future Oncol 13:1–3

Declerck P, Bakalos G, Zintzaras E, Barton B, Schreitmüller T (2018) Monoclonal antibody biosimilars in oncology: critical appraisal of available data on switching. Clin Ther 40:798–809.e2

Finn RS, Qin S, Ikeda M, Galle PR, Ducreux M, Kim TY et al (2020) Atezolizumab plus bevacizumab in unresectable hepatocellular carcinoma. N Engl J Med 382(20):1894–1905

Garcia J, Hurwitz HI, Sandler AB, Miles D, Coleman RL, Deurloo R et al (2020) Bevacizumab (Avastin®) in cancer treatment: a review of 15 years of clinical experience and future outlook. Cancer Treat Rev 86:102017

García JJ, Raez LE, Rosas D (2020) A narrative review of biosimilars: a continued journey from the scientific evidence to practice implementation. Transl Lung Cancer Research 9(5):2113

Greene L, Singh RM, Carden MJ et al (2019) Strategies for overcoming barriers to adopting biosimilars and achieving goals of the Biologics Price Competition and Innovation Act: a survey of managed care and specialty pharmacy professionals. J Manag Care Spec Pharm 25:904–912

He K, Chen H, Gwise T, Casak S, Lemery S, Keegan P et al (2016) Statistical considerations in evaluating a biosimilar product in an oncology clinical study. Clin Cancer Res 22(21):5167–5170. https://doi.org/10.1158/1078-0432.CCR-16-1010

Herndon K, Braithwaite J, Berry B et al (2021) Biosimilar perceptions among healthcare professionals and commercial medical benefit policy analysis in the United States. BioDrugs 35:103–112
Huang HY, Liu CC, Yu Y, Wang L, Wu DW, Guo LW, Wang SH, Fang H, Bai Y, Fang Y, Fan Q (2020) Pharmacoeconomic evaluation of cancer biosimilars worldwide: a systematic review. Front Pharmacol 11:572569
Jacobs I, Singh E, Sewell KL et al (2016) Patient attitudes and understanding about biosimilars: an international cross-sectional survey. Patient Prefer Adherence 10:937–948
Jacobs I, Ewesuedo R, Lula S, Zacharchuk C (2017) Biosimilars for the treatment of cancer: a systematic review of published evidence. BioDrugs 31:1–36
Jin R, Mahtani RL, Accortt N, Lawrence T, Sandschafer D, Loaiza-Bonilla A (2021) Clinical and treatment characteristics of patients treated with the first therapeutic oncology biosimilars bevacizumab-awwb and trastuzumab-anns in the US. Ther Adv Med Oncol 13:17588359211041961
Kaida-Yip F, Deshpande K, Saran T et al (2018) Biosimilars: review of current applications, obstacles, and their future in medicine. World J Clin Cases 6:161–166
Ledón N, Lage A (2017) Biosimilars and the real world. MEDICC Rev 19:9
Leonard E, Wascovich M, Oskouei S et al (2019) Factors affecting health care provider knowledge and acceptance of biosimilar medicines: a systematic review. J Manag Care Spec Pharm 25:102–112
Li E, Subramanian J, Anderson S et al (2015) Development of biosimilars in an era of oncologic drug shortages. Drug Des Devel Ther 9:3247–3255
Luo X, Liu Q, Zhou Z, Yi L, Peng L, Wan X et al (2022) Cost-effectiveness of bevacizumab biosimilar LY01008 combined with chemotherapy as first-line treatment for Chinese patients with advanced or recurrent nonsquamous non-small cell lung cancer. Front Pharmacol 13:832215. https://doi.org/10.3389/fphar.2022.832215
Luo X, Du X, Li Z, Liu J, Lv X, Li H, Guo Q, Wang C, Xue X, Le K, Jiang X (2023) Clinical benefit, price, and uptake for cancer biosimilars vs reference drugs in China: a systematic review and meta-analysis. JAMA Netw Open 6(10):e2337348
Lyman GH, Zon R, Harvey RD et al (2018a) Rationale, opportunities, and reality of biosimilar medications. N Engl J Med 378:2036–2044
Lyman GH, Balaban E, Diaz M et al (2018b) American Society of Clinical Oncology statement: biosimilars in oncology. J Clin Oncol 36:1260–1265
Ma JT, Guo YJ, Song J, Sun L, Zhang SL, Huang LT et al (2021) Rational application of first-line EGFR-TKIs combined with antiangiogenic inhibitors in advanced EGFR-mutant non-small-cell lung cancer: a systematic review and meta-analysis. Biomed Res Int 2021:8850256. https://doi.org/10.1155/2021/8850256
Markets and Markets: biosimilars market by product (recombinant non-glycosylated proteins (insulin, filgrastim, interferons, rHGH), glycosylated (monoclonal antibodies, EPO), peptides (glucagon, calcitonin)) & application (oncology, blood disorders): Global forecast to 2020. http://www.marketsandmarkets.com/Market-Reports/biosimilars-40.html
Mellstedt H, Niederwieser D, Ludwig H (2008) The challenge of biosimilars. Ann Oncol 19(3):411–419
Monk BJ, Huh WK, Rosenberg JA, Jacobs I (2017) Will bevacizumab biosimilars impact the value of systemic therapy in gynecologic cancers? Gynecol Oncol Res Pract 4:7. https://doi.org/10.1186/s40661-017-0045-x
Nabhan C, Parsad S, Mato AR, Feinberg BA (2018) Biosimilars in oncology in the United States: a review. JAMA Oncol 4(2):241–247
Ngo D, Chen J (2021) A clinical review of biosimilars approved in oncology. Ann Pharmacother 55(3):362–377
Niederwieser D, Schmitz S (2011) Biosimilar agents in oncology/haematology: from approval to practice. Eur J Haematol 86:277–288

Niho S, Kunitoh H, Nokihara H, Horai T, Ichinose Y, Hida T et al (2012) Randomized phase II study of first-line carboplatin and paclitaxel with or without bevacizumab in Japanese patients with advanced non-squamous non-small-cell lung cancer. Lung Cancer 76(3):362–367

Press MF, Lenz HJ (2007) EGFR, HER2 and VEGF pathways: validated targets for cancer treatment. Drugs 67:2045–2075

Reck M, Luft A, Bondarenko I, Shevnia S, Trukhin D, Kovalenko NV, Vacharadze K, Andrea F, Hontsa A, Choi J, Shin D (2020) A phase III, randomized, double-blind, multicenter study to compare the efficacy, safety, pharmacokinetics, and immunogenicity between SB8 (proposed bevacizumab biosimilar) and reference bevacizumab in patients with metastatic or recurrent nonsquamous non-small cell lung cancer. Lung Cancer 146:12–18

Reinmuth N, Bryl M, Bondarenko I, Syrigos K, Vladimirov V, Zereu M, Bair AH, Hilton F, Liau K, Kasahara K (2019) PF-06439535 (a bevacizumab biosimilar) compared with reference bevacizumab (Avastin®), both plus paclitaxel and carboplatin, as first-line treatment for advanced non-squamous non-small-cell lung cancer: a randomized, double-blind study. BioDrugs 33:555–570

Saleem T, Qurashi H, Jamali M, Gomez JC, Kanderi T (2020) Biosimilars as a future, promising solution for financial toxicity: a review with emphasis on bevacizumab. Cureus 12(7):e9300

Sandler A, Gray R, Perry MC, Brahmer J, Schiller JH, Dowlati A et al (2006) Paclitaxel-carboplatin alone or with bevacizumab for non-small-cell lung cancer. N Engl J Med 355(24):2542–2550

Santos SB, Lobo JM, Silva AC (2019) Biosimilar medicines used for cancer therapy in Europe: a review. Drug Discov Today 24(1):293–299

Shukla NA, Yan MN, Hanna N (2020) The story of angiogenesis inhibitors in non-small-cell lung cancer: the past, present, and future. Clin Lung Cancer 21(4):308–313. https://doi.org/10.1016/j.cllc.2020.02.024

Syrigos KN (2013) Bevacizumab in lung cancer: impact and exceptional cases. Focus on Health Ltd., Athens, pp 9–11

Syrigos K, Abert I, Andric Z, Bondarenko IN, Dvorkin M, Galic K, Galiulin R, Kuchava V, Sriuranpong V, Trukhin D, Zhavrid E (2021) Efficacy and safety of bevacizumab biosimilar FKB238 versus originator bevacizumab: results from AVANA, a phase III trial in patients with non-squamous non-small-cell lung cancer (non-sq-NSCLC). BioDrugs 35:417–428

Thatcher N, Goldschmidt JH, Thomas M, Schenker M, Pan Z, Paz-Ares Rodriguez L, Breder V, Ostoros G, Hanes V (2019a) Efficacy and safety of the biosimilar ABP 215 compared with bevacizumab in patients with advanced nonsquamous non–small cell lung cancer (MAPLE): a randomized, double-blind, phase III study. Clin Cancer Res 25(7):2088–2095

Thatcher N, Goldschmidt JH, Thomas M et al (2019b) Efficacy and safety of the biosimilar ABP 215 compared with bevacizumab in patients with advanced nonsquamous non-small cell lung cancer (MAPLE): a randomized, double-blind, phase III study. Clin Cancer Res 25:2088–2095. https://doi.org/10.1158/1078-0432.Ccr-18-2702

Trukhin D, Poddubskaya E, Andric Z, Makharadze T, Bellala RS, Charoentum C, Yanez Ruiz EP, Fulop A, Hyder Ali IA, Syrigos K, Katgi N (2021) Efficacy, safety and immunogenicity of MB02 (bevacizumab biosimilar) versus reference bevacizumab in advanced non-small cell lung cancer: a randomized, double-blind, phase III study (STELLA). BioDrugs 35(4):429–444

US Food and Drug Administration (n.d.) FDA and FTC announce new efforts to further deter anticompetitive business practices, support competitive market for biological products to help Americans. https://www.fda.gov/newsevents/pressannouncementsfdaandftcannouncenewefforts furtherdeteranticompetitivebusinesspracticessupport

Vaddepally R, Doddamani R, Sodavarapu S, Madam NR, Katkar R, Kutadi AP, Mathew N, Garje R, Chandra AB (2022) Review of immune-related adverse events (irAEs) in non-small-cell lung cancer (NSCLC)—their incidence, management, multiorgan irAEs, and rechallenge. Biomedicines 10(4):790

Xiao X, Zhang G, Sun B, Wang C, Wang X, Kong F, Jia Y (2022) Comparison of efficacy and safety of bevacizumab biosimilar and original bevacizumab in non-squamous non-small cell lung cancer: a systematic review and meta-analysis. Transl Cancer Res 11(6):1472

Xu X, Zhang S, Xu T, Zhan M, Chen C, Zhang C (2022) Efficacy and safety of bevacizumab biosimilars compared with reference biologics in advanced non-small cell lung cancer or metastatic colorectal cancer patients: a network meta-analysis. Front Pharmacol 13:880090

Xue X, Truong B, Qian J (2023) Adverse event reporting of marketed biosimilar and biological monoclonal antibody cancer treatments in the United States. Expert Opin Biol Ther 23(8):841–849

Yang J, Yu S, Yang Z, Yan Y, Chen Y, Zeng H, Ma F, Shi Y, Shi Y, Zhang Z, Sun F (2019) Efficacy and safety of supportive care biosimilars among cancer patients: a systematic review and meta-analysis. BioDrugs 33:373–389

Zhao Z, Zhao L, Xia G, Lu J, Shen B, Zhou G, Wu F, Hu X, Feng J, Yu S (2023) Efficacy and safety of bevacizumab biosimilar compared with reference bevacizumab in locally advanced and advanced non-small cell lung cancer patients: a retrospective study. Front Oncol 12:1036906

A Recent Overview of the Growing Applications of Biosimilars in Pancreatic Cancer Management: Current Picture and Future Perspectives

9

Dhruv Sanjay Gupta, Daksh Sanjay Gupta, and Saritha R. Shetty

Abstract

With an annual growth rate in incidence of 1.1%, high lethality and difficulties in early detection, pancreatic cancer is projected to become the second leading cause of cancer-linked deaths by 2030, and is a cause for alarm. Given the steadily increasing incidence and mortality associated with this type of cancer, and poor prognosis resulting in mounting pressure on healthcare systems, there is a need to explore alternate strategies that promise safety, efficacy and cost-effectiveness. Rising rates of chemoresistance have prompted discovery in the domain of immunotherapy, with monoclonal antibodies (MABs) occupying a spot of prominence in this domain. However, a major hindrance to their usage has been the high cost of therapy. This has further led to the screening of biosimilar candidates, to drive down costs and improve patient access. Biosimilars are medications with substantial similarity to licensed biologics, established through laboratory testing focused on gauging similarity in pharmacodynamic and pharmacokinetic profiles. The authors of this chapter aim to offer an overview of the growing disease burden of pancreatic cancer and its pathogenesis, currently employed diagnostic and management strategies, as well as newer avenues such as the utilization of biosimilars of some key biologics. In addition to this, an insight into recently undertaken clinical trials has been offered to the readers. Any discussion on an emerging frontier would remain incomplete without a focus on lacunae and challenges, such as cost–benefit analysis, the need for

D. S. Gupta · S. R. Shetty (✉)
Department of Pharmaceutics, Shobhaben Pratapbhai Patel School of Pharmacy and Technology Management, SVKM's Narsee Monjee Institute of Management Studies (NMIMS) Deemed-to-University, Mumbai, Maharashtra, India
e-mail: saritha.shetty@nmims.edu

D. S. Gupta
Vivekanand Education Society's College of Pharmacy, Mumbai, Maharashtra, India

S. Bhatt et al. (eds.), *Biosimilars for Cancer Treatment*,
https://doi.org/10.1007/978-981-97-3001-8_9

better-designed clinical protocols and collaboration among key stakeholders, to expedite market access. Finally, given the unprecedented growth in the domain of artificial intelligence and machine learning, a future perspective on these tools and their applicability in the domain of pancreatic cancer has been briefly discussed, as well as implications on expanding the usage of biosimilars.

Keywords

Pancreatic cancer · Biosimilars · Biologics · Genetic mutations · Oncogenes · Oncology

9.1 Introduction

Biologics occupy a spot of prominence in the management of chronic conditions, chiefly those with an immune underpinning (Hsiao et al. 2020). Since the landmark approval of human insulin as a first-of-its-kind therapy, there has been a significant focus on the development of biologics. However, this process is rigorous and cost-intensive, adding incrementally to the ever-increasing burden on healthcare systems, accounting for nearly one-third of the yearly expenditure on medications, in certain European countries (Grieshaber-Bouyer and Lorenz 2020).

A biosimilar medication is one that is substantially similar to, but not identical to, a licensed biological product (the reference or original product) in contrast to small-molecule generic pharmaceuticals. Biopharmaceutical giants across the world are shifting their focus to providing cost-effective management strategies for diseases with complex etiologies, as originator biologics reach the patent cliff. However, with the introduction of biosimilars, it is essential for clinicians to gain familiarity with this modality, chiefly with respect to the development of these agents, areas of similarity and variation as compared to originator moieties, as well as guidelines with respect to integrating these agents into specialized therapeutic regimens (Thill et al. 2019).

While there has been an overall global reduction in cancer-related mortality since 2020, owing to a reduction in smoking, early-stage detection and advancements in adjuvant therapies, there has been a marked increase in the incidence of pancreatic cancer (Siegel et al. 2024). With an annual projected increase of 1% globally, high rates of morbidity and mortality associated with the disease are found in developed as well as low and middle-income countries (LMICs), further highlighting disparities in healthcare delivery models (Ilic and Ilic 2016). In terms of its classification, pancreatic cancer may be broadly understood under two headings: those that are parenchymal in origin (encompassing exocrine and endocrine tumours), and stromal tumours (Zalatnai 2006). Exocrine tumours account for a majority of cases, with cystadenomas being presented primarily as benign tumours, and ductal adenocarcinomas forming a major chunk of aggressive, malignant cases of the disease (Collisson et al. 2019). While surgery, radiotherapy and conventional chemotherapy have yielded promising results, their usage is complicated by a number of factors, such as poor progression-free survival times, and an increased possibility of disease recurrence (Ferrantella et al. 2019). This has further led to the exploration of biologics and subsequently biosimilars, to manage treatment costs.

The global market for oncology is expected to shift towards biosimilars, with a 30–45% reduction in healthcare expenditure predicted, following their adoption (Vizgirda and Jacobs 2017). Despite the initial levels of scepticism and apprehension from clinicians and patients alike with respect to the safety and efficacy of these agents, global awareness and education programmes have been undertaken to make the adoption process significantly smoother. In addition to this, other efforts to bring about standardization, that are being undertaken, include standardization in terms of nomenclature, harmonization of regulatory requirements and the implementation of pharmacovigilance programmes (Camacho 2017).

This chapter aims to provide readers with a comprehensive understanding of pancreatic cancer as a disease condition, dabbling in its risk factors and pathogenesis, screening and diagnostic methodologies, as well as currently accepted treatment strategies. As the chapter progresses, readers are introduced to the concept of biosimilars in cancer management, delving further into the domain of pancreatic cancer as the area of focus. This discussion is followed by a bird's-eye view of the current clinical picture, certain lacunae that must be addressed to drive usage and market access, as well as future perspectives pertaining to the usage of biosimilars.

9.2 Risk Factors and Pathogenesis of Pancreatic Cancer

As discussed briefly, the incidence of pancreatic cancer has steadily been on the rise over the last few decades, accounting for nearly 5% of global malignancies (Goral 2015). With a predominance in men, chiefly those over 40 years of age, it begins as an asymptomatic malignancy and eventually takes on an aggressive course (Loveday et al. 2019). Owing to the nature of the ailment, detection is complicated, and coupled with swift metastases and a propensity to spread to surrounding organs, the survival rate remains dismal (Halbrook et al. 2023). On the basis of the functions of the pancreas, malignancies of the organ may be broadly divided into non-endocrine and endocrine tumours. Benign, non-endocrine tumours broadly include adenomas, lipomas, neuromas and cystadenomas. Malignant tumours, on the other hand, differ in their histopathological characteristics and may be broadly grouped as ductal adenocarcinomas (accounting for a majority of cases), cystadenocarcinomas, and other types of malignant tumours, which include sarcomas (Galli et al. 2021). Continued research efforts have indicated that pancreatic cancer has a strong genetic basis, with family history and metabolic conditions such as diabetes mellitus increasing the likelihood of its onset (Roy et al. 2021). The genetic underpinnings of pancreatic malignancies shall be discussed in greater detail, further on.

In addition to a hereditary predisposition, factors such as age, long-term cigarette smoking and alcohol consumption, prolonged exposure to radiotherapy and continued consumption of a high-fat diet are prominent risk factors (Klein 2021). Of these, it is crucial to highlight smoking as a major contributor. This may be chiefly attributed to an increased number of cancerous mutations in smokers, as compared to non-smokers. Chronic pancreatitis is an additional risk factor, increasing the likelihood of disease occurrence to the tune of 2.7-fold (Del Chiaro et al. 2007). The progression of non-cancerous, pancreatic epithelial cells to malignant cells is also influenced by a number of germline and acquired mutations in oncogenes, tumour suppressor genes, and those responsible for regulating the cell cycle and apoptosis (Mizrahi et al. 2020). Hallmarks of early-stage cancer include telomerase shortening, mutations of the p16 gene and KRAS, as well as reactive oxygen species (ROS) mediated inflammation. An increased cellular turnover, in turn, drives genomic instability (Hata et al. 2016). Later in the sequence of pathogenesis, loss of the p53 gene as well as inactivation of the SMAD4 oncogene cause significant disruption of the TGF-β signalling pathway (Kamisawa et al. 2016).

Certain conditions may also act as precursors to pancreatic malignancies and are commonly observed to originate from cancer stem cells (CSCs) (Ishiwata et al. 2018). These include intraepithelial neoplasms (concentrated near the pancreatic duct, and associated with invasive carcinomas) (Zhao et al. 2017), intraductal papillary neoplasms (typically smaller in size and usually asymptomatic, with mucin production being a primary distinguishing feature between subtypes), and mucinous cystic neoplasms (characterized by a thickened, fibrotic wall and greater incidence in females) (McGuigan et al. 2018). This may be coupled with mechanisms of oncogene mutations, such as point mutation and amplification, resulting in their activation. While a mutation in the KRAS gene is observed in a vast majority of pancreatic cancers, other prominent genes include TP53, SMAD4 and CDKN2A (Kleeff et al. 2016). Such genetic mutations are observed to result in the formation of precancerous lesions, which remain focused around the pancreatic duct. In addition to this, various signalling pathways are also directly influenced by such oncogene mutations. These include the P13K-AKT pathway (affecting cellular circulation) the ERK 1/2 (Yan et al. 2019) and mitogen-activated extracellular signal-regulated kinase (MEK) pathways (responsible for regulating cellular proliferation, phases of the cell cycle, angiogenesis and cellular apoptosis), as well as the NOTCH signalling pathway. KRAS mutation, in particular, also has a bearing on the STAT3 and hedgehog pathways, significantly amplifying metastases (Datta et al. 2022). It is also essential to factor in the impact of microRNAS (miRNAS) in tumour progression. Some major uncoded RNA moieties influencing gene expression include miRNA-196a (observed to be upregulated in pancreatic neuroendocrine cancer cell lines), miRNA-190 and miRNA-21 (playing a major role in cellular stemness) (Yonemori et al. 2017). Lastly, the dysregulation of tumour suppressor genes (TSGs) also plays a role in the manifestation of tumours. Of these, p53 mutation is observed in the early three fourth of pancreatic cancer cases. Other TSG mutations include that of liver kinase B1 (LKB1) and mitogen-activated protein kinase 4 (MKK4) (Bhattacharya et al. 2021).

Epigastric pain remains a leading symptom in pancreatic cancer, with a sharp pain that may subsequently radiate to the back. This pain is exacerbated by simple, daily activities and locomotion (Stevens and Lisanti 2024). In addition to this, sudden weight loss, jaundice, diarrhoea or steatorrhea may also be observed. Symptoms may also be specific to the location of the tumour. For instance, cancers originating at the head of the pancreas are characterized by dark jaundice, owing to bile duct obstruction. In summation, persistent pain, unexplained reduction in appetite, and jaundice are key symptoms to be monitored clinically (De La Cruz et al. 2014).

Taking the example of Pancreatic ductal adenocarcinoma (PDAC), genetic mutations result in molecular alterations in cancer cells. Notable phenomena include kataegis (resulting in clustered substitutions of nucleotides) and chromothripsis (characterized by genomic rearrangements in a small number of chromosomes) (Australian Pancreatic Cancer Genome Initiative, ICGC Breast Cancer Consortium, ICGC MMML-Seq Consortium, ICGC PedBrain, et al. 2013). In addition to DNA alterations, other mutations observed include epigenetic modulations and acetylation to histones (Zhou et al. 2022). With a growing focus on the tumour microenvironment (TME) in a number of cancers, PDAC is characterized by an immunosuppressive TME, and recruitment of regulatory T-cells, myeloid-derived suppressor cells and macrophages, resulting in interleukin secretion (Ho et al. 2020). Cancer-associated fibroblasts (CAFs) have also been observed to play a role in microenvironmental regulation, promoting the growth and survival of malignant cells through lipid enrichment. A major influencer of CAF activity has been identified to be Netrin G1, providing further insights into the crosstalk between cells in the TME (Francescone et al. 2021). Figure 9.1 offers an insight into the progression of pancreatic cancer, the classification of tumours, precursors of the disease, as well as major genetic and microenvironmental changes.

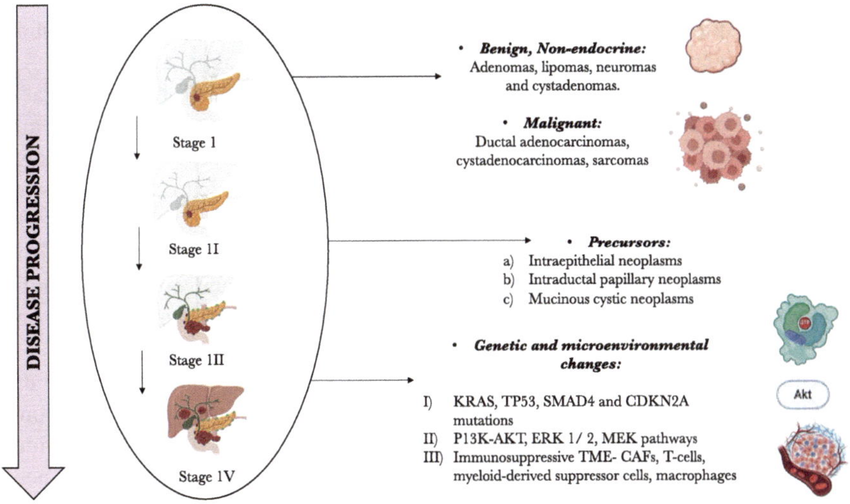

Fig. 9.1 Overview of progression, precursors and major physiological changes accompanying the pathogenesis of pancreatic cancer (*Created using Biorender*)

9.3 Screening Measures and Diagnostics of Pancreatic Cancer

While pancreatic cancer has a lower incidence in comparison with other cancer types, it is associated with high mortality owing to its complex etiology (Zhang et al. 2018). This may be attributed to localized nature of the ailment in a very small population of patients, with most cases being metastatic at the time of diagnosis, and rapidly spreading (Huang et al. 2021). Thus, screening has emerged as a vital tool to enable the potential detection of curable pancreatic malignancies. Undertaking screening greatly influences the overall survival time of patients and takes into consideration chiefly the size of the tumour and the disease stage (Cai et al. 2021). While population-based screening is not widely undertaken owing to the low incidence of this cancer type in comparison with those such as lung, breast and colorectal cancers, targeting of high-risk subgroups has been identified as the way forward (Lindquist et al. 2018). Screening studies are designed primarily considering those with genetic predispositions, and patients with metabolic disorders such as diabetes mellitus (Andersen et al. 2017). Measures undertaken to detect lesions as a part of screening studies include endoscopic ultrasound (EUS), computer tomography-based screening (CT scans) and magnetic resonance imaging (MRI) (Lee 2014). In addition to this, factoring in the linkage of pancreatic cancers with metabolic dysregulation, periodic screening of blood glucose levels is also essential. Given the underlying genetic basis, genetic testing is also recommended, in combination with image-based detection techniques (Daly et al. 2021). Surveillance programs aim to improve 5-year survival times, and offering treatment alternatives such as surgeries to patients at early stages of the disease. However, a number of challenges cloud the usage of conventional screening methods, such as a high cost, certain techniques being invasive in nature, as well as low detection specificity. This may be combined with the possibilities for false negatives and false positives, prompting the adoption of more advanced and contemporary screening measures (Singhi et al. 2019).

Progress in research has enabled tapping into biomarkers, for detection at a microscopic biological level. Utilizing biomarkers both in the pancreatic juice and blood, including exosomes, circulating tumour DNA and mutant genes, early-stage detection is facilitated (Yang et al. 2021). Antibody tests specific to cancer subtypes are utilized, and these assays have assisted in the identification of various biomarkers associated with blood coagulation, hormonal regulation and tissue metabolism. Emerging strategies, employing artificial intelligence (AI) and machine learning (ML) are increasingly gaining popularity in biomarker detection (Hameed and Krishnan 2022).

With respect to diagnostic evaluation, lab tests such as a complete blood count (CBC), liver function tests and electrolyte monitoring are advised, alongside monitoring biomarker levels such as that of CA 19-9 (Scarà et al. 2015). This provides useful insights into prognosis and the stage of the disease, and the adoption of neo-adjuvant therapeutic approaches. While a CT scan of the abdomen remains to be the imaging technique of choice, it may be combined with other tests such as MRI, to further enhance the detection of lesions. Other potentially useful tests include positron emission tomography (PET) scans, and are particularly useful in case of elevated CA 19-9 levels, to further confirm the diagnosis of pancreatic cancer (Chu et al. 2017).

With respect to the staging of tumours, physicians utilize a four-category system of classification, in order to understand disease progression and suitably design treatment regimens. This classification broadly includes resectable (showing no arterial or venous involvement or angiogenesis, as well as a limited spread), borderline resectable tumours (showing a greater venous contact and thrombosis), locally advanced stage of the disease (associated with unreconstructable venous involvement) and metastasis (Perri et al. 2020).

9.4 Currently Available Treatment Strategies

With a survival time of merely 5–6 months in case of untreated pancreatic cancer and a median survival ranging from 10 to 12 months following treatment (Wood et al. 2022), obtaining a bird's-eye view of currently available management strategies is of the essence. The first-line therapy of pancreatic cancer comprises a combination of 5-fluorouracil (5-FU), leucovorin, irinotecan and oxaliplatin (abbreviated as FOLFIRONOX) (Zhang et al. 2021). In addition to this, gemcitabine in combination with albumin-bound paclitaxel has also provided promising clinical benefits (Von Hoff et al. 2013). However, it is important to note that other drug combinations have not matched up either in terms of a prolongation of survival time and arresting tumour progression (efficacy) or a satisfactory toxicity profile associated with long-term usage (safety). Keeping in mind the premise of this chapter, advancements in research with respect to the TME and other molecular targets have enabled the designing of personalized therapy and biologics for certain populations (Krepline and Tsai 2019).

Treatment approaches may be broadly designed on the basis of pancreatic cancer sub-type, or the stage of the disease. As previously discussed at length, the contribution of genetic alterations in pathogenesis cannot be overlooked. In tumours showing BRCA mutations, PARP inhibitors such as Olaparib have yielded promising results, owing to the prevention of single-strand break repair, resulting in apoptosis (Zhu et al. 2020). This, in turn, led to its FDA approval in the year 2019, as a maintenance therapy for pancreatic cancer patients (Parsels et al. 2021). In tumours showcasing microsatellite instability (MSI), pembrolizumab, an immune checkpoint protein PD-1 (programmed cell death protein-1), has been approved for usage in clinical settings. It primarily functions by checking the proliferation of T cells, and triggering cellular apoptosis, thereby increasing the elicitation of innate tumour responses (Marabelle et al. 2020). In patients with KRAS mutations, breakthroughs in therapy have been marked by the FDA approval of inhibitors such as Sotorasib and Adagrasib (Bannoura et al. 2022; Strickler et al. 2023).

Directing our focus on treatment regimens designed on the basis of the stage of the disease, partial removal through surgery remains the gold standard for resectable tumours (Karunakaran and Barreto 2021). This is typically followed by adjuvant chemotherapy administration. The 'Whipple procedure', a commonly used term for pancreaticoduodenectomy surgery, involves the removal portion of the organ. In some cases, total pancreatectomy may be required (Masiak-Segit et al. 2018). In the case of locally advanced tumours, an additional component of the regimen is neo-adjuvant chemotherapy (either independently or in combination with radiotherapy), prior to surgery. Following recovery from surgery, adjuvant chemotherapy may be undertaken (Springfeld et al. 2023). The selection for the most appropriate treatment strategy is made taking into consideration the overall health status of the patient.

In the case of nonresectable tumours and cases of metastasis, the most commonly utilized standard of care is FOLFIRINOX administration. Second-line therapy involves drug combinations, as independent usage of chemotherapeutic agents may not yield satisfactory results (Paluri et al. 2020). Taking into consideration maintenance therapy, Olaparib has recently been approved for usage in patients with BRCA mutations. In order to enhance long-term prognosis, a number of clinical trials are now taking into account maintenance therapy (Kindler et al. 2022). Although the therapeutic strategies aforementioned have shown a certain degree of therapeutic benefits, major problems restricting their usage include high costs associated with formulation as well as the possibility of drug resistance (Zeng et al. 2019). This necessitates the development of intricately coordinated, large-scale efforts focused on reducing treatment costs and improving the overall impact on the survival and well-being of pancreatic patients. The subsequent sections offer an insight into how this may be achieved through the increased adoption of biosimilars.

9.5 Biosimilars in Oncology

Biologics have gained a spot of prominence in the domain of oncology, as therapeutic agents as well as strategies intended for supportive care (Zinzani et al. 2019). As previously highlighted, cost-intensive product development and manufacturing limit accessibility to a vast majority of patients. The lens has, therefore, shifted to biosimilars, in an attempt to drive down treatment costs and enhance market access (Joshi et al. 2022). With malignancies accounting for one in six deaths globally, annual cancer-related deaths are poised at growing upwards of the ten million mark, with low- and medium-income countries (LMICs) being especially vulnerable (Calamia and Abraham 2023). This may be attributed to an acute lack of resources and the current structure of healthcare systems. Owing to an inverted population pyramid in various nations across the globe, it is estimated that medical and treatment costs directly associated with malignancies are expected to increase by 34% by 2030, when compared with the previous decade (Bennett et al. 2018). With various trends observed such as the growing focus on precision medicine and personalized treatment modalities, cost-effectiveness has taken centre stage, as it is assumed that such specialized therapies would be expensive (Ramaswami et al. 2018).

However, it is observed that several barriers impede the integration of biosimilars into oncology treatment plans. A significant barrier is the perception of prescribers as well as patients, with results from recent surveys indicating a considerable gap in knowledge (Shubow et al. 2023). A few major areas that must be addressed include generating greater awareness with respect to the regulatory and approval process for these agents, possibilities for substitution in treatment plans, as well as their efficacy and safety profiles (Barbier et al. 2020). In order to equip physicians with crucial information, organizations such as the American Society of Clinical Oncology (ASCO), as well as the FDA, provide guidelines and training through educational webinar programs (Chopra and Lopes 2017). In addition to this, several patient forums and advocacy groups have sprung up to further aid the dissemination of knowledge from physicians to patients, enabling clinicians to offer tailor-made material to patients. This, in turn, is expected to boost the understanding and in turn, acceptability of biosimilars (Vandenplas et al. 2021).

9.6 Major Drug Classes for which Biosimilars Have Been Explored

As previously discussed, biosimilars play a role in cancer management under two broad headings: disease management (biosimilars of monoclonal antibodies like Rituximab, Bevacizumab and Trastuzumab), and supportive agents (which include pegfligrastim, filgrastrim and epoetin) (Jacobs et al. 2017). Filgrastim is a recombinant granulocyte colony-stimulating factor (G-CSF), regulating neutrophil production. It is majorly indicated for reducing the incidence of neutropenia, in cancer patients receiving myelosuppressive medications (Mehta et al. 2015). Zarxio® (filgrastim-sndz) was the first biosimilar to be approved and marketed in the United States, in 2015 (Dale et al. 2018). Pegfilgrastim comprises an added polyethylene glycol unit and offers the benefit of an extended duration of action (Yang and Kido 2011). Following binding with G-CSF receptors, they work on enhancing the proliferation as well as maturation of neutrophils (Hoy 2019). Currently, there are eight approved Pegfilgrastim biosimilars in Europe, and four such products in the United States (Botteri et al. 2018). With reference to epoetins (EPOs), they are used primarily for the management of anaemia induced as a result of chemotherapy. With similarity to the hormone erythropoietin, they are responsible for stimulating erythrocyte production, chiefly in the bone marrow (Link 2019). These agents act on the JAK–STAT signalling pathway, and are gaining popularity as supporting agents (Muller and Baribeault 2007). In Europe, three biosimilars of EPO-α and two of EPO-ζ have been approved, and only one agent, Retacrit®, is marketed in the United States (Laribi et al. 2022).

With respect to monoclonal antibodies (mABs), Bevacizumab is a humanized mAb, inhibiting angiogenesis, and thereby arresting tumour growth (Lazurko et al. 2023). Mvasi® (bevacizumab-awwb) (Khachigian et al. 2023) and Zirabev® (bevacizumab-bvzr) are two Bevacizumab biosimilars available in the United States

(Yang et al. 2022). With the recent patent expiry in Europe in 2022, it is expected that a greater number of biosimilars of this agent are expected to reach the market in the near future (Harter et al. 2022). The reference product for Rituximab is Rituxan®, and is observed to trigger cellular apoptosis by upregulating interleukin levels (chiefly IL-10) and B-cell lymphoma-2 (Bcl-2) (Poletto et al. 2022). While it is primarily indicated for blood malignancies, it has been increasingly adopted in combination with conventional chemotherapeutic agents, such as cyclophosphamide (Atisha-Fregoso et al. 2021). Some biosimilars approved by the USFDA and EMA include Truxima® (rituximab-abbs), Riabni® (rituximab-arrx) and Ritemvia® (Salles et al. 2017). Lastly, Trastuzumab has been steadily gaining research attention. With respect to its mechanism of action, it binds to the human epidermal growth factor receptor 2 protein (HER2), and has been used independently as well as in combination with paclitaxel for the management of HER2-positive breast cancer, as well as gastric cancer (Sarosiek and Morawski 2018). Ogivri® (trastuzumab-dkst) was the first trastuzumab biosimilar approved by the USFDA, following the patent expiration of Herceptin® in 2019 (Rathore 2018). Following this, four more biosimilar candidates have been approved in the United States. Zercepac® was approved by the EMA recently, for the management of early-stage breast malignancies (Liu et al. 2022).

9.7 Clinical Picture

A simulation-modelling study undertaken by MacDonald et al., in 2500 patients receiving FOLFIRINOX therapy, aimed to understand the impact of switching to a biosimilar of pegfilgrastim. The primary parameters evaluated included cost-effectiveness and patient access. It was concluded that conversion to a prophylactic biosimilar moiety resulted in significant cost reduction, with thousands of dollars saved over multiple therapy cycles. In addition to an improvement in the number of doses provided, a budget-neutral improvement in access to chemotherapeutic medication was envisioned (MacDonald et al. 2021). In a retrospective cohort study, the effectiveness of pegfilgrastim biosimilars, in comparison with the innovator product, was mapped. This study was undertaken with a special focus on the prevention of febrile neutropenia across a variety of cancers, including breast, lung, colorectal, prostate, ovarian and pancreatic cancer. It was concluded that the biosimilar candidates demonstrated a similar safety and efficacy profile to the reference product, encouraging wider clinical adoption. However, it was recommended that further studies must be undertaken, to obtain a better understanding of the tolerability and efficacy of these agents (Wang et al. 2023). Another comparative study was conducted to assess the benefits offered by a biosimilar of low-molecular-weight enoxaparin for its anti-thrombotic activity, in the presence of pancreatic cancer cells BXPC3. The cancer cells were observed to modulate the inhibitory efficacy of the biosimilar candidate, creating a clear point of differentiation from the innovator product. It was, therefore, concluded that malignancies may have the capacity to significantly alter the therapeutic capabilities of biosimilars, prompting more focused studies on a global level (Gerotziafas et al. 2015). A cross-sectional study undertaken by Wang et al. aimed to evaluate the impact of the introduction of biosimilars and novel drug delivery systems on the uptake of G-CSFs, over a 5-year window. Patients with a variety of cancers, including gastric, pancreatic, and ovarian cancers were a part of this study. While it was observed that the utilization of G-CSFs had increased in patients with a risk of developing febrile neutropenia, there is a need to direct a greater focus towards preventive treatment, to improve therapeutic outcomes and lower the increasing disease burden of pancreatic cancer (Wang et al. 2021).

9.8 Challenges Associated with the Usage of Biosimilars for Pancreatic Cancer Management

With respect to the usage of biosimilars for the management of a variety of cancers, there are a number of barriers restricting equitable access, including varying regulations across markets, limited market opportunities and challenges with respect to nomenclature using non-proprietary names, on a global scale. This divide is even more pronounced in the case of breast and colorectal cancer (Teran et al. 2022). In addition to this, there remain several unaddressed questions with respect to interchangeability in terms of usage, as well as pharmacovigilance, which must be tackled through coordination mediated by key stakeholders such as regulatory bodies, policy makers, pharmaceutical companies, and research organizations. Another challenge that must be discussed is that of switching over from reference products to biosimilars. This would require carefully planned training and upskilling of healthcare staff, as well as provision of information through patient advocacy programs (Hester et al. 2020). As previously discussed, the need for a prudently undertaken cost-to-benefit analysis is also essential, through economic evaluation. This, in turn, must be mapped with health determinants such as quality-adjusted life years (QALY) and disability-adjusted life years (DALY), to drive up to threshold of willingness to pay from the end of patients, on understanding the value offered by these regimens (Deng et al. 2022). A study undertaken by Karanyotova et al. in Bulgaria concluded that long timelines with respect to patient access to targeted oncology therapy were a major impediment in the uptake of biosimilars, alongside a framework for pricing and reimbursement that is still in its nascent stages (Karanyotova et al. 2023).

While biologics focused on the domain of oncology is one of the fastest growing segments in the arena of pharmaceutical product development, there is a need to address the gap between regulatory approval and marketing of approved products. Of the five FDA-approved trastuzumab biosimilars by the end of 2019, only two have been marketed. While regulations such as the Biologics Price Competition and Innovation Act of 2009 are catalysts for innovation and shrinking product development timelines, there is a pressing need for a more robust regulatory framework, in order to address concerns with respect to cost, efficacy and safety, from healthcare providers and patients (Whalen 2020).

9.9 Future Perspectives and Conclusion

While the impact of AI and ML has been briefly touched upon over the course of this chapter, various models have been designed and tested in recent times, with applications ranging from pancreatic cancer detection, biomarker identification to monitoring of therapeutic responses (Fyfe 2023). As the range of applications of such technologies in cancer management continues to grow, it is expected that sustained growth will be observed in the domains of disease risk predictions and directing attention towards emerging sub-types and concurrent survival periods (Huang et al. 2022; Kenner et al. 2021). Notwithstanding the tremendous progress made in the domain, there is a significant amount of work that remains to be done with respect to integrating biologics and biosimilars into these models. The authors of this chapter envision significant strides being made in the personalization of therapies and subsequent review of therapeutic responses in patient subgroups, following such efforts. As discussed previously, the chronic and terminal nature of malignancies has encouraged a growth in focus on biosimilars, to pave the way for considerable cost reduction and improved market access (Buske et al. 2017). PDAC malignancies form a considerable share of pancreatic cancer cases, however, biosimilars must be explored for the management of other cancer subtypes as well, owing to a poor prognosis observed across the board following conventional therapy (Yamamoto et al. 2023). Figure 9.2 encapsulates the key benefits of biosimilars in pancreatic cancer as discussed over the course of the chapter, some challenges that remain to be addressed, and exciting future prospects.

While offering readers a comprehensive overview of the effort currently underway and the clinical picture of biosimilar candidates for pancreatic cancer

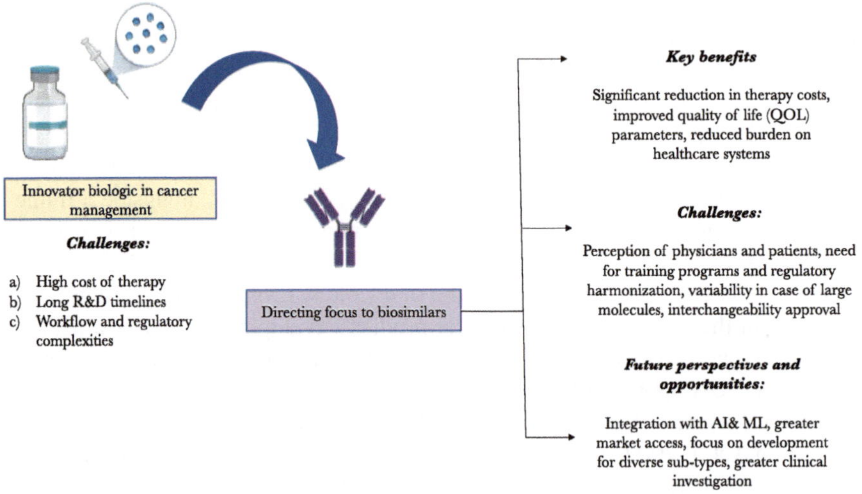

Fig. 9.2 Benefits, challenges and future opportunities surrounding the usage of biosimilars in pancreatic cancer management (*Created using Biorender*)

amelioration, the authors have identified the need for a greater number of clinical trials, in diverse patient groups and populations. There is a pressing need for clinical protocols to be designed with a special focus on the pharmacokinetic and pharmacodynamic profiles of proposed candidates, as well as due consideration to immunogenicity and possible toxicity (Lucchesi et al. 2018). It has been observed that the progress in the arena of biologics and biosimilars in one cancer type may prompt further research in other disease types, thereby resulting in an expansion of indications following regulatory authorization. This, in turn, would be beneficial for clinicians and prospective patients alike. In summary, the key takeaways from this chapter broadly encompass the pressing necessity for focused research and discovery of potential biosimilar candidates, a comparison of their performance and safety profiles in comparison with reference/innovator products, careful attention to designing comprehensive clinical protocols, reforms in HTA to improve market access and pricing regulations in order provide timely, cost-effective intervention to patients in need.

References

Andersen DK, Korc M, Petersen GM, Eibl G, Li D, Rickels MR, Chari ST, Abbruzzese JL (2017) Diabetes, pancreatogenic diabetes, and pancreatic cancer. Diabetes 66(5):1103–1110. https://doi.org/10.2337/db16-1477

Atisha-Fregoso Y, Malkiel S, Harris KM, Byron M, Ding L, Kanaparthi S, Barry WT, Gao W, Ryker K, Tosta P, Askanase AD, Boackle SA, Chatham WW, Kamen DL, Karp DR, Kirou KA, Sam Lim S, Marder B, McMahon M et al (2021) Phase II randomized trial of rituximab plus cyclophosphamide followed by Belimumab for the treatment of lupus nephritis. Arthritis Rheumatol 73(1):121–131. https://doi.org/10.1002/art.41466

Australian Pancreatic Cancer Genome Initiative, ICGC Breast Cancer Consortium, ICGC MMML-Seq Consortium, ICGC PedBrain, Alexandrov LB, Nik-Zainal S, Wedge DC, Aparicio SAJR, Behjati S, Biankin AV, Bignell GR, Bolli N, Borg A, Børresen-Dale A-L, Boyault S, Burkhardt B, Butler AP, Caldas C, Davies HR et al (2013) Signatures of mutational processes in human cancer. Nature 500(7463):415–421. https://doi.org/10.1038/nature12477

Bannoura SF, Khan HY, Azmi AS (2022) KRAS G12D targeted therapies for pancreatic cancer: has the fortress been conquered? Front Oncol 12:1013902. https://doi.org/10.3389/fonc.2022.1013902

Barbier L, Simoens S, Vulto AG, Huys I (2020) European stakeholder learnings regarding Biosimilars: part I—improving biosimilar understanding and adoption. BioDrugs 34(6):783–796. https://doi.org/10.1007/s40259-020-00452-9

Bennett JE, Stevens GA, Mathers CD, Bonita R, Rehm J, Kruk ME, Riley LM, Dain K, Kengne AP, Chalkidou K, Beagley J, Kishore SP, Chen W, Saxena S, Bettcher DW, Grove JT, Beaglehole R, Ezzati M (2018) NCD countdown 2030: worldwide trends in non-communicable disease mortality and progress towards sustainable development goal target 3.4. Lancet 392(10152):1072–1088. https://doi.org/10.1016/S0140-6736(18)31992-5

Bhattacharya A, Santhoshkumar A, Kurahara H, Harihar S (2021) Metastasis suppressor genes in pancreatic cancer: an update. Pancreas 50(7):923–932. https://doi.org/10.1097/MPA.0000000000001853

Botteri E, Krendyukov A, Curigliano G (2018) Comparing granulocyte colony-stimulating factor filgrastim and pegfilgrastim to its biosimilars in terms of efficacy and safety: a meta-analysis of randomised clinical trials in breast cancer patients. Eur J Cancer 89:49–55. https://doi.org/10.1016/j.ejca.2017.10.034

Buske C, Ogura M, Kwon H-C, Yoon SW (2017) An introduction to biosimilar cancer therapeutics: definitions, rationale for development and regulatory requirements. Future Oncol 13(15s):5–16. https://doi.org/10.2217/fon-2017-0153

Cai J, Chen H, Lu M, Zhang Y, Lu B, You L, Zhang T, Dai M, Zhao Y (2021) Advances in the epidemiology of pancreatic cancer: trends, risk factors, screening, and prognosis. Cancer Lett 520:1–11. https://doi.org/10.1016/j.canlet.2021.06.027

Calamia M, Abraham I (2023) The economics of biosimilars and challenges to biosimilar adoption in low- and middle-income countries. Expert Opin Biol Ther 23(8):653–657. https://doi.org/10.1080/14712598.2023.2247972

Camacho LH (2017) Current status of biosimilars in oncology. Drugs 77(9):985–997. https://doi.org/10.1007/s40265-017-0743-z

Chopra R, Lopes G (2017) Improving access to cancer treatments: the role of biosimilars. J Glob Oncol 3(5):596–610. https://doi.org/10.1200/JGO.2016.008607

Chu LC, Goggins MG, Fishman EK (2017) Diagnosis and detection of pancreatic cancer. Cancer J 23(6):333–342. https://doi.org/10.1097/PPO.0000000000000290

Collisson EA, Bailey P, Chang DK, Biankin AV (2019) Molecular subtypes of pancreatic cancer. Nat Rev Gastroenterol Hepatol 16(4):207–220. https://doi.org/10.1038/s41575-019-0109-y

Dale DC, Crawford J, Klippel Z, Reiner M, Osslund T, Fan E, Morrow PK, Allcott K, Lyman GH (2018) A systematic literature review of the efficacy, effectiveness, and safety of filgrastim. Support Care Cancer 26(1):7–20. https://doi.org/10.1007/s00520-017-3854-x

Daly MB, Pal T, Berry MP, Buys SS, Dickson P, Domchek SM, Elkhanany A, Friedman S, Goggins M, Hutton ML, CGC, Karlan BY, Khan S, Klein C, Kohlmann W, CGC, Kurian AW, Laronga C, Litton JK et al (2021) Genetic/familial high-risk assessment: breast, ovarian, and pancreatic, version 2.2021, NCCN clinical practice guidelines in oncology. J Natl Compr Cancer Netw 19(1):77–102. https://doi.org/10.6004/jnccn.2021.0001

Datta J, Dai X, Bianchi A, De Castro Silva I, Mehra S, Garrido VT, Lamichhane P, Singh SP, Zhou Z, Dosch AR, Messaggio F, Ban Y, Umland O, Hosein PJ, Nagathihalli NS, Merchant NB (2022) Combined MEK and STAT3 inhibition uncovers stromal plasticity by enriching for cancer-associated fibroblasts with mesenchymal stem cell-like features to overcome immunotherapy resistance in pancreatic cancer. Gastroenterology 163(6):1593–1612. https://doi.org/10.1053/j.gastro.2022.07.076

De La Cruz MSD, Young AP, Ruffin MT (2014) Diagnosis and management of pancreatic cancer. Am Fam Physician 89(8):626–632

Del Chiaro M, Zerbi A, Falconi M, Bertacca L, Polese M, Sartori N, Boggi U, Casari G, Longoni BM, Salvia R, Caligo MA, Di Carlo V, Pederzoli P, Presciuttini S, Mosca F (2007) Cancer risk among the relatives of patients with pancreatic ductal adenocarcinoma. Pancreatology 7(5–6):459–469. https://doi.org/10.1159/000108963

Deng W, Hu J, Li M, Yang S, Xie Z, Chen J (2022) Trastuzumab biosimilar HLX02 versus reference trastuzumab in patients with recurrent or metastatic HER2-positive breast cancer: a model-based economic evaluation for China. Expert Rev Pharmacoecon Outcomes Res 22(7):1117–1126. https://doi.org/10.1080/14737167.2022.2107506

Ferrantella A, Dudeja V, Are C (2019) Radiotherapy as an adjunct to surgery for pancreatic cancer: where are we after more than 30 years of research and trials? Ann Surg Oncol 26(13):4166–4167. https://doi.org/10.1245/s10434-019-07787-w

Francescone R, Barbosa Vendramini-Costa D, Franco-Barraza J, Wagner J, Muir A, Lau AN, Gabitova L, Pazina T, Gupta S, Luong T, Rollins D, Malik R, Thapa RJ, Restifo D, Zhou Y, Cai KQ, Hensley HH, Tan Y, Kruger WD et al (2021) Netrin G1 promotes pancreatic tumorigenesis through cancer-associated fibroblast–driven nutritional support and immunosuppression. Cancer Discov 11(2):446–479. https://doi.org/10.1158/2159-8290.CD-20-0775

Fyfe I (2023) AI predicts pancreatic cancer risk. Nat Rev Gastroenterol Hepatol 20(7):413–413. https://doi.org/10.1038/s41575-023-00802-2

Galli R, Schmutz-Kober K, Kühnel J, Maak M, Rosenberg R (2021) Pankreaskarzinom. Ther Umsch 78(10):605–613. https://doi.org/10.1024/0040-5930/a001317

Gerotziafas GT, Rouseau A, Mbemba E, Khartechi A, Van Dreden P, Walenga J, Fareed J, Elalamy I (2015) Characterization of the antithrombotic fingerprint of the branded and copies of the low-molecular-weight enoxaparin using thrombin generation assay. Clin Appl Thromb Hemost 21(8):697–704. https://doi.org/10.1177/1076029615587355

Goral V (2015) Pancreatic cancer: pathogenesis and diagnosis. Asian Pac J Cancer Prev 16(14):5619–5624. https://doi.org/10.7314/APJCP.2015.16.14.5619

Grieshaber-Bouyer R, Lorenz H-M (2020) Biosimilars—Chancen und Risiken. Der Internist 61(5):522–529. https://doi.org/10.1007/s00108-020-00784-2

Halbrook CJ, Lyssiotis CA, Pasca Di Magliano M, Maitra A (2023) Pancreatic cancer: advances and challenges. Cell 186(8):1729–1754. https://doi.org/10.1016/j.cell.2023.02.014

Hameed BS, Krishnan UM (2022) Artificial intelligence-driven diagnosis of pancreatic cancer. Cancers (Basel) 14(21):5382. https://doi.org/10.3390/cancers14215382

Harter P, Mouret-Reynier MA, Pignata S, Cropet C, González-Martín A, Bogner G, Fujiwara K, Vergote I, Colombo N, Nøttrup TJ, Floquet A, El-Balat A, Scambia G, Guerra Alia EM, Fabbro M, Schmalfeldt B, Hardy-Bessard A-C, Runnebaum I, Pujade-Lauraine E, Ray-Coquard I (2022) Efficacy of maintenance olaparib plus bevacizumab according to clinical risk in patients with newly diagnosed, advanced ovarian cancer in the phase III PAOLA-1/ENGOT-ov25 trial. Gynecol Oncol 164(2):254–264. https://doi.org/10.1016/j.ygyno.2021.12.016

Hata T, Ishida M, Motoi F, Yamaguchi T, Naitoh T, Katayose Y, Egawa S, Unno M (2016) Telomerase activity in pancreatic juice differentiates pancreatic cancer from chronic pancreatitis: a meta-analysis. Pancreatology 16(3):372–381. https://doi.org/10.1016/j.pan.2016.01.007

Hester A, Gaß P, Fasching PA, Krämer AK, Ettl J, Diessner J, Wöckel A, Egger T, Stock K, Redlin J, Andraschko M, Harbeck N, Würstlein R (2020) Trastuzumab Biosimilars in the therapy of breast cancer—"real world" experiences from four Bavarian University Breast Centres. Geburtshilfe Frauenheilkd 80(9):924–931. https://doi.org/10.1055/a-1226-6666

Ho WJ, Jaffee EM, Zheng L (2020) The tumour microenvironment in pancreatic cancer—clinical challenges and opportunities. Nat Rev Clin Oncol 17(9):527–540. https://doi.org/10.1038/s41571-020-0363-5

Hoy SM (2019) Pegfilgrastim-jmdb/MYL-1401H: a pegfilgrastim biosimilar. BioDrugs 33(1):117–120. https://doi.org/10.1007/s40259-019-00334-9

Hsiao B, Khan A, Kang I (2020) Vaccinations and biologics. Infect Dis Clin North Am 34(2):425–450. https://doi.org/10.1016/j.idc.2020.02.012

Huang J, Lok V, Ngai CH, Zhang L, Yuan J, Lao XQ, Ng K, Chong C, Zheng Z-J, Wong MCS (2021) Worldwide burden of, risk factors for, and trends in pancreatic cancer. Gastroenterology 160(3):744–754. https://doi.org/10.1053/j.gastro.2020.10.007

Huang B, Huang H, Zhang S, Zhang D, Shi Q, Liu J, Guo J (2022) Artificial intelligence in pancreatic cancer. Theranostics 12(16):6931–6954. https://doi.org/10.7150/thno.77949

Ilic M, Ilic I (2016) Epidemiology of pancreatic cancer. World J Gastroenterol 22(44):9694–9705. https://doi.org/10.3748/wjg.v22.i44.9694

Ishiwata T, Matsuda Y, Yoshimura H, Sasaki N, Ishiwata S, Ishikawa N, Takubo K, Arai T, Aida J (2018) Pancreatic cancer stem cells: features and detection methods. Pathol Oncol Res 24(4):797–805. https://doi.org/10.1007/s12253-018-0420-x

Jacobs I, Ewesuedo R, Lula S, Zacharchuk C (2017) Biosimilars for the treatment of cancer: a systematic review of published evidence. BioDrugs 31(1):1–36. https://doi.org/10.1007/s40259-016-0207-0

Joshi D, Khursheed R, Gupta S, Wadhwa D, Singh TG, Sharma S, Porwal S, Gauniyal S, Vishwas S, Goyal S, Gupta G, Eri RD, Williams KA, Dua K, Singh SK (2022) Biosimilars in oncology: latest trends and regulatory status. Pharmaceutics 14(12):2721. https://doi.org/10.3390/pharmaceutics14122721

Kamisawa T, Wood LD, Itoi T, Takaori K (2016) Pancreatic cancer. Lancet 388(10039):73–85. https://doi.org/10.1016/S0140-6736(16)00141-0

Karanyotova S, Topova B, Petrova E, Doychev P, Kapitanska E, Petrova G, Mitkova Z, Dimitrova M (2023) Treatment patterns, adherence to international guidelines, and financial mechanisms of the market access of advanced breast cancer therapy in Bulgaria. Front Public Health 11:1073733. https://doi.org/10.3389/fpubh.2023.1073733

Karunakaran M, Barreto SG (2021) Surgery for pancreatic cancer: current controversies and challenges. Future Oncol 17(36):5135–5162. https://doi.org/10.2217/fon-2021-0533

Kenner B, Chari ST, Kelsen D, Klimstra DS, Pandol SJ, Rosenthal M, Rustgi AK, Taylor JA, Yala A, Abul-Husn N, Andersen DK, Bernstein D, Brunak S, Canto MI, Eldar YC, Fishman EK, Fleshman J, Go VLW, Holt JM et al (2021) Artificial intelligence and early detection of pancreatic cancer: 2020 summative review. Pancreas 50(3):251–279. https://doi.org/10.1097/MPA.0000000000001762

Khachigian LM, Liew G, Teo KYC, Wong TY, Mitchell P (2023) Emerging therapeutic strategies for unmet need in neovascular age-related macular degeneration. J Transl Med 21(1):133. https://doi.org/10.1186/s12967-023-03937-7

Kindler HL, Hammel P, Reni M, Van Cutsem E, Macarulla T, Hall MJ, Park JO, Hochhauser D, Arnold D, Oh D-Y, Reinacher-Schick A, Tortora G, Algül H, O'Reilly EM, Bordia S, McGuinness D, Cui K, Locker GY, Golan T (2022) Overall survival results from the POLO trial: a phase III study of active maintenance Olaparib versus placebo for germline BRCA-mutated metastatic pancreatic cancer. J Clin Oncol 40(34):3929–3939. https://doi.org/10.1200/JCO.21.01604

Kleeff J, Korc M, Apte M, La Vecchia C, Johnson CD, Biankin AV, Neale RE, Tempero M, Tuveson DA, Hruban RH, Neoptolemos JP (2016) Pancreatic cancer. Nat Rev Dis Primers 2(1):16022. https://doi.org/10.1038/nrdp.2016.22

Klein AP (2021) Pancreatic cancer epidemiology: understanding the role of lifestyle and inherited risk factors. Nat Rev Gastroenterol Hepatol 18(7):493–502. https://doi.org/10.1038/s41575-021-00457-x

Krepline A, Tsai S (2019) Has personalized medicine for pancreatic cancer arrived? Adv Surg 53:103–115. https://doi.org/10.1016/j.yasu.2019.04.006

Laribi K, Spaeth D, Scotte F, Ray-Coquard I (2022) Retacrit®, biosimilar of Epoetin alfa, in chemotherapy-induced anemia in routine practice: impact of iron supplementation. Oncology 100(10):519–528. https://doi.org/10.1159/000522069

Lazurko C, Linder R, Pulman K, Lennox G, Feigenberg T, Fazelzad R, May T, Zigras T (2023) Bevacizumab treatment for low-grade serous ovarian cancer: a systematic review. Curr Oncol 30(9):8159–8171. https://doi.org/10.3390/curroncol30090592

Lee ES (2014) Imaging diagnosis of pancreatic cancer: a state-of-the-art review. World J Gastroenterol 20(24):7864. https://doi.org/10.3748/wjg.v20.i24.7864

Lindquist CM, Miller FH, Hammond NA, Nikolaidis P (2018) Pancreatic cancer screening. Abdom Radiol (NY) 43(2):264–272. https://doi.org/10.1007/s00261-017-1308-z

Link H (2019) Hematopoietic growth factors in the management of anemia and febrile neutropenia. Breast Care (Basel) 14(2):93–98. https://doi.org/10.1159/000497408

Liu Z, Guan Y, Yao Y, Zhang W, Zhuang X, Zhang Y, Zhu T (2022) Effectiveness and safety of Zercepac and reference Trastuzumab in the neoadjuvant setting for early-stage breast cancer: a retrospective cohort study. J Oncol 2022:9998114. https://doi.org/10.1155/2022/9998114

Loveday BP, Lipton L, Thomson BN (2019) Pancreatic cancer: an update on diagnosis and management. Aust J Gen Pract 48(12):826–831. https://doi.org/10.31128/AJGP-06-19-4957

Lucchesi S, Marcianò I, Panagia P, Intelisano R, Randazzo MP, Sgroi C, Altavilla G, Santarpia M, Adamo V, Franchina T, Ferraù F, Reitano P, Trifirò G (2018) Prevalence of use and cost of biological drugs for cancer treatment: a 5-year picture from southern Italy. Clin Drug Investig 38(3):269–278. https://doi.org/10.1007/s40261-017-0591-3

MacDonald K, Alrawashdh N, McBride A, Abraham I (2021) Conversion to biosimilar pegfilgrastim-cbqv enables budget-neutral access to FOLFIRINOX treatment for metastatic pancreatic cancer. Future Oncol 17(33):4561–4570. https://doi.org/10.2217/fon-2021-0718

Marabelle A, Le DT, Ascierto PA, Di Giacomo AM, De Jesus-Acosta A, Delord J-P, Geva R, Gottfried M, Penel N, Hansen AR, Piha-Paul SA, Doi T, Gao B, Chung HC, Lopez-Martin J, Bang Y-J, Frommer RS, Shah M, Ghori R et al (2020) Efficacy of pembrolizumab in patients with noncolorectal high microsatellite instability/mismatch repair–deficient cancer: results from the phase II KEYNOTE-158 study. J Clin Oncol 38(1):1–10. https://doi.org/10.1200/JCO.19.02105

Masiak-Segit W, Rawicz-Pruszyński K, Skórzewska M, Polkowski WP (2018) Surgical treatment of pancreatic cancer. Pol Przegl Chir 90(2):45–53. https://doi.org/10.5604/01.3001.0011.7493

McGuigan A, Kelly P, Turkington RC, Jones C, Coleman HG, McCain RS (2018) Pancreatic cancer: a review of clinical diagnosis, epidemiology, treatment and outcomes. World J Gastroenterol 24(43):4846–4861. https://doi.org/10.3748/wjg.v24.i43.4846

Mehta HM, Malandra M, Corey SJ (2015) G-CSF and GM-CSF in neutropenia. J Immunol 195(4):1341–1349. https://doi.org/10.4049/jimmunol.1500861

Mizrahi JD, Surana R, Valle JW, Shroff RT (2020) Pancreatic cancer. Lancet 395(10242):2008–2020. https://doi.org/10.1016/S0140-6736(20)30974-0

Muller RJ, Baribeault D (2007) Extended-dosage-interval regimens of erythropoietic agents in chemotherapy-induced anemia. Am J Health Syst Pharm 64(24):2547–2556. https://doi.org/10.2146/ajhp070018

Paluri RK, Kasi A, Young C, Posey JA (2020) Second-line treatment for metastatic pancreatic cancer. Clin Adv Hematol Oncol 18(2):106–115

Parsels LA, Engelke CG, Parsels J, Flanagan SA, Zhang Q, Tanska D, Wahl DR, Canman CE, Lawrence TS, Morgan MA (2021) Combinatorial efficacy of olaparib with radiation and ATR inhibitor requires PARP1 protein in homologous recombination–proficient pancreatic cancer. Mol Cancer Ther 20(2):263–273. https://doi.org/10.1158/1535-7163.MCT-20-0365

Perri G, Prakash L, Katz MHG (2020) Defining and treating borderline resectable pancreatic cancer. Curr Treat Options in Oncol 21(9):71. https://doi.org/10.1007/s11864-020-00769-1

Poletto S, Novo M, Paruzzo L, Frascione PMM, Vitolo U (2022) Treatment strategies for patients with diffuse large B-cell lymphoma. Cancer Treat Rev 110:102443. https://doi.org/10.1016/j.ctrv.2022.102443

Ramaswami R, Bayer R, Galea S (2018) Precision medicine from a public health perspective. Annu Rev Public Health 39(1):153–168. https://doi.org/10.1146/annurev-publhealth-040617-014158

Rathore AS (2018) Approval of Ogivri. PDA J Pharm Sci Technol 72(1):1–1. https://doi.org/10.5731/pdajpst.2018.001105

Roy A, Sahoo J, Kamalanathan S, Naik D, Mohan P, Kalayarasan R (2021) Diabetes and pancreatic cancer: exploring the two-way traffic. World J Gastroenterol 27(30):4939–4962. https://doi.org/10.3748/wjg.v27.i30.4939

Salles G, Barrett M, Foà R, Maurer J, O'Brien S, Valente N, Wenger M, Maloney DG (2017) Rituximab in B-cell hematologic malignancies: a review of 20 years of clinical experience. Adv Ther 34(10):2232–2273. https://doi.org/10.1007/s12325-017-0612-x

Sarosiek T, Morawski P (2018) [Trastuzumab and its biosimilars]. Pol Merkur Lekarski 44(263):253–257

Scarà S, Bottoni P, Scatena R (2015) CA 19-9: biochemical and clinical aspects. In: Scatena R (ed) Advances in cancer biomarkers, vol 867. Springer, Dordrecht, pp 247–260. https://doi.org/10.1007/978-94-017-7215-0_15

Shubow S, Sun Q, Nguyen Phan AL, Hammell DC, Kane M, Lyman GH, Gibofsky A, Lichtenstein GR, Bloomgarden Z, Cross RK, Yim S, Polli JE, Wang Y (2023) Prescriber perspectives on biosimilar adoption and potential role of clinical pharmacology: a workshop summary. Clin Pharmacol Ther 113(1):37–49. https://doi.org/10.1002/cpt.2765

Siegel RL, Giaquinto AN, Jemal A (2024) Cancer statistics, 2024. CA Cancer J Clin 74(1):12–49. https://doi.org/10.3322/caac.21820

Singhi AD, Koay EJ, Chari ST, Maitra A (2019) Early detection of pancreatic cancer: opportunities and challenges. Gastroenterology 156(7):2024–2040. https://doi.org/10.1053/j.gastro.2019.01.259

Springfeld C, Ferrone CR, Katz MHG, Philip PA, Hong TS, Hackert T, Büchler MW, Neoptolemos J (2023) Neoadjuvant therapy for pancreatic cancer. Nat Rev Clin Oncol 20(5):318–337. https://doi.org/10.1038/s41571-023-00746-1

Stevens KJ, Lisanti C (2024) Pancreas imaging. In: StatPearls. StatPearls Publishing. http://www.ncbi.nlm.nih.gov/books/NBK547725/

Strickler JH, Satake H, George TJ, Yaeger R, Hollebecque A, Garrido-Laguna I, Schuler M, Burns TF, Coveler AL, Falchook GS, Vincent M, Sunakawa Y, Dahan L, Bajor D, Rha S-Y, Lemech C, Juric D, Rehn M, Ngarmchamnanrith G et al (2023) Sotorasib in *KRAS* p.G12C–mutated advanced pancreatic cancer. N Engl J Med 388(1):33–43. https://doi.org/10.1056/NEJMoa2208470

Teran E, Gomez H, Hannois D, Lema M, Mantilla W, Rico-Restrepo M, McElwee E, Castro Sanchez N, Valdivieso N, Espinoza MA (2022) Streamlining breast cancer and colorectal cancer biosimilar regulations to improve treatment access in Latin America: an expert panel perspective. Lancet Oncol 23(7):e348–e358. https://doi.org/10.1016/S1470-2045(22)00121-8

Thill M, Thatcher N, Hanes V, Lyman GH (2019) Biosimilars: what the oncologist should know. Future Oncol 15(10):1147–1165. https://doi.org/10.2217/fon-2018-0728

Vandenplas Y, Simoens S, Van Wilder P, Vulto AG, Huys I (2021) Informing patients about biosimilar medicines: the role of European patient associations. Pharmaceuticals 14(2):117. https://doi.org/10.3390/ph14020117

Vizgirda V, Jacobs I (2017) Biosimilars: considerations for oncology nurses. Clin J Oncol Nurs 21(2):E54–E60. https://doi.org/10.1188/17.CJON.E54-E60

Von Hoff DD, Ervin T, Arena FP, Chiorean EG, Infante J, Moore M, Seay T, Tjulandin SA, Ma WW, Saleh MN, Harris M, Reni M, Dowden S, Laheru D, Bahary N, Ramanathan RK, Tabernero J, Hidalgo M, Goldstein D et al (2013) Increased survival in pancreatic cancer with nab-paclitaxel plus gemcitabine. N Engl J Med 369(18):1691–1703. https://doi.org/10.1056/NEJMoa1304369

Wang C-Y, Heldermon CD, Vouri SM, Park H, Wheeler SE, Ramnaraign BH, Dang NH, Brown JD (2021) Trends in use of granulocyte colony-stimulating factor following introduction of biosimilars among adults with cancer and commercial or Medicare insurance from 2014 to 2019. JAMA Netw Open 4(11):e2133474. https://doi.org/10.1001/jamanetworkopen.2021.33474

Wang C-Y, Vouri SM, Park H, Heldermon CD, Brown JD (2023) Comparative effectiveness of pegfilgrastim biosimilars vs originator for prevention of febrile neutropenia: a retrospective cohort study. J Manag Care Spec Pharm 29(2):119–127. https://doi.org/10.18553/jmcp.2023.29.2.119

Whalen J (2020) Recognizing and addressing challenges to the adoption of trastuzumab biosimilars and HER2-targeted therapies. Am J Manag Care 26(Suppl 2):S23–S31. https://doi.org/10.37765/ajmc.2020.42899

Wood LD, Canto MI, Jaffee EM, Simeone DM (2022) Pancreatic cancer: pathogenesis, screening, diagnosis, and treatment. Gastroenterology 163(2):386–402.e1. https://doi.org/10.1053/j.gastro.2022.03.056

Yamamoto Y, Kasashima H, Fukui Y, Tsujio G, Yashiro M, Maeda K (2023) The heterogeneity of cancer-associated fibroblast subpopulations: their origins, biomarkers, and roles in the tumor microenvironment. Cancer Sci 114(1):16–24. https://doi.org/10.1111/cas.15609

Yan Z, Ohuchida K, Fei S, Zheng B, Guan W, Feng H, Kibe S, Ando Y, Koikawa K, Abe T, Iwamoto C, Shindo K, Moriyama T, Nakata K, Miyasaka Y, Ohtsuka T, Mizumoto K, Hashizume M, Nakamura M (2019) Inhibition of ERK1/2 in cancer-associated pancreatic stellate cells suppresses cancer–stromal interaction and metastasis. J Exp Clin Cancer Res 38(1):221. https://doi.org/10.1186/s13046-019-1226-8

Yang B-B, Kido A (2011) Pharmacokinetics and pharmacodynamics of pegfilgrastim. Clin Pharmacokinet 50(5):295–306. https://doi.org/10.2165/11586040-000000000-00000

Yang J, Xu R, Wang C, Qiu J, Ren B, You L (2021) Early screening and diagnosis strategies of pancreatic cancer: a comprehensive review. Cancer Commun 41(12):1257–1274. https://doi.org/10.1002/cac2.12204

Yang J, Kelton JM, Thompson J, Alvir JMJ, Maculaitis MC, Shelbaya A (2022) Real-world usage of bevacizumab-bvzr biosimilar in US oncology practice. Am J Manag Care 28(4):160–166. https://doi.org/10.37765/ajmc.2022.88831

Yonemori K, Kurahara H, Maemura K, Natsugoe S (2017) MicroRNA in pancreatic cancer. J Hum Genet 62(1):33–40. https://doi.org/10.1038/jhg.2016.59

Zalatnai A (2006) Molecular aspects of stromal-parenchymal interactions in malignant neoplasms. Curr Mol Med 6(6):685–693. https://doi.org/10.2174/156652406778195053

Zeng S, Pöttler M, Lan B, Grützmann R, Pilarsky C, Yang H (2019) Chemoresistance in pancreatic cancer. Int J Mol Sci 20(18):4504. https://doi.org/10.3390/ijms20184504

Zhang L, Sanagapalli S, Stoita A (2018) Challenges in diagnosis of pancreatic cancer. World J Gastroenterol 24(19):2047–2060. https://doi.org/10.3748/wjg.v24.i19.2047

Zhang B, Zhou F, Hong J, Ng DM, Yang T, Zhou X, Jin J, Zhou F, Chen P, Xu Y (2021) The role of FOLFIRINOX in metastatic pancreatic cancer: a meta-analysis. World J Surg Oncol 19(1):182. https://doi.org/10.1186/s12957-021-02291-6

Zhao C, Gao F, Weng S, Liu Q (2017) Pancreatic cancer and associated exosomes. Cancer Biomarkers 20(4):357–367. https://doi.org/10.3233/CBM-170258

Zhou Y, Jin X, Yu H, Qin G, Pan P, Zhao J, Chen T, Liang X, Sun Y, Wang B, Ren D, Zhu S, Wu H (2022) HDAC5 modulates PD-L1 expression and cancer immunity via p65 deacetylation in pancreatic cancer. Theranostics 12(5):2080–2094. https://doi.org/10.7150/thno.69444

Zhu H, Wei M, Xu J, Hua J, Liang C, Meng Q, Zhang Y, Liu J, Zhang B, Yu X, Shi S (2020) PARP inhibitors in pancreatic cancer: molecular mechanisms and clinical applications. Mol Cancer 19(1):49. https://doi.org/10.1186/s12943-020-01167-9

Zinzani PL, Dreyling M, Gradishar W, Andre M, Esteva FJ, Boulos S, González Barca E, Curigliano G (2019) Are biosimilars the future of oncology and haematology? Drugs 79(15):1609–1624. https://doi.org/10.1007/s40265-019-01193-y

Biosimilars in Breast Cancer

10

Dhara Bhatt

Abstract

Breast cancer is a significant health concern globally, leading to maximum mortalities amongst women worldwide, which is 11.6% of all cancers accounting for 2.3 million cases. The commonly used treatments for breast cancer include chemotherapy, surgery, and also radiotherapy. The selection of the treatment relies on various factors, including the disease's grade, stage, and molecular subtype. This ensures that the treatment is tailored to be effective and secure. Targeted or personalised therapies are the results of cutting-edge cancer treatments that have been made possible by significant advancements in our understanding of cancer biology. In the past 20 years, the biologics, also known as the biologic drugs, have become an essential and valuable tool in treating diseases such as cancer and other therapeutic areas. In the field of breast cancer treatment, biosimilars have been approved for use in both the disease itself and as supportive drugs for cancer patients. The reduction in the treatment cost using trastuzumab biosimilars is approximately 20–30% lower in comparison to the costs using the reference product. The supportive drugs include haematopoietic growth factors like erythropoietin and filgrastim. These drugs are used to manage the patients clinically. The impact of biosimilars on reducing healthcare budgets relies on their acceptance by regulatory agencies, prescribers, chemists, and patients. So far, there have been no significant differences in terms of effectiveness, safety, and immune response between the original product and the biosimilar version. The present chapter focuses mainly the use of biosimilars in breast cancer with the target-based mechanisms and its clinical implications.

D. Bhatt (✉)
Medisynth Chemicals Pvt. Ltd., Navi Mumbai, Maharashtra, India

S. Bhatt et al. (eds.), *Biosimilars for Cancer Treatment*,
https://doi.org/10.1007/978-981-97-3001-8_10

Keywords

Oncology · Breast cancer · Biosimilars · Monoclonal antibody · Trastuzumab

10.1 Introduction

Cancer is considered to be one of the deadliest diseases in today's time, causing several mortalities every year. The impact of the disease varies in different parts throughout the world, depending on the availability of the healthcare facilities, disease management, and several socio-economic factors related to the disease (Chhikara and Parang 2023). The figures for the burden of cancer globally have been released by the cancer agency, International Agency for Research on Cancer (IARC) of the World Health Organization (WHO). The most commonly occurring cancer in the world is lung cancer, comprising 12.4% of all cancers worldwide, with total of 2.5 million new cases. The second most frequently occurring cancer is the breast cancer amongst female, which is 11.6% of all cancers accounting for 2.3 million cases. It is then, followed by colorectal cancer (9.6%, 1.9 million cases), prostate cancer (7.3%, 1.5 million cases), and stomach cancer (4.9%, 970,000 cases) (International Agency for Research on Cancer (IARC) 2024). The cancer statistics in India are recorded by Indian Council of Medical Research's (ICMR) institute, the National Cancer Registry Programme (NCRP) of the National Centre for Disease Informatics and Research (NCDIR) located in Bengaluru. The registry of ICMR collects the data through approximately 272 different population-based registries and the cancer registries of the hospitals across the country. In the year 2020, the cancer cases in India were approximately, 1,392,179, i.e. near to ~1.4 million, leading to nearly 850,000 (~0.85 million) deaths due to cancer (Mathur et al. 2020). By the end of the year 2040, the cancer cases are projected to rise at 47% in comparison to 2020, leading to a total of 28.4 million cancer cases (Sung et al. 2021).

Breast cancer is a foremost cause of mortalities amongst women on a global scale and a substantial concern for public health. In Indian females, the breast cancer, cervical cancer, and uterine cancer are highly prevalent. There has been a consistent rise in the occurrence of breast cancer in India, especially in the urban regions of the country (Sung et al. 2021). In India, the prevalence of breast cancer is very frequent, at the rate of 25.8 cases /10,000 people. In the year 2020, it represented 13.6% cancer cases of all types of cancer cases and resulted in approximately 13.3 deaths/100,000 individuals (Kerketta et al. 2023).

The most common therapies used to treat breast cancer are surgery, chemotherapy, and radiotherapy. Additionally, there are innovative approaches such as oncogene inactivation, genetic therapies, cell-target suicide, tumour suppressor gene augmentation, immunomodulation (such as expression of ectopic cytokine, immune enhancement), virus mediated oncolysis, and nanotechnology. The focused treatment involves endocrine therapy primarily for hormone receptor (HR) -positive breast cancer (HR+) anti-HER2 therapy mainly for HER2 receptor positive (HER2+) breast cancer. Regrettably, there is currently no specific treatment option for

Triple-Negative Breast Cancer (TNBC) subtype, where the above-mentioned receptors are absent. In the realm of treating metastatic breast cancer, systemic therapies play a significant part in managing the progression of the tumour (Harbeck et al. 2019).

Targeted or personalised therapies are the results of cutting-edge cancer treatments that have been made possible by significant advancements in our understanding of cancer biology. In the past 20 years, biologics have become an essential and valuable tool in treating diseases such as cancer and other therapeutic areas (Howard et al. 2015). According to estimates, over 250 biological antibodies have been produced by recombinant DNA technology and are presently licenced for use in medicine. Nevertheless, the high cost of such medicines has made it difficult for doctors to prescribe them (Lee et al. 2012). Biosimilars are becoming more and more common in clinical practice as a potential way to lower the cost of monoclonal antibody therapy as an alternative to increasing the usage of biologic medicines. Biosimilars are replicas of the original biologic drugs that have shown similarities to a benchmark product through extensive comparative analysis and evaluations conducted during preclinical and clinical trials (Hirsch and Lyman 2011). The current chapter is mainly focused on the use of the biosimilars used in the treatment and the clinical management of the breast cancer.

10.2 Biosimilars in Oncology

With the advancement in the science and technology, we have in-depth knowledge of cancer biology, which has ultimately led to newer and advanced therapies in cancer. The use of the biologic drugs in the past few years has been an integral part of the cancer treatment. Although, it should be noted that the costs of available healthcare have been steadily increasing (Howard et al. 2015). The active ingredients of biologics or biological medicinal products are obtained from living organisms, tissues, or cells, such as humans, animals, or microorganisms. Proteins come in a wide range of sizes and complex structures, from simple ones like insulin to the more intricate ones like monoclonal antibodies or coagulation factors. They are the main components of biologics. When it comes to chemical drugs, their composition is distinct, whereas the biologic's composition can only be defined up to some extent. Furthermore, it should be noted that the successive batches of identical biological medicines may exhibit a certain level of inherent minor variability, known as microheterogeneity. It is crucial for these slight variations to remain within a range that is acceptable, in order to maintain consistent safety and efficacy levels (EMA and EU 2023). After a biologic drug is approved, there is a period of market protection to guarantee reimbursement for research and development expenses. After a certain period of time (typically around 10 years), the biosimilars are authorised for its use clinically (Weise et al. 2012).

10.2.1 Definition of Biosimilars

The "biosimilar" or the "similar biological medicinal products" are defined by EMA (European Medicines Agency) as "products that contain a version of the active substance of an already authorised original biological medicinal product (reference medicinal product)", being similar "in terms of quality characteristics, biological activity, safety and efficacy based on a comprehensive comparability exercise" (EMA and EU 2023). Similarly, the "biosimilars" are defined by the United States Food and Drug Administration (U.S. FDA), as "biological products that are highly similar to and have no clinically meaningful differences from an existing FDA-approved reference product in terms of safety, purity, and potency (safety and effectiveness)" (US FDA n.d.).

10.2.2 Biosimilars Vs Reference Medicine

Biosimilars are biological products that closely resemble their biological reference monoclonal antibody. While there may be slight variations in non-active substances, the sequence of amino acid remains identical. However, the biosimilars differ from generics because they are not exactly similar to their reference product, whereas generics are the copies of the branded drugs. The difference in the biosimilars is because of the use of distinctive cell lines and their process of manufacturing, leading to slight structural differences between them (Li et al. 2015). Due to the potential impact of these structural variations on clinical response, it is necessary to conduct comparative analysis on safety, efficacy, pharmacodynamics, pharmacokinetics, and the immunogenicity data in order to establish the similarity of a biosimilar product with its reference or original product. For the medicines to be classified as biosimilars, they must meet additional criteria to guarantee that they produce the same clinical outcome as the original product in every patient (Glassy and Gupta 2014). Minor adjustments in manufacturing processes can lead to variations in protein stability, which includes changes in post-translational modifications like glycosylation. Such type of alterations could potentially impact the biological function, as well as the overall effectiveness, safety, and immune response (Laslop 2013).

Unlike the original or the reference biological product, that has proven to be safe and effective against different types of tumours and diseases like adjuvant or metastatic disease, biosimilars are being developed through studies to demonstrate their comparable safety and efficacy to their original/reference medicine. While developing the biosimilars, the pharmacological models are mainly focused on the analytical comparability in both in vitro and the clinical settings. The last stage of the biosimilar's development involves a phase III confirmatory study for the establishment of the comparable effectiveness, safety profile, and immunogenicity with the reference or original product. The simpler biologics uses sensitive pharmacodynamics markers to measure clinical efficacy (e.g. haemoglobin levels in case of erythropoietin treatment), whereas, for cancer monoclonal antibodies, this type of surrogate measures are not available. As a result, in order to show their effectiveness, cancer

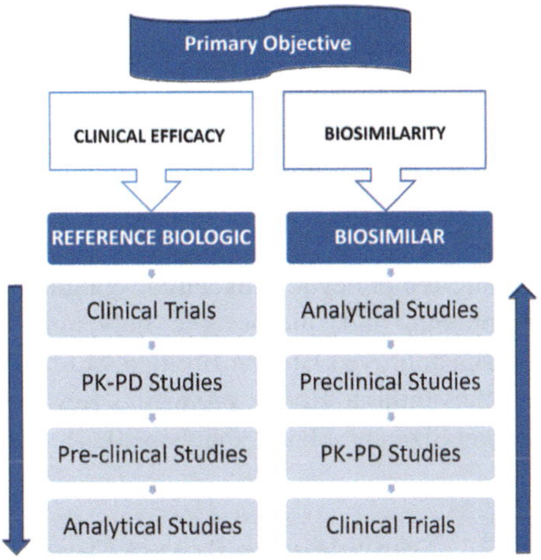

Fig. 10.1 Primary objective of reference biologic vs biosimilar development (Weise et al. 2012)

monoclonal antibodies are dependent on the clinical endpoints in the disease like progression-free or disease-free survival, and/or overall survival of the patients. For the development of the biologics, the animal studies are essential, but in case of biosimilars, they are only necessary if additional qualifications are needed to address any remaining differences among the biosimilars and their original or reference product (Migliavacca Zucchetti et al. 2019).

Biosimilars have distinct characteristics compared to reference drugs, such as variations in the active substance's size, complexity, and its process of manufacturing. The thorough examination of physicochemical as well as the functional properties of the biosimilars in relation to their original/reference product forms the basis of biosimilar development (Fig. 10.1) (Weise et al. 2012). Therefore, it is necessary to establish the similarity to the reference product (European Medicines Agency 2014). If bio-similarity has been proven, it may be possible to extrapolate the efficacy and safety data to other uses of the original/reference product, as long as the mechanism of action is consistent across various uses and the biosimilar's safety profile is well-defined and free from any concerning immune reactions (Weise et al. 2012; Migliavacca Zucchetti et al. 2019). By avoiding the need for redundant clinical trials conducted with the reference medicine, the development costs for biosimilars can be significantly reduced. This, in turn, allows for the availability of these medications at a more affordable rate. In addition, the entry of biosimilars in the European market resulted in price reduction for both biosimilars and the original reference medicines (Migliavacca Zucchetti et al. 2019; Cornes 2012).

10.2.3 Global Market of the Biosimilars

The size of the global biosimilars market is estimated to be USD 35.47 billion in the year 2024 and USD 82.27 billion by the year 2029, with a CAGR (compound annual growth rate) of 18.32% from year 2024 to 2029. The increased prevalence of cancer globally is the main reason for the increase in the research and development of biosimilars over the projected period. The oncology accounts for the remarkable market share of biosimilars. The expansion of the examined market is also anticipated to be growing because of the focus on research and development in oncology by major players of industry and an increase in regulatory body approvals. For example, Amgen's biosimilar to Rituxan (rituximab), RIABNI (rituximab-arrx), was granted approval by the US FDA in December 2020 to treat adult patients with chronic lymphocytic leukaemia, granulomatosis with polyangiitis, non-Hodgkin's lymphoma, and microscopic polyangiitis. Furthermore, in May 2022, Health Canada approved Biocon Biologics Ltd. and Viatris Inc.'s Abevmy (Bevacizumab), as a biosimilar to Roche's Avastin's Bevacizumab, for cancer treatment (Biosimilars Market Size and Share Analysis 2024).

10.3 Overview on Breast Cancer

10.3.1 Breast Cancer Types Based on Molecular Targets

There are various breast cancer types, each with distinct characteristics based on behaviour, appearance, and response to treatment (Weigelt et al. 2010). Based on the histological findings, breast cancers are primarily classified as invasive or non-invasive types of cancer (Miyazaki et al. 2022). There are four main molecular subtypes of invasive breast cancer that can be identified using immune histological technique. This technique relies on the expression of specific receptors, including the oestrogen receptor (ER), the progesterone receptor (PR), and also the human epidermal growth factor receptor 2 (HER2) (Joshi and Press 2018). The selection of the treatment relies on various factors, including the disease's stage, grade, and molecular subtype. This ensures that the treatment is tailored to be effective and secure. The grade of the tumour refers to the visual characteristics of cancer cells in relation to normal cells, including tubule differentiation, mitotic count, and nuclear pleomorphism (Elston and Ellis 1993). The stage of the tumour is a metric used to categorise the scope of the cancer within the body. The TNM staging system is utilised, taking into account factors such as tumour size, involvement of lymph node, and metastases (Amin et al. 2017).

Diagnosing breast cancer involves analysing tissue samples using established criteria. The invasive ductal carcinoma is the most prevalent breast cancer type found in 50–75% patients. The second most common type is invasive lobular carcinoma, occurring in 5–15% patients. The mixed ductal or lobular carcinomas or other less common histologies are found in rest of the patients (Schnitt 2000). The two primary molecular targets are identified for breast cancer pathogenesis. About 70% cases of the invasive breast cancers show the presence of oestrogen receptor alpha (ERα). ERα is a transcription factor and also a steroid hormone receptor, which upon activation, activates the pathways responsible for the growth of cancer in the breast cells. ERα signalling is marked by an expression of progesterone receptor (PR), which is a steroid hormone (Joshi and Press 2018). The HR+, i.e. hormone receptor positive tumours are the tumours that shows the presence of either receptors, ER or PR, or both in minimum 1% tumour cells. Utilising endocrine therapy to regulate ER signalling is the main approach for treating breast cancers that are ER+ or PR+ (Hammond 2010).

The second key molecular target is the epidermal growth factor 2 (ERBB2, which was formerly known as HER2/neu or HER2). It is a transmembrane receptor: tyrosine kinase of epidermal growth factor receptor family. It is overexpressed in around 20% of breast cancers. It is linked to a negative prognosis when systemic therapy is not administered. Tumours that have overexpression or amplification of the ERBB2 genes are classified as ERBB2+ (Wolff et al. 2014). ERBB2-targeted treatments like anti-ERBB2 antibodies such as trastuzumab and pertuzumab or tyrosine kinase inhibitors such as lapatinib and neratinib, which are small molecules, could be beneficial to patients diagnosed with overexpressed or amplified ERBB2.

Triple-negative breast cancer (TNBC), accounts for about 15% of the breast tumours, is identified by the absence of molecular target receptors, i.e. oestrogen, progesterone, or ERBB2. Such tumours are associated with a significant likelihood of distant relapse within the initial 3–5 years after diagnosis. The molecular pathophysiology of TNBC is still an area to be understood (Denkert et al. 2017; Foulkes et al. 2010).

10.3.2 Treatment Options Available for Breast Cancer

Several treatment options are available for breast cancer treatment, including surgery, radiotherapy, chemotherapy, personalised treatments, and the new and emerging treatments targeting breast cancer as described in Fig. 10.2.

10.3.2.1 Surgery

There are two common approaches to breast surgery. One is mastectomy (complete breast removal), often followed by breast reconstruction. The other approach is breast-conserving surgery. A lumpectomy involves partial breast removal along with tumour. The “recommended margins status” is described as “no ink on tumour”, indicating the absence of any tumour cells near the tissue edge. Research indicates that lumpectomy and the total mastectomy together with the radiation have similar outcomes in terms of relapse-free and also overall survival (OS) (Li et al. 2015).

10.3.2.2 Radiotherapy

The radiation therapy has been playing a crucial tool in cancer treatment (Grubbé 1933). Intense radiations are used to treat either the entire breast or a specific breast area (following breast-conserving surgery), the chest wall (post mastectomy), and nearby lymph nodes (Boyages 2017). Research findings indicate that radiation therapy may provide more benefits to the high-risk breast cancer patients, who have undergone conservative surgery. However, patients with small, low-grade tumours may not necessarily require radiation therapy (Kwong et al. 2014). Radiation

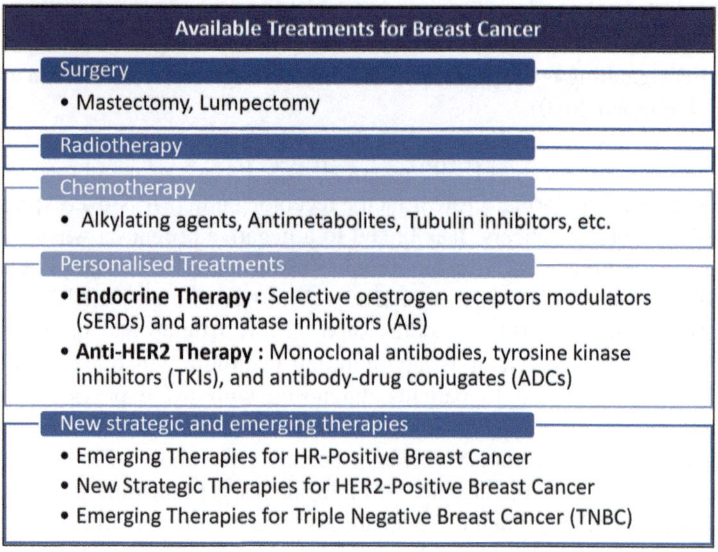

Fig. 10.2 Treatments available for breast cancer

therapy is a crucial treatment for all types of breast cancer, but it holds particular significance for TNBC because of unavailability of the personalised treatments available for this subtype. Research has demonstrated the beneficial effects of radiotherapy on TNBC patients, whether they undergo conserving surgery or mastectomy (He et al. 2018).

10.3.2.3 Chemotherapy

It is a treatment method comprising several classes of the cytotoxic drugs. While chemotherapy can be challenging and may cause side effects, it is efficacious in fighting against breast cancer. Chemotherapy consists of various families of cytotoxic drugs, such as antimetabolites, alkylating agents, and tubulin inhibitors (Nabholtz and Gligorov 2005). Cyclophosphamide is potent alkylating agent inducing DNA strand breakage (Penel et al. 2012). Anthracyclines such as doxorubicin, daunorubicin, epirubicin, and idarubicin work by intercalating with DNA, which effectively hinders macromolecular biosynthesis (Gewirtz 1999). Taxanes, such as docetaxel and paclitaxel, bind to microtubules, halting their disassembly. This ultimately results into arrest of cell cycle and apoptosis. Chemotherapy can be given before or after surgery, as well as for treating breast cancer metastases (Moos and Fitzpatrick 1998).

10.3.2.4 Personalised Treatments for Breast Cancer

(a) **Endocrine Therapy**: The primary approach used to treat hormone receptor positive invasive (HR+) breast cancer is endocrine therapy. This therapy directly targets the ER, either through selective oestrogen receptors degraders and modulators or by inhibiting oestrogen synthesis with aromatase inhibitors. There are several commonly used types of medications in the field of scientific research. These include: i. selective oestrogen receptor degraders (SERDs) and ii. aromatase inhibitors (AIs) (Howlader et al. 2014; El Sayed et al. 2019).

(b) **Anti-HER2 Therapy**: HER2 overexpression has been linked to low survival outcome when compared to HR+ or HER2-negative breast cancer. Therefore, it is necessary to have therapies that specifically target HER2 in order to effectively treat HER2-positive breast cancer. Current therapies for HER2 involve the use of antibodies that focus on particular HER2 epitopes, antibody-drug conjugates (ADCs), and tyrosine kinase inhibitors (TKIs) (Slamon et al. 1987; Press et al. 1993; Goutsouliak et al. 2020).

10.3.2.5 New Strategies/Challenges: Breast Cancer Treatment

(a) **Emerging Therapies (HR+ Breast Cancer)**: The primary causes of resistance to current endocrine therapy involve the mTOR/PI3K/Akt signalling pathways and also the pathway CDK4/6 (pathway for cell cycle progression). Thus, new treatments for HR+ breast cancer primarily focus on these pathways to overcome oestrogen-independent cell survival (Brufsky and Dickler 2018).

(b) **New Strategic Therapies (HER2+ Breast Cancer)**: The HER2+ breast cancer is typically treated using tyrosine kinase inhibitors (TKIs) or targeted antibodies. A more recent addition to the treatment options is TDM-1, which is an ADC. The advancements in treatment have significantly enhanced the patient survival rates with HER2+ breast cancer. Nonetheless, a considerable number of patients having HER2+ breast cancer experience resistance to anti-HER2 therapy. Therefore, there are emerging therapeutic strategies, including antibodies that target HER2 receptors, new TKIs, inhibitors for PI3K/mTOR pathway and CDK4/6 pathways, and also vaccines (Escrivá-de-Romaní et al. 2018).

(c) **Emerging Therapies (TNBC)**: The most aggressive subtypes of the breast cancer are TNBC. The unique characteristic of TNBC is lack of three receptors (ER, PR, and HER2 overexpression), along with its significant heterogeneity, and has posed challenges in the development of effective treatments. Therefore, various strategic therapies have been formulated to address the different subtypes of TNBC. Some of the advancements in this field involve use of targeted therapy, conjugated antibodies, and immunotherapy (Bianchini et al. 2016).

10.4 Biosimilars Approved for Breast Cancer Treatment

For treatment of breast cancer, biosimilars are approved for use in both, the disease itself and as supportive drugs for cancer patients. These supportive drugs include haematopoietic growth factors like erythropoietin and filgrastim. There has been a major development in the market with the introduction of biosimilars of growth factors. These biosimilars were first approved by EMA in 2008 (EMA and EU 2023) and now extensively utilised to address the side effects of cancer chemotherapy, such as neutropenia and anaemia.

In terms of monoclonal antibodies, trastuzumab has the potential to alter the progression of HER2 disease especially for the treatment of HER2+ breast cancer. It is also indicated in metastatic gastric cancer with HER2 amplification (European Medicines Agency 2014). In addition to trastuzumab's biosimilar, bevacizumab (targeting VEGF, i.e. vascular endothelial growth factor) has been approved in Europe. It is used as a first-line treatment, with paclitaxel, for metastatic breast cancer in adults (Cornes 2012). While the approval of bevacizumab's biosimilar is undoubtedly a significant advancement in ensuring that more cancer patients have access to monoclonal antibodies treatment, it is worth mentioning that it is approved by EMA for use in breast cancer.

10.4.1 Trastuzumab Biosimilars

Trastuzumab has a remarkable effect on the progression of HER2-amplified breast cancer. Trastuzumab along with cytotoxic chemotherapy has markedly enhanced the response, and survival, both overall and progression-free survival of the patients with HER2+ metastatic breast cancer. Additionally, there have been studies that show enhanced survival rates in initial stages of HER2 breast cancer, when compared to chemotherapy alone (Perez et al. 2014; Piccart-Gebhart et al. 2005; Slamon et al. 2011). The drug has been considered revolutionary, but its high cost has limited its widespread use. However, with the expiration of the European patent in the year 2014, and the expiration of the US patent in the year 2019, made several manufacturers to apply for approval of trastuzumab biosimilars (Blackwell et al. 2018). The arrival of trastuzumab's biosimilars is viewed as a potential way to enhance HER2 blockage in the treatment of breast cancer. For the approval of the biosimilars, majority of the studies compared the safety and efficacy of the biosimilars to the original trastuzumab, combined with a cytotoxic drugs, usually a taxane (Migliavacca Zucchetti et al. 2019).

10.4.2 Mechanism of Action of Antibodies Targeting HER2

Trastuzumab (Herceptin; USFDA approval 1998) was the first antibody that was developed which targeted HER2 receptor (Slamon et al. 2001). Trastuzumab specifically focuses on subdomain IV, which is a part of HER2 extracellular domain. However, there is still much to learn about the mechanism behind trastuzumab's therapeutic effect. The mechanism of action of antibodies targeting HER2 is described in Fig. 10.3. Several hypotheses have been put forward on the basis of scientific studies to explain the mechanism of action of trastuzumab. As an example, trastuzumab has the ability to prevent HER2-HER3 heterodimer formation, which is recognised as pair that is the most oncogenic among the HER family (Junttila et al. 2009). Additionally, it has the potential to hinder the active creation of p95HER2 fragment by inhibiting the cleavage of the HER2 extracellular domain (Molina et al. 2001). There is a potential way to combat tumours by stimulating

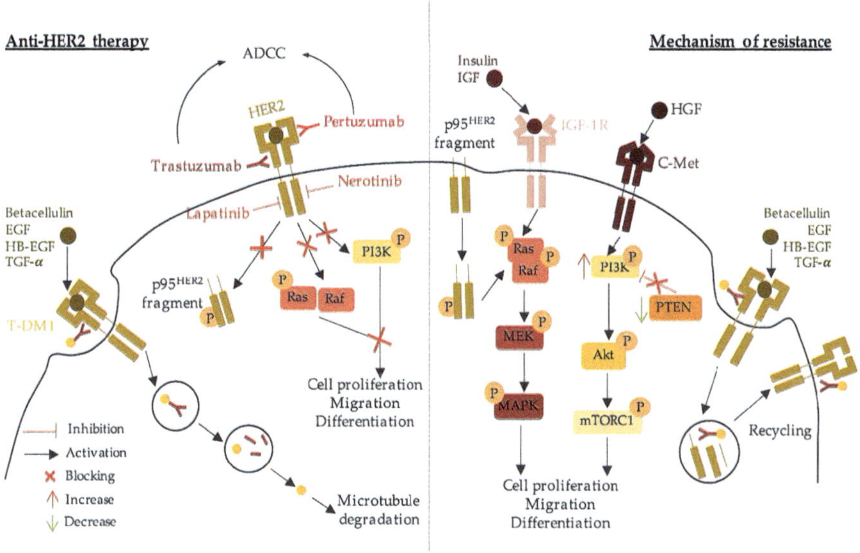

Fig. 10.3 Mechanisms of action and resistance of anti-HER2 therapy (Burguin et al. 2021). The left part of the figure describes the mechanism of action of anti-HER2 therapy through anti-HER2 antibody (trastuzumab and pertuzumab), tyrosine kinase inhibitors (lapatinib and neratinib), and trastuzumab emtansine (T-DM1). The right part of the figure describes the mechanism of resistance to anti-HER2 therapy through constitutive active p95HER2 fragment, activation of other signalling pathways, and rapid recycling of HER2-T-DM1. *ADCC* antibody-dependent cellular cytotoxicity, *HER2* human epidermal growth factor receptor 2, *EGF* epidermal growth factor, *HB-EGF* heparin-binding EGF-like growth factor, *TGF-_* transforming growth factor alpha, *T-DM1* trastuzumab emtansine, *IGF-1R* insulin growth factor receptor 1, *IGF* insulin growth factor, *HGF* hepatocyte growth factor, *MEK/MAPK* mitogen activated protein kinase, *PI3K* phosphoinositide 3-kinase, *mTOR* mammalian target of rapamycin, *PTEN* phosphatase and tensin homolog. (Source: https://www.researchgate.net/figure/Anti-HER2-therapy-mechanisms-of-action-and-resistance-The-left-part-of-the-figure_fig2_354011807; Licence link: Canonical URL https://creativecommons.org/licenses/by/4.0/)

ADCC, Antibody-Dependent Cellular Cytotoxicity, through interaction with Fc receptors situated on the immune effector cells (Clynes et al. 2000).

At first, trastuzumab was authorised in the treatment of metastatic HER2+ breast cancer. It highly increased the clinical advantages of first-line treatment (Slamon et al. 2001). It has also shown its safety and efficacy in the early stages HER2+ breast cancer. It is commonly used simultaneously along with other anti-HER2 therapies and/or chemotherapy (Perez et al. 2011a, b). The suggested dose of trastuzumab is 4 mg/kg intravenously (IV) initially, followed by 2 mg/kg/week, for a duration of 1 year as an adjuvant therapy in early stages of HER2+ breast cancer. The therapy should continue up till disease-free progression for metastatic HER2+ breast cancer (Iqbal and Iqbal 2014). Pertuzumab also known as Perjeta is an antibody that specifically focuses on the HER2 extracellular domain, specifically binding to subdomain II. Pertuzumab effectively blocks the formation of HER2 heterodimers with other members of the HER family, by binding to HER2. This, in turn, inhibits the activation of downstream signalling pathways (Capelan et al. 2013). Similar to trastuzumab, pertuzumab also has an indirect antitumor effect by activating the ADCC pathway (Scheuer et al. 2009).

Although trastuzumab and pertuzumab have had significant positive effects on HER2+ breast cancer treatment, only one-third of HER2+ breast cancer patients experience the benefits of anti-HER2 antibodies (Rimawi et al. 2015). The probable hypotheses which explain this resistance involve changes to the structure of HER2, which make it more difficult for antibodies to bind. Through alternative splicing, a truncated isoform may arise that lacks the extracellular domain. This leads to formation of a constitutive active p95HER2 fragment (Zagozdzon et al. 2011). The signalling pathways which are mediated by HER2 could be bypassed by the overexpression of different tyrosine kinases. Research has demonstrated that cells with high levels of IGF-1R can bypass arrest of cell cycle by boosting CDK2 kinase activity (Lu et al. 2001). In addition, the increased expression of c-Met, a receptor for hepatic growth factor, works together with HER2 signalling to provide resistance against anti-HER2 antibodies. It is worth noting that c-Met has a physical interaction with HER2, and depleting c-Met makes cells more responsive to trastuzumab (Shattuck et al. 2008; Agarwal et al. 2009). One possible explanation for resistance to anti-HER2 antibodies is changed which occur within the HER2 downstream signalling pathways. The receptor HER2 leads to activation of PI3K/Akt signalling pathway. This pathway is regulated by PTEN- Phosphatase and tensin homolog, an inhibitor of this pathway (Davis et al. 2014). Tumours that lose PTEN function and/or constitutive activation of PI3K because of genetic mutations tend to have poorer therapeutic outcomes when treated with trastuzumab (Berns et al. 2007; Kataoka et al. 2010).

10.4.3 Approved Trastuzumab Biosimilars

(a) **Ogivri**: It is trastuzumab-dkst, which is manufactured by Mylan GmbH. It was the initial trastuzumab biosimilar to receive FDA approval. The Heritage trial, conducted by Rugo H. et al. and published in JAMA in 2017, provided confirmation of the study's findings. Furthermore, a presentation on the long-term follow-up of trastuzumab-dkst maintenance monotherapy along with taxane was made at the 2018, ASCO Annual Meeting. This presentation further supported the efficacy and safety profiles, which were found to be similar to the originator trastuzumab (Rugo et al. 2017; Manikhas et al. 2018).

(b) **Ontruzant**: It is in the market as SB3, manufactured by Samsung Bioepis UK limited (SBUK). In November 2017, Ontruzant became the first trastuzumab biosimilar to be approved by EMA. It was also the first of its kind to be launched in the United Kingdom. It was approved by the regulatory body after a Phase III clinical trial. The study was randomised and double-blinded study that was aimed to compare the effectiveness, safety, and immune response of Ontruzant SB3 with the original trastuzumab in patients who received neoadjuvant therapy in the initial breast cancer stages with HER2 overexpression (Pivot et al. 2018).

(c) **Herzuma**: It was created and manufactured by collaboration between Celltrion Health Care Hungary Kft and TEVA pharmaceuticals in the name of CT-P6 (Trastuzumab-pkrb). The treatment received approval from the EMA in February, 2018. In the year 2018 December, US FDA approved the biosimilar. In the patients with overexpression of HER2, it is used as adjuvant therapy and neoadjuvant therapy. It is a first-line therapy, combined with paclitaxel or is also administered on its own in the patients having metastatic diseases, who have undergone single cycle or more chemotherapy regimens. Additionally, it is authorised to be used conjunction with chemotherapy as an initial treatment for gastric cancer that overexpresses the HER2 protein. The approval was granted after a thorough examination of the Phase III confirmatory study. This study evaluated drug's safety and effectiveness in HER2-positive patients as an adjuvant and neoadjuvant therapy for a duration of 1 year in HER2 positive breast cancer patient (EMA and EU 2023; Moore et al. 2021; Esteva et al. 2018).

(d) **Trazimera**: Trazimera is fourth biosimilar Pfizer and the first one in the oncology segment to be granted European approval in the name of PF-05280014. In July 2018, EMA approved it as adjuvant therapy and also as a therapy for metastatic HER2-overexpressed breast cancer. Along with breast cancer, it also received approval as a first-line treatment for of metastatic gastric cancer with HER2-overexpression. The European approval was granted on the basis of the exhaustive data showcasing a strong resemblance between Trazimera and trastuzumab. The data was inclusive of the results of REFLECTIONS B327_02, which was a comparative clinical study which used Trazimera for first-line therapy in patients with HER2 overexpression. The study results demonstrated that Trastuzumab and Trazimera were clinically equivalent (Migliavacca Zucchetti et al. 2019).

(e) **Kanjinti**: It is manufactured by Amgen Europe B.V. as ABP 980. The biosimilar Kanjinti, was approved by EMA as an adjuvant and neoadjuvant therapy for early breast cancer with HER2 overexpression, metastatic breast cancer, and also for metastatic gastric cancer with amplified HER2 in May 2018. However, the FDA did not grant its approval. The approval from EMA was granted after conducting an equivalence study comparing Kanjinti to the original trastuzumab (von Minckwitz et al. 2018).

10.5 Biosimilar: Implications for Clinical Practice

After over a decade of availability in Europe, biosimilars are now starting to receive regulatory approval from the US FDA and are making their way into the market in the United States. These recent approvals highlight the increasing acceptance of biosimilars by regulatory agencies. The process of approval has also streamlined and shortened. However, getting a marketing licence is only the initial challenge for a biosimilar. Overcoming the hurdle of gaining market penetration and gaining widespread acceptance from physicians, chemists, patients, and other healthcare professionals is yet an obstacle to overcome.

(a) **Pharmacy and supply chain management**: Pharmacy and the supply chain management will present significant challenges in integration and adoption of the biosimilars in the health care systems.
(b) **Cost**: The consideration of cost is a crucial factor for determining the level of adoption of biosimilars. Cost reduction could be achieved either through direct savings or by encouraging competition among the branded product manufacturers. However, while generic drugs typically have a significantly lower cost compared to their corresponding branded products, it is projected that the discounts on biosimilars will be smaller, ranging from 15% to 35% in comparison their reference biologics owing to their higher research and development costs.
(c) **Acceptance of biosimilars amongst patients and physicians**: Patient acceptance of biosimilars will depend on their level of awareness, knowledge, and perception. This will play a significant role in determining the extent and speed at which the biosimilars would be integrated into the clinical practice. While it is important to educate patients directly, it is equally crucial for physicians to have a strong understanding of biosimilars in order to effectively communicate with patients and address any concerns they may have. According to the studies, it appears that physicians would benefit from further education regarding the necessary criteria for biosimilars to gain regulatory approval. The findings from studies showed that a reference biologic and its biosimilar can be safely switched during the treatment.
(d) **Safety aspect**: As the biosimilars are becoming more accessible, ensuring safety is a top priority for both patients and clinicians. However, there is a common misconception that the low costs of biosimilars may compromise it safety. The regulatory guidelines for the approval by the FDA state that a candidate for biosimilar must show similarity without significant difference in safety, potency, and purity, as compared to the reference biological products. The confirmatory trails are of short duration as compared to long-term studies and hence the long-term unknown side effects could arise. However, through post marketing pharmacovigilance and registry studies, any such abnormality would be identified (Rifkin and Peck 2017).

10.6 How Biosimilars Play a Key Role in Changing the Trajectory of Breast Cancer Research and Treatment?

The comprehensive examination of the clinical advancement of breast cancer biosimilars has opened up new possibilities for research in oncology, providing valuable insights. Trastuzumab is a vital medication recognised by the World Health Organisation for breast cancer treatment. It is the only biologic included in the WHO's essential medicine list for this specific purpose, whether in the early stages or advanced stages of the disease. The implementation of clinical trials should be in line with national priorities and tailored to the capacity and structure of the health system. It is crucial to meet the needs of the population while focusing on patient-centred care.

In order to revolutionise breast cancer care on a global scale, it is crucial to focus on the scientific advancements in drug development, particularly in the field of biosimilars. By prioritising research in clinical practice, we can not only improve patient outcomes but also reap economic and financial rewards. By prioritising population public health, we can ensure that biosimilar development is integrated into the larger healthcare landscape. This will guarantee that breast cancer patients have timely access to the most effective medicines, while also guiding the allocation of health funding and resources. Ultimately, it allows us to establish clear priorities throughout the entire continuum of care (Trapani and Curigliano 2020).

10.7 Conclusion

It is evident that the biosimilar's integration in the field of oncology sector is on the horizon because of the extremely high expenses associated with anticancer monoclonal antibodies. The rising costs have become a significant challenge for both public and private healthcare system in the present day. The use of biosimilars can help control rising healthcare expenses while also improving patient accessibility. The effect of biosimilars on reducing treatment costs relies on how the biosimilars are accepted by regulatory agencies, doctors, chemists, and patients. So far, there have been no variations in terms of safety, effectiveness, and immune response between the original product and the biosimilar version (Migliavacca Zucchetti et al. 2019).

Over the time, with increasing use of biosimilars, there has been a greater focus on post marketing surveillance to better understand any uncertainties surrounding their use. However, for exploration of full potential of the biosimilars, several challenges need to be addressed. Various factors, including distribution channels, formulary decisions, pricing, and reimbursement, will have an impact on the adoption of this. Enhancing the knowledge of everyone involved, including doctors, chemists, healthcare administrators, and patients, will be crucial in fostering overall acceptance. These efforts in creating awareness will focus on the importance of understanding regulatory standards for their approval, the necessary preclinical/clinical data, and the significance of similarity in terms of safety and efficacy (Rifkin and Peck 2017).

To make the biosimilars more affordable and accessible, the World Health Organisation has initiated a pilot programme for manufacturers, for inclusion of trastuzumab biosimilars in the List of Essential Medicines. The costs of the trastuzumab biosimilars are estimated to be 20–30% lower than the reference product. More such initiatives are needed for the cost reduction of biosimilars in future, and also the boost in the field of the research is definitely a way ahead.

References

Agarwal S, Zerillo C, Kolmakova J, Christensen J, Harris L, Rimm D et al (2009) Association of constitutively activated hepatocyte growth factor receptor (Met) with resistance to a dual EGFR/Her2 inhibitor in non-small-cell lung cancer cells. Br J Cancer 100(6):941–949

Amin MB, Greene FL, Edge SB, Compton CC, Gershenwald JE, Brookland RK et al (2017) The eighth edition AJCC cancer staging manual: continuing to build a bridge from a population-based to a more "personalized" approach to cancer staging. CA Cancer J Clin 67(2):93–99

Berns K, Horlings HM, Hennessy BT, Madiredjo M, Hijmans EM, Beelen K et al (2007) A functional genetic approach identifies the PI3K pathway as a major determinant of trastuzumab resistance in breast cancer. Cancer Cell 12(4):395–402

Bianchini G, Balko JM, Mayer IA, Sanders ME, Gianni L (2016) Triple-negative breast cancer: challenges and opportunities of a heterogeneous disease. Nat Rev Clin Oncol 13(11):674–690

Biosimilars Market Size & Share Analysis—Growth trends & forecasts (2024–2029). 2024. https://www.mordorintelligence.com/industry-reports/global-biosimilars-market-industry

Blackwell K, Gligorov J, Jacobs I, Twelves C (2018) The global need for a trastuzumab biosimilar for patients with HER2-positive breast cancer. Clin Breast Cancer 18(2):95–113

Boyages J (2017) Radiation therapy and early breast cancer: current controversies. Med J Aust 207(5):216–222

Brufsky AM, Dickler MN (2018) Estrogen receptor-positive breast cancer: exploiting signaling pathways implicated in endocrine resistance. Oncologist 23(5):528–539

Burguin A, Diorio C, Durocher F (2021) Breast cancer treatments: updates and new challenges. J Pers Med 11(8):808

Capelan M, Pugliano L, De Azambuja E, Bozovic I, Saini K, Sotiriou C et al (2013) Pertuzumab: new hope for patients with HER2-positive breast cancer. Ann Oncol 24(2):273–282

Chhikara BS, Parang K (2023) Global cancer statistics 2022: the trends projection analysis. Chem Biol Lett 10(1):451

Clynes RA, Towers TL, Presta LG, Ravetch JV (2000) Inhibitory Fc receptors modulate in vivo cytoxicity against tumor targets. Nat Med 6(4):443–446

Cornes P (2012) The economic pressures for biosimilar drug use in cancer medicine. Target Oncol 7:57–67

Davis NM, Sokolosky M, Stadelman K, Abrams SL, Libra M, Candido S et al (2014) Deregulation of the EGFR/PI3K/PTEN/Akt/mTORC1 pathway in breast cancer: possibilities for therapeutic intervention. Oncotarget 5(13):4603

Denkert C, Liedtke C, Tutt A, von Minckwitz G (2017) Molecular alterations in triple-negative breast cancer—the road to new treatment strategies. Lancet 389(10087):2430–2442

El Sayed R, El Jamal L, El Iskandarani S, Kort J, Abdel Salam M, Assi H (2019) Endocrine and targeted therapy for hormone-receptor-positive, HER2-negative advanced breast cancer: insights to sequencing treatment and overcoming resistance based on clinical trials. Front Oncol 9:510

Elston EW, Ellis IO (1993) Method for grading breast cancer. J Clin Pathol 46(2):189–190

EMA and EU (2023) Biosimilars in the EU. Information guide for healthcare professionals. European Medicines Agency. https://www.ema.europa.eu/en/documents/leaflet/biosimilars-eu-information-guide-healthcare-professionals_en.pdf

Escrivá-de-Romaní S, Arumí M, Bellet M, Saura C (2018) HER2-positive breast cancer: current and new therapeutic strategies. Breast 39:80–88

Esteva FJ, Saeki T, Kim H, Stebbing J (2018) Efficacy and safety of the trastuzumab biosimilar candidate CT-P6. Future Oncol 14(19):1909–1919

European Medicines Agency (2014) Guideline on similar biological medicinal products. https://www.ema.europa.eu/en/documents/scientific-guideline/guideline-similar-biological-medicinal-products-rev1_en.pdf

Foulkes WD, Smith IE, Reis-Filho JS (2010) Triple-negative breast cancer. N Engl J Med 363(20):1938–1948

Gewirtz D (1999) A critical evaluation of the mechanisms of action proposed for the antitumor effects of the anthracycline antibiotics adriamycin and daunorubicin. Biochem Pharmacol 57(7):727–741

Glassy MC, Gupta R (2014) Technical and ethical limitations in making human monoclonal antibodies (an overview). In: Human monoclonal antibodies: methods and protocols, pp 9–36

Goutsouliak K, Veeraraghavan J, Sethunath V, De Angelis C, Osborne CK, Rimawi MF et al (2020) Towards personalized treatment for early stage HER2-positive breast cancer. Nat Rev Clin Oncol 17(4):233–250

Grubbé EH (1933) Priority in the therapeutic use of X-rays. Radiology 21(2):156–162

Hammond MEH (2010) American Society of Clinical Oncology/College of American Pathologists guideline recommendations for immunohistochemical testing of estrogen and progesterone receptors in breast cancer. J Oncol Pract 6(4):195–197

Harbeck N, Penault-Llorca F, Cortes J, Gnant M, Houssami N, Poortmans P et al (2019) Breast cancer. Nat Rev Dis Primers 5(1):66

He MY, Rancoule C, Rehailia-Blanchard A, Espenel S, Trone J-C, Bernichon E et al (2018) Radiotherapy in triple-negative breast cancer: current situation and upcoming strategies. Crit Rev Oncol Hematol 131:96–101

Hirsch BR, Lyman GH (2011) Biosimilars: are they ready for primetime in the United States? J Natl Compr Cancer Netw 9(8):934–943

Howard DH, Bach PB, Berndt ER, Conti RM (2015) Pricing in the market for anticancer drugs. J Econ Perspect 29(1):139–162

Howlader N, Altekruse SF, Li CI, Chen VW, Clarke CA, Ries LA et al (2014) US incidence of breast cancer subtypes defined by joint hormone receptor and HER2 status. J Natl Cancer Inst 106(5):dju055

International Agency for Research on Cancer (IARC) (2024) Global cancer burden growing, amidst mounting need for services. World Health Organization, Geneva. https://www.who.int/news/item/01-02-2024-global-cancer-burden-growing%2D%2Damidst-mounting-need-for-services

Iqbal N, Iqbal N (2014) Human epidermal growth factor receptor 2 (HER2) in cancers: overexpression and therapeutic implications. Mol Biol Int 2014:852748

Joshi H, Press MF (2018) Chapter 22—Molecular oncology of breast cancer. In: Bland KI, Copeland EM, Klimberg VS, Gradishar WJ (eds) The breast, 5th edn. Elsevier, pp 282–307.e5

Junttila TT, Akita RW, Parsons K, Fields C, Phillips GDL, Friedman LS et al (2009) Ligand-independent HER2/HER3/PI3K complex is disrupted by trastuzumab and is effectively inhibited by the PI3K inhibitor GDC-0941. Cancer Cell 15(5):429–440

Kataoka Y, Mukohara T, Shimada H, Saijo N, Hirai M, Minami H (2010) Association between gain-of-function mutations in PIK3CA and resistance to HER2-targeted agents in HER2-amplified breast cancer cell lines. Ann Oncol 21(2):255–262

Kerketta ZH, Kujur A, Kumari N, Sagar V, Pushpa F, Kerketta Z et al (2023) A cross-sectional study on the epidemiology of newly diagnosed breast cancer patients attending tertiary care hospitals in a tribal preponderant state of India: regression analysis. Cureus 15(6):e40489

Kwong D, McGale P, Taylor C, Correa C, Cutter D, Duane F et al (2014) Effect of radiotherapy after mastectomy and axillary surgery on 10-year recurrence and 20-year breast cancer mortality: meta-analysis of individual patient data for 8135 women in 22 randomised trials. Lancet 383(9935):2127–2135

Laslop A (2013) Biosimilar monoclonal antibodies—challenges and opportunities in Europe. GaBI J 2:110–111

Lee JF, Litten JB, Grampp G (2012) Comparability and biosimilarity: considerations for the healthcare provider. Curr Med Res Opin 28(6):1053–1058

Li E, Ramanan S, Green L (2015) Pharmacist substitution of biological products: issues and considerations. J Manag Care Spec Pharm 21(7):532–539

Lu Y, Zi X, Zhao Y, Mascarenhas D, Pollak M (2001) Insulin-like growth factor-I receptor signaling and resistance to trastuzumab (Herceptin). J Natl Cancer Inst 93(24):1852–1857

Manikhas A, Pennella EJ, Bondarenko I, Mukhametshina G, Abesamis-Tiambeng MLT, Akewanlop C et al (2018) Biosimilar trastuzumab-dkst monotherapy versus trastuzumab

monotherapy after combination therapy: toxicity, efficacy, and immunogenicity from the phase 3 heritage trial. J Clin Oncol 36(15)
Mathur P, Sathishkumar K, Chaturvedi M, Das P, Sudarshan KL, Santhappan S et al (2020) Cancer statistics, 2020: report from national cancer registry programme, India. JCO Glob Oncol 6:1063–1075
Migliavacca Zucchetti B, Nicolò E, Curigliano G (2019) Biosimilars for breast cancer. Expert Opin Biol Ther 19(10):1015–1021
Miyazaki Y, Shimizu J, Kubo Y, Tabata N, Aso T (2022) Quantitative classification of invasive and noninvasive breast cancer using dynamic magnetic resonance imaging of the mammary gland. J Clin Imaging Sci 12:45
Molina MA, Codony-Servat J, Albanell J, Rojo F, Arribas J, Baselga J (2001) Trastuzumab (herceptin), a humanized anti-Her2 receptor monoclonal antibody, inhibits basal and activated Her2 ectodomain cleavage in breast cancer cells. Cancer Res 61(12):4744–4749
Moore TJ, Mouslim MC, Blunt JL, Alexander GC, Shermock KM (2021) Assessment of availability, clinical testing, and US Food and Drug Administration review of biosimilar biologic products. JAMA Intern Med 181(1):52–60
Moos PJ, Fitzpatrick FA (1998) Taxanes propagate apoptosis via two cell populations with distinctive cytological and molecular traits. Cell Growth Differ 9(8):687
Nabholtz J-M, Gligorov J (2005) The role of taxanes in the treatment of breast cancer. Expert Opin Pharmacother 6(7):1073–1094
Penel N, Adenis A, Bocci G (2012) Cyclophosphamide-based metronomic chemotherapy: after 10 years of experience, where do we stand and where are we going? Crit Rev Oncol Hematol 82(1):40–50
Perez EA, Romond EH, Suman VJ, Jeong J-H, Davidson NE, Geyer CE Jr et al (2011a) Four-year follow-up of trastuzumab plus adjuvant chemotherapy for operable human epidermal growth factor receptor 2–positive breast cancer: joint analysis of data from NCCTG N9831 and NSABP B-31. J Clin Oncol 29(25):3366
Perez EA, Suman VJ, Davidson NE, Gralow JR, Kaufman PA, Visscher DW et al (2011b) Sequential versus concurrent trastuzumab in adjuvant chemotherapy for breast cancer. J Clin Oncol 29(34):4491
Perez EA, Romond EH, Suman VJ, Jeong J-H, Sledge G, Geyer CE Jr et al (2014) Trastuzumab plus adjuvant chemotherapy for human epidermal growth factor receptor 2–positive breast cancer: planned joint analysis of overall survival from NSABP B-31 and NCCTG N9831. J Clin Oncol 32(33):3744
Piccart-Gebhart MJ, Procter M, Leyland-Jones B, Goldhirsch A, Untch M, Smith I et al (2005) Trastuzumab after adjuvant chemotherapy in HER2-positive breast cancer. N Engl J Med 353(16):1659–1672
Pivot X, Bondarenko I, Nowecki Z, Dvorkin M, Trishkina E, Ahn J-H et al (2018) Phase III, randomized, double-blind study comparing the efficacy, safety, and immunogenicity of SB3 (trastuzumab biosimilar) and reference trastuzumab in patients treated with neoadjuvant therapy for human epidermal growth factor receptor 2–positive early breast cancer. J Clin Oncol 36(10):968–974
Press MF, Pike MC, Chazin VR, Hung G, Udove JA, Markowicz M et al (1993) Her-2/neu expression in node-negative breast cancer: direct tissue quantitation by computerized image analysis and association of overexpression with increased risk of recurrent disease. Cancer Res 53(20):4960–4970
Rifkin RM, Peck SR (2017) Biosimilars: implications for clinical practice. J Oncol Pract 13(9_suppl):24s–31s
Rimawi MF, Schiff R, Osborne CK (2015) Targeting HER2 for the treatment of breast cancer. Annu Rev Med 66:111–128
Rugo HS, Barve A, Waller CF, Hernandez-Bronchud M, Herson J, Yuan J et al (2017) Effect of a proposed trastuzumab biosimilar compared with trastuzumab on overall response rate in patients with ERBB2 (HER2)–positive metastatic breast cancer: a randomized clinical trial. JAMA 317(1):37–47

Scheuer W, Friess T, Burtscher H, Bossenmaier B, Endl J, Hasmann M (2009) Strongly enhanced antitumor activity of trastuzumab and pertuzumab combination treatment on HER2-positive human xenograft tumor models. Cancer Res 69(24):9330–9336

Schnitt SJ (2000) Pathology of invasive breast cancer. Diseases of the breast, 9:349

Shattuck DL, Miller JK, Carraway KL III, Sweeney C (2008) Met receptor contributes to trastuzumab resistance of Her2-overexpressing breast cancer cells. Cancer Res 68(5):1471–1477

Slamon DJ, Clark GM, Wong SG, Levin WJ, Ullrich A, McGuire WL (1987) Human breast cancer: correlation of relapse and survival with amplification of the HER-2/neu oncogene. Science 235(4785):177–182

Slamon DJ, Leyland-Jones B, Shak S, Fuchs H, Paton V, Bajamonde A et al (2001) Use of chemotherapy plus a monoclonal antibody against HER2 for metastatic breast cancer that overexpresses HER2. N Engl J Med 344(11):783–792

Slamon D, Eiermann W, Robert N, Pienkowski T, Martin M, Press M et al (2011) Adjuvant trastuzumab in HER2-positive breast cancer. N Engl J Med 365(14):1273–1283

Sung H, Ferlay J, Siegel RL, Laversanne M, Soerjomataram I, Jemal A et al (2021) Global cancer statistics 2020: GLOBOCAN estimates of incidence and mortality worldwide for 36 cancers in 185 countries. CA Cancer J Clin 71(3):209–249

Trapani D, Curigliano G (2020) How can biosimilars change the trajectory of breast cancer therapy? Expert Rev Anticancer Ther 20(5):325–328

US FDA (n.d.) Biological product definitions. https://www.fda.gov/media/108557/download

von Minckwitz G, Colleoni M, Kolberg H-C, Morales S, Santi P, Tomasevic Z et al (2018) Efficacy and safety of ABP 980 compared with reference trastuzumab in women with HER2-positive early breast cancer (LILAC study): a randomised, double-blind, phase 3 trial. Lancet Oncol 19(7):987–998

Weigelt B, Geyer FC, Reis-Filho JS (2010) Histological types of breast cancer: how special are they? Mol Oncol 4(3):192–208

Weise M, Bielsky M-C, De Smet K, Ehmann F, Ekman N, Giezen TJ et al (2012) Biosimilars: what clinicians should know. Blood 120(26):5111–5117

Wolff AC, Hammond MEH, Hicks DG, Dowsett M, McShane LM, Allison KH et al (2014) Recommendations for human epidermal growth factor receptor 2 testing in breast cancer: American Society of Clinical Oncology/College of American Pathologists clinical practice guideline update. Arch Pathol Lab Med 138(2):241–256

Zagozdzon R, Gallagher WM, Crown J (2011) Truncated HER2: implications for HER2-targeted therapeutics. Drug Discov Today 16(17–18):810–816

Biosimilars in Hepatocellular Carcinoma 11

Seema Yadav, Abhishek Singh, Narhari N. Palei,
Arghya Kusum Dhar, and Subas Chandra Dinda

Abstract

Hepatocellular carcinoma (HCC) stands as a major global health concern with limited therapeutic options. Biosimilars are becoming popular in oncological settings as more affordable options to biologic medications. The present state of biosimilar development and use in the context of HCC is examined in detail in this study. This chapter discussed HCC, highlighting its rising frequency and the difficulties associated with the current treatment options. The study then explores the molecular targets and pathways associated with the advancement of HCC, highlighting the possibility of using biosimilars to target various pathways. The regulatory environment controlling biosimilar approval and market access is critically examined, illuminating the difficulties and factors particular to the development of biosimilars in the oncology sector. In addition, the study compared reference biologics with biosimilars in order to assess the clinical evidence regarding the safety and efficacy of biosimilars in the treatment of HCC. The goal of synthesizing the current data and identifying knowledge gaps is to help policymakers, researchers, and clinicians make well-informed decisions about the inclusion of biosimilars in HCC treatment regimens. This chapter demonstrated the biosimilar in hepatocellular carcinoma, which may be utilized to treat carcinoma diseases, and discussed their related available biosimilar treatment.

Keywords

Biosimilars · HCC · Lenvatinib · Nivolumab · TKIs

S. Yadav · A. Singh · N. N. Palei
Amity Institute of Pharmacy, Amity University, Lucknow, Uttar Pradesh, India

A. K. Dhar (✉) · S. C. Dinda
School of Pharmacy, The Neotia University, Sarisha, (South) West Bengal, India

S. Bhatt et al. (eds.), *Biosimilars for Cancer Treatment*,
https://doi.org/10.1007/978-981-97-3001-8_11

11.1 Introduction

The most common primary liver cancer and the one with the third-highest death rate globally is hepatocellular carcinoma (HCC). Since 2009, patients with advanced hepatocellular carcinoma (aHCC) have been advised to begin treatment with the multi-tyrosine kinase inhibitor (mTKI) sorafenib (Giannini et al. 2018). A lot of work has gone into expanding the range of long-lasting and efficient systemic treatment options to help with aHCC prognosis (Ren et al. 2021). Many monoclonal antibodies and tyrosine kinase inhibitors (TKIs) that target immunological checkpoints and angiogenesis have shown therapeutic benefit in phase III trials recently, most notably the IMbrave 150 study. With 830,000 projected deaths from cancer-related causes worldwide in 2020, primary liver cancer ranks third in terms of cancer-related deaths worldwide (Sun et al. 2022). Hepatocellular carcinoma (HCC), 75–85% of cases, is the most common. Patients with liver cirrhosis (LC) are at high risk of dying from hepatocellular carcinoma (HCC). The majority of tumors are not eligible for curative treatments because only a small percentage of HCCs are found during surveillance in the Western world, even though all hepatology and oncology scientific societies recommend ultrasound surveillance for HCC in patients who are at risk, with or without an alpha-fetoprotein (AFP) determination (Jalil et al. 2023). Indeed, up to 30–35% of patients in nations lacking national HCC surveillance systems present with extrahepatic spread (EHS) and/or macrovascular invasion (MVI), or with a poor performance status (PS) as a result of the tumor's progression. These individuals are categorized as having "advanced" HCC (BCLC C) in the Barcelona Clinic Liver Cancer (BCLC) system, and sorafenib systemic therapy is the only antitumoral treatment that is advised (Zhang et al. 2023). The hepatitis B virus is the primary cause of HCC (HBV). Approximately, half of all new instances of HBV infection worldwide are from China. The prognosis for people with HCC is dismal, and the disease is highly prevalent in China (Zhang et al. 2023). As the first-line conventional treatment for advanced HCC, sorafenib provides limited benefit in terms of survival and moderate effectiveness. Recent advancements in research have indicated that a viable first-line therapy for HCC could involve the combination of anti-programmed cell death-1 (PD-1) antibodies and antiangiogenic medicines (Liu et al. 2023). It is an aggressive cancer, often diagnosed in later stages when treatment options are limited, and 5-year survival rates are low at around 18%. With the high-cost burden of current targeted therapies for HCC, there has been growing interest in the potential for more affordable biosimilar versions of these treatments to improve accessibility and outcomes (Ouyang et al. 2023). Biosimilars are highly similar versions of approved originator biologic medications, offering the same clinical benefits but potentially at lower costs. They can provide more affordable and accessible treatment options once originators lose patent exclusivities. Biosimilars are expected to play an important role in oncology treatment, with projected cost savings that could enable access for more patients. Several originator biologics have shown survival benefits for HCC, but high prices limit their reach. Biosimilars may expand access (Xie et al. 2023).

In HCC, the original monoclonal antibody ramucirumab has shown notable safety and effectiveness improvements (Fong et al. 2023). In the second-line situation following sorafenib, ramucirumab compared to placebo improved overall and progression-free survival in a statistically significant and clinically meaningful way in the REACH-2 trial. However, many patients and payers find the monthly expense of $10,000+ to be unaffordable. If significant cost reductions are realized, biosimilar versions may be able to get past the affordability barrier (Liu et al. 2022). Bevacizumab has also shown promising efficacy in HCC, improving outcomes for patients with advanced disease. In the MOA study, patients who received bevacizumab plus metronomic capecitabine had a median overall survival of 13.3 months compared to 6.5 months with capecitabine alone (Zhang et al. 2022a). But again, the over $5000 monthly cost limits access. Biosimilars may enable broader usage through lower pricing. Several bevacizumab biosimilars have already been approved for other cancer types in Europe and the US (Zhang et al. 2022b).

Another expensive immunotherapy, nivolumab, has been approved for use in HCC based on survival improvements; however, it is only available to a limited subset of patients with unresectable HCC who have not responded to sorafenib. The FDA expedited the approval of Checkmate 040 with objective response rates of about 20% as a result of the experiment. However, considering the restricted patient eligibility, the $15,000 monthly fee is incredibly exorbitant (Dawkins and Webster 2019). Biosimilars like nivolumab may lessen this financial burden. In recent studies, there are limitations on patient eligibility and payer coverage because of the high cost and high efficacy of the several originator biologics that have been licensed for the treatment of HCC. The release of biosimilar versions of these medications, however, presents a genuine chance for more affordable access (Rizzo and Ricci 2022). A major goal should be achieving significant cost savings to enhance utilization and results. It will be crucial to keep a careful eye on the ongoing development of biosimilars and their approvals in HCC as originators risk losing their exclusivity. The full influence that biosimilars may have on HCC patient survival in the future will depend on how well they uphold comparable clinical results to reference drugs while enabling broader treatment accessibility through biosimilar uptake (Booth and Pilz 2022).

Biologics play an important role in the treatment of Hepatocellular carcinoma. The prognosis has changed with the introduction of systemic biologic treatments that can block tumor development pathways. It is now feasible for the median survival to last longer than a year (Li et al. 2022). Biologics that target the signaling cascades that fuel the progression of HCC have changed the therapeutic paradigm, leading to small but significant increases in survival. The incorporation of biologics is transforming clinical practice quickly and could result in new combinatorial techniques that can prolong life even more, with several late-stage trials still in progress. Their appearance has sparked fresh optimism (Sun et al. 2023). The biosimilar for HCC treatments is depicted in Fig. 11.1.

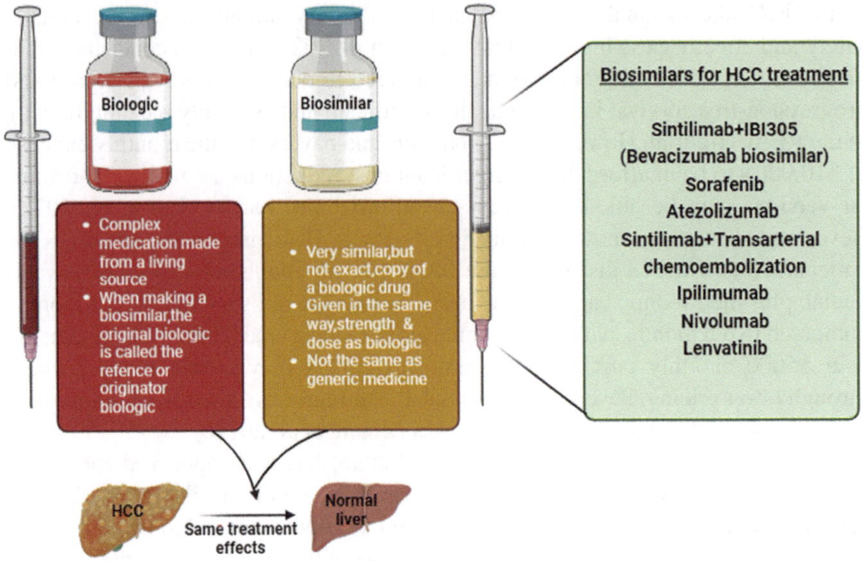

Fig. 11.1 Biosimilars for HCC treatments

11.2 Biologics and Biosimilars

Biosimilars are biologic that are very similar to a previously FDA-approved biologic (known as the original biologic). An overview of biologics and biosimilars is shown in Table 11.1. Minor variations across batches of the very same drug are both expected and normal for both original biologics and biosimilars. This means that biosimilars are not identical as their original biologic because biologics cannot be precisely duplicated (Weik 2000).

11.3 Biologics

Biologics comprise complex molecules derived from living organisms such as protein molecules, antibodies (Abs), nucleic acids, and cellular therapies. They are intended to target specific chemicals, pathways, or cells in disease processes. Stem cell therapies, monoclonal antibodies, recombinant proteins, gene therapies, and transplant tissue are a few examples of biologic medications. They are frequently used to treat a wide range of serious and life-threatening disorders (Patel et al. 2015).

In the context of pharmaceuticals, this lends itself to the term biologic, which refers to any of the class of medicines in which the active pharmaceutical ingredient (API) is derived from a living organism. This includes substances derived from plants and animals, as well as microbes like yeast and bacteria. It would also include human blood transfusions and tissue transplants. The field of biology emerged in

Table 11.1 Overview of biologics and biosimilars

Category	Biologics	Biosimilars
Definition	Original complex molecules such as proteins, hormones, enzymes, and antibodies, that come from living things	Highly equivalent copies of approved biologics with identical quality, efficacy, and safety
Development	Developed through clinical trials, original research, and regulatory approval procedures	Developed following the expiration of reference biologics' patents, requiring comparative analytical, non-clinical, and clinical studies (Blandizzi et al. 2017)
Regulatory guidelines	Standards for safety, effectiveness, and quality are ensured by stringent regulatory oversight and approval processes (such as those provided by the FDA and EMA)	Strict FDA and EMA regulatory pathways and criteria ensuring medicinal equivalency, safety, and quality
Interchangeability	Not always interchangeable with other biological substances	Has the potential to attain interchangeability status, meaning there would be no apparent clinical difference from the reference biologic
Immunogenicity	Due to structural complexity and variability, there is a chance for immunogenicity and undesirable reactions	Immunogenicity profiles exhibited homology to reference biologics (Windisch 2015)
Cost	Higher production costs result in higher medicine prices	Generally, more affordable, and accessible when compared to reference biologics
Market entry	Limited competition during a patent-protected exclusivity period	Market entry following patent expiration or settlement of patent dispute
Examples	Rituxan, Herception, and Humira	Herceptin, Rituxan, and Humira biosimilars (such as a biosimilar version of Adalimumab) (Johnson 2016)

the twentieth century. In 1921, Frederick Banting and Charles Best isolated the first-ever biologic, the hormone insulin, in Toronto (Revers and Furczon 2010).

The production of biologics involves fermentation, r-DNA technology, living cells, and sophisticated manufacturing methods. Because of the complexities involved, precision production and quality control are required for ensuring purity, consistency, and safety.

Biologics are governed by strict regulatory bodies such as the European Medicines Agency (EMA) and the United States Food and Drug Administration (USFDA). Comprehensive data proving safety, consistency, quality, and efficacy by exacting analytical, non-clinical, and clinical assessments are needed for approval.

The production method, regulatory restrictions, structural complexity, and material sources of biologics are different from those of traditional pharmaceuticals (Giannini et al. 2018). As a result, biological products can sometimes be more expensive, creating limitations on accessibility for some patients. Biosimilars were developed to improve patient outcomes by increasing access to biological medicines.

The use of biologics has transformed the treatment of many diseases, such as cancer (using therapies such as monoclonal antibodies (MAbs) and targeted therapies), autoimmune diseases (using TNF inhibitors and interleukin inhibitors), and rare health conditions (using enzyme replacement therapies) (Ingrasciotta et al. 2018).

11.4 Biosimilars

Biosimilars are a category of drug that is intended to improve access for people who require biologic medicine treatment. Biosimilars are defined by the USFDA as "a biological product that is highly similar to and has no clinically meaningful differences from an existing FDA-approved reference product" in terms of safety, purity, and potency. In simple terms, biosimilars are the same as the biologic drugs they were designed to mimic (Weik 2000).

To develop a biosimilar, it is necessary to show that it is similar to the reference biologic using functional tests, comparative clinical trials, non-clinical research, and analytical characterization. In order to facilitate the development, review, and approval of biosimilars, regulatory bodies have set up procedures and guidelines that emphasize the importance of comprehensive comparability analyses (Rawla et al. 2018).

Biosimilars have the potential to become interchangeable, meaning that a prescribing healthcare provider does not need to take any further steps to replace the reference biologic. To get interchangeability, other regulatory requirements must be fulfilled, such as proving the same clinical results in any patient as the reference biologic.

There are a number of issues with biosimilars, including post-market surveillance, pharmacovigilance, physician and patient acceptability, regulatory complexity, reimbursement policies, pricing strategies, and market dynamics. Stakeholders from the healthcare sector, patients, regulators, and healthcare professionals of healthcare must work together to address these issues.

By providing cost-effective alternatives, biosimilars have the potential to enhance patient access to biologic medicines. They are used in a variety of therapeutic fields, including oncology, rheumatology, gastrointestinal, dermatology, and endocrinology.

The Biologics Price Competition and Innovation (BPCI) Act of 2009 established a regulatory system for biosimilar approval. BPCI was established with the goal of increasing treatment options, boost medicine access, and promote competition for reduced healthcare costs.

In certain circumstances, pharmaceutical companies may apply to the FDA for an additional designation for a biosimilar medicine to make sure it can be used interchangeably in place of the original, or a reference, medicine (Sun et al. 2022). According to Mark Latymer, Senior Director, Biosimilars Portfolio, I&I Medical

Affairs, NMD at Pfizer, biosimilars are not required to have this extra designation; however, some drug manufacturers apply for it in order to facilitate the substitution of a biosimilar, based on state laws, for a reference medication at the pharmacy (Rawla et al. 2018).

Though their safety and efficacy requirements are comparable to biologics, there are some significant variations between biologics and biosimilars.

Regulatory Path: Prior to approval, every medicine is subjected to extensive testing. The approval procedure associated with biosimilars is simpler than that for biologics (Liu et al. 2023). The development route does not imply that biosimilars tend to be less safe, less efficient, or subject to lower standards. The data submitted during the biosimilar approval process is intended to demonstrate clinical equivalence between it and the reference biologic, whereas the data submitted during the biologic approval process is intended to demonstrate its safety and effectiveness (Ouyang et al. 2023).

Cost: As biosimilars go through a more simplified development procedure than biologics, they might be less expensive (Xie et al. 2023). The Association for Accessible Medicines (AAM) states that the rivalry between biosimilars has reduced the cost of reference products as well as their biosimilars, saving $7.9 billion in 2020 and more than $12.6 billion over the previous 10 years (Chan and Chan 2017).

In the changing field of HCC treatment, biologics and biosimilars are essential because they provide immune-based, targeted, and combination treatments that enhance patient outcomes. While biologics have been shown to be effective in advanced HCC, biosimilars offer the potential to improve affordability, availability, and healthcare sustainability. Collaboration among doctors, researchers, regulators, politicians, industry stakeholders, including patient groups is critical for properly navigating the complexities, problems, and opportunities related to biologics and biosimilars in HCC treatment. For patients with HCC to receive the best possible care, unfulfilled requirements must be addressed, and their quality of life must be enhanced, all of which depend on ongoing research, education, innovation, and evidence-based practice (Jalil et al. 2023).

11.5 Role of Biosimilars in HCC

Recently approved molecularly targeted medicines and immune checkpoint inhibitors have introduced new therapy options for patients with metastatic HCC. In the Guidelines of the Chinese Society of Clinical Oncology Hepatocellular Carcinoma 2020, targeted medications like lenvatinib are advised as first-line therapy plans. Furthermore, in the Chinese Society of Clinical Oncology Immune Checkpoint Inhibitor Clinical Practice 2021 Guidelines, sintilimab (a programmed cell death protein-1 inhibitor) plus bevacizumab or biosimilar (a vascular endothelial growth factor inhibitor) combination therapy has been upgraded as a first-line treatment (grade II) recommendation for advanced HCC (Wang et al. 2021).

The significance of biosimilars in HCC care is currently growing as an important factor in cancer treatment, with potential benefits in terms of accessibility, price, and healthcare sustainability. The following summarises the role of biosimilars in HCC:

11.5.1 Increasing Access and Affordability

Biosimilars offer more affordable options to reference biologics, which may lower medical costs and increase accessibility for payers, healthcare systems, and patients. In HCC, biosimilars offer possibilities to improve therapeutic access, increase treatment options, and meet unfulfilled medical requirements, especially in situations with limited resources or in areas with inequities in healthcare.

11.5.2 Healthcare Sustainability

Facilitating Healthcare Sustainability Biosimilars help to ensure the long-term sustainability of healthcare by optimizing resource allocation, increasing cost-effectiveness, and expanding access to novel medicines for HCC and other cancers. Biosimilars are consistent with healthcare policy goals such as increasing patient access, lowering costs, creating competition, and supporting value-based treatment in cancer treatment.

11.5.3 Improving Market Competition and Dynamics

In the biologics markets for HCC and other cancers, biosimilars encourage competition in the market, which in turn spurs innovation, price wars, and cost-containment measures. Biosimilars have the potential to drive down the cost of biologics by encouraging value-based pricing models, competitive pricing techniques, and affordability considerations for cancer treatment.

11.5.4 Regulatory Issues and Approval Processes

Biosimilars in HCC follow strict regulatory requirements, which include quality, effectiveness, safety, and immunogenicity studies in comparison to reference biologics. Regulatory organizations, such as the EMA and the USFDA, have set up specific approval processes for biosimilars in order to guarantee post-approval monitoring and patient safety (Dawkins and Webster 2019).

11.6 FDA Approval of Biosimilars for Hepatocellular Carcinoma

The most prevalent and lethal type of liver cancer is hepatocellular carcinoma (HCC), which has a substantial impact on world health. Having accessible and efficient treatment options is critical as the incidence of HCC keeps rising (Rizzo et al. 2020; Llovet et al. 2022). The FDA has provided patients with more therapeutic options in recent years by allowing biosimilars to treat HCC. This is a significant turning point in improving the accessibility and affordability of cutting-edge medicines. The USFDA has approved biosimilars for the treatment of HCC, a common and deadly form of liver cancer. This clearance represents a major step forward in tackling the worldwide health burden associated with HCC. Biosimilars are biological products that are very similar to reference goods that the FDA has approved. Biosimilars are biological products that bear a striking resemblance to an FDA-approved reference product currently on the market, with no discernible clinical differences. Biosimilars require stringent development and approval procedures because they are complex compounds made in living cells, in contrast to generic medications, which are exact replicas of their reference equivalents. They have gone through a stringent approval procedure that guarantees their quality, safety, and efficacy (Finn et al. 2020a; Galle et al. 2021). The authorization of biosimilars is significant in the context of HCC since it provides an affordable substitute for reference biologics. In addition to relieving the financial burden on patients and healthcare systems, this greater affordability also makes cutting-edge treatments more accessible, eventually improving patient outcomes. The pharmaceutical industry now faces healthy competition thanks to the approval of biosimilars, which may spur additional hepatocellular carcinoma research and development. This rivalry creates a dynamic environment that pushes for improvements in the knowledge and treatment of the disease. Nivolumab and pembrolizumab are two immunotherapy medications that have been licensed for the treatment of hepatocellular cancer. Immune checkpoint inhibitors (ICIs) have shown promise in treating a variety of tumor forms, and patients with HCC are seeing a similar progression. First, after the early phase CheckMate 040 and KEYNOTE-224 trials, the USFDA approved two PD-1 inhibitors: pembrolizumab and nivolumab. Nevertheless, the primary objectives of the two confirmatory phase III trials, KEYNOTE-240 and CheckMate 459, which evaluated pembrolizumab with placebo as second-line therapy and nivolumab versus sorafenib as first-line treatment, were not met (Yau et al. 2022; Finn et al. 2020b). Clinical trials assessing combinatorial techniques incorporating ICIs and other anticancer drugs, particularly antiangiogenic agents, have yielded more compelling findings, signaling a new era in HCC care if ICIs alone as a first-line treatment have not provided the desired impact (Ouyang et al. 2022; Rizzo et al. 2021, 2022; Wen et al. 2023; Llovet et al. 2021; Rizzo 2022).

11.7 Clinical Trial

Efforts to produce biosimilars for several therapeutic drugs used in the treatment of hepatocellular carcinoma (HCC) are underway (Table 11.2).

Table 11.2 Clinical trials using biosimilars for the treatment of HCC

S. no	Drug	Study design	Phase	Effect	References
1.	Sintilimab + IBI305 (bevacizumab biosimilar) and sorafenib	Randomized, open-label, multi-center study	Phase 2, 3	When compared to sorafenib in the first-line context, sintilimab plus IBI305 showed a significant advantage in both overall survival and progression-free survival. This combined regimen may offer a potential therapy option for people with HCC	Ren et al. (2021)
2.	Atezolizumab-bevacizumab and sintilimab-bevacizumab biosimilar	Randomized controlled trial	Phase 3	Except for tremelimumab-durvalumab, anti-programmed-death and anti-VEGF pathway inhibitor monoclonal antibodies (Anti-PD-(L)1/VEGF Ab) exhibited an overall survival benefit over all other medications	Fong et al. (2023)
3.	Sintilimab plus transarterial chemoembolization	Prospective, open-label	Phase 2	Sintilimab plus drug-eluting bead transarterial chemoembolization (DEB-TACE) before surgery showed good efficacy and safety in patients with HCC of Barcelona Clinic Liver Cancer (BCLC) stage A exceeding the Milan criteria or BCLC stage B	Guo et al. (2023)
4.	Ipilimumab and nivolumab	Single-arm, open-label study	Phase 1 Phase 2	With an acceptable percentage of reversible grade 3–4 toxicities (27.1%) and a low rate of discontinuation (2%) in patients with HCC, nivolumab/ipilimumab co-administration proved to be safe and preliminarily effective when used in advanced HCC patients	Pinato et al. (2021)

(continued)

Table 11.2 (continued)

S. no	Drug	Study design	Phase	Effect	References
5.	Sintilimab and lenvatinib	Single-arm, single-center, not-randomized, open-label	Phase 2	Sintilimab in combination with lenvatinib is a safe and feasible therapy option for intermediate to locally advanced HCC that was previously unsuitable for surgical resection	Wang et al. (2023)

11.8 Conclusion and Future Prospective

Biosimilars offer a revolutionary approach to the treatment of hepatocellular carcinoma (HCC), improving patient access and affordability while providing more affordable alternatives to reference biologics. Although biologics have transformed the treatment of HCC, biosimilars promote market competition, foster innovation, and tackle healthcare issues. To ensure patient safety, optimize use, improve anticancer outcomes in HCC and worldwide practice, and advance biosimilar development, regulatory harmonization, clinical research, stakeholder collaboration, and policy making are crucial. Prospects for the future include sustained innovation, international cooperation, evidence-based practice, and policy modification to fully utilize biosimilars, improve healthcare equity, and revolutionize cancer and HCC patient care.

Biosimilars are more affordable alternatives to biologics, which are widely used in cancer treatment around the world. The proposed biosimilar medication is put through a rigorous testing process and needs to pass common safety and efficacy standards before receiving approval from the FDA (or similar international organizations). When it comes to the proper use of biosimilars, efforts should be taken to raise awareness and educate physicians and other healthcare professionals. For those suffering from advanced HCC, bevacizumab plus atezolizumab combination therapy has now become the standard first-line treatment. The FDA has approved three bevacizumab biosimilars so far and more clinical trials must be conducted to investigate alternative biosimilar possibilities, considering the critical role bevacizumab plays in the management of HCC and many other malignancies.

References

Blandizzi C, Meroni PL, Lapadula G (2017) Comparing originator biologics and biosimilars: a review of the relevant issues. Clin Ther 39(5):1026–1039

Booth JP, Pilz J (2022) Retrospective indication-matched cohort study of reference product and biosimilar: bevacizumab versus bevacizumab-awwb. Hosp Pharm 57(4):455–461

Chan JCN, Chan ATC (2017) Biologics and biosimilars: what, why and how? ESMO Open 2(1):e000180

Dawkins J, Webster RM (2019) The hepatocellular carcinoma market. Nat Rev Drug Discov 18(1):13–14

Finn RS, Qin S, Ikeda M, Galle PR, Ducreux M, Kim TY et al (2020a) Atezolizumab plus bevacizumab in unresectable hepatocellular carcinoma. N Engl J Med 382(20):1894–1905

Finn RS, Ryoo BY, Merle P, Bouattour M, Lim HY, Breder V et al (2020b) Pembrolizumab as second-line therapy in patients with advanced hepatocellular carcinoma in KEYNOTE-240: a randomized, double-blind, phase III trial. J Clin Oncol 38:193

Fong KY, Zhao JJ, Sultana R, Lee JJX, Lee SY, Chan SL et al (2023) First-line systemic therapies for advanced hepatocellular carcinoma: a systematic review and patient-level network meta-analysis. Liver Cancer 12(1):7–18

Galle P, Finn RS, Qin S, Ikeda M, Ducreux M, Kim TY et al (2021) IMbrave150: updated overall survival (OS) data from a global, randomized, open-label phase III study of atezolizumab (atezo) plus bevacizumab (bev) vs sorafenib (sor) in patients (pts) with unresectable hepatocellular carcinoma (HCC). Oncol Res Treat 44:34

Giannini EG, Bucci L, Garuti F, Brunacci M, Lenzi B, Valente M et al (2018) Patients with advanced hepatocellular carcinoma need a personalized management: a lesson from clinical practice. Hepatology 67(5):1784–1796

Guo C, Zhang J, Huang X, Chen Y, Sheng J, Huang X et al (2023) Preoperative sintilimab plus transarterial chemoembolization for hepatocellular carcinoma exceeding the Milan criteria: a phase II trial. Hepatol Commun 7(3):e0054

Ingrasciotta Y, Cutroneo PM, Marcianò I, Giezen T, Atzeni F, Trifirò G (2018) Safety of biologics, including biosimilars: perspectives on current status and future direction. Drug Saf 41:1013–1022

Jalil A, Wert J, Jain AG, Ahmad S (2023) Emerging role of biosimilars: focus on bevacizumab and hepatocellular carcinoma. In: Computational methods in drug discovery and repurposing for cancer therapy, pp 387–402

Johnson JA (2016) Biologics and biosimilars: background and key issues. In: Congressional Research Service, Library of Congress

Li L, Yang S, Chen Y, Tian L, He Y, Wu B et al (2022) Immune checkpoint inhibitors plus an anti-VEGF antibody as the first-line treatment for unresectable hepatocellular carcinoma: a network meta-analysis and cost-effectiveness analysis. Front Pharmacol 13:891008

Liu D, Mu H, Liu C, Zhang W, Cui Y, Wu Q et al (2022) Hepatic artery infusion chemotherapy (HAIC) combined with sintilimab and bevacizumab biosimilar (IBI305) for initial unresectable hepatocellular carcinoma (HCC): a prospective, single-arm phase II trial. American Society of Clinical Oncology. J Clin Oncol 40:4073

Liu K, Zhu Y, Zhu H (2023) Immunotherapy or targeted therapy as the first-line strategies for unresectable hepatocellular carcinoma: a network meta-analysis and cost-effectiveness analysis. Front Immunol 13:1103055

Llovet J, Kelley R, Villanueva A, Singal A, Pikarsky E, Roayaie S et al (2021) Hepatocellular carcinoma. Nat Rev Dis Primers 7(1):6

Llovet JM, Castet F, Heikenwalder M, Maini MK, Mazzaferro V, Pinato DJ et al (2022) Immunotherapies for hepatocellular carcinoma. Nat Rev Clin Oncol 19(3):151–172

Ouyang T, Kan X, Zheng C (2022) Immune checkpoint inhibitors for advanced hepatocellular carcinoma: monotherapies and combined therapies. Front Oncol 12:898964

Ouyang J, Yang Y, Zhou Y, Ye F, Wang Z, Li Q et al (2023) The MAPS-CRAFITY score: a novel efficacy predictive tool for unresectable hepatocellular carcinoma treated with targeted therapy plus immunotherapy. Hepatol Int 17(6):1519–1531

Patel PK, King CR, Feldman SR (2015) Biologics and biosimilars. J Dermatolog Treat 26(4):299–302

Pinato DJ, Cortellini A, Sukumaran A, Cole T, Pai M, Habib N et al (2021) PRIME-HCC: phase Ib study of neoadjuvant ipilimumab and nivolumab prior to liver resection for hepatocellular carcinoma. BMC Cancer 21(1):301

Rawla P, Sunkara T, Raj JP (2018) Role of biologics and biosimilars in inflammatory bowel disease: current trends and future perspectives. J Inflamm Res 11:215–226

Ren Z, Xu J, Bai Y, Xu A, Cang S, Du C et al (2021) Sintilimab plus a bevacizumab biosimilar (IBI305) versus sorafenib in unresectable hepatocellular carcinoma (ORIENT-32): a randomised, open-label, phase 2-3 study. Lancet Oncol 22(7):977–990

Revers L, Furczon E (2010) An introduction to biologics and biosimilars. Part I: biologics: what are they and where do they come from? Can Pharm J 143(3):134–139

Rizzo A (2022) Locoregional treatments plus immunotherapy in hepatocellular carcinoma: where do we stand? Future Oncol 18:1665–1668

Rizzo A, Ricci AD (2022) Challenges and future trends of hepatocellular carcinoma immunotherapy. Int J Mol Sci 23:11363

Rizzo A, Nannini M, Novelli M, Dalia Ricci A, Di Scioscio V, Pantaleo MA (2020) Dose reduction and discontinuation of standard-dose regorafenib associated with adverse drug events in cancer patients: a systematic review and meta-analysis. Ther Adv Med Oncol 12:1758835920936932

Rizzo A, Ricci AD, Di Federico A, Frega G, Palloni A, Tavolari S et al (2021) Predictive biomarkers for checkpoint inhibitor-based immunotherapy in hepatocellular carcinoma: where do we stand? Front Oncol 11:803133

Rizzo A, Cusmai A, Gadaleta-Caldarola G, Palmiotti G (2022) Which role for predictors of response to immune checkpoint inhibitors in hepatocellular carcinoma? Expert Rev Gastroenterol Hepatol 16(4):333–339

Sun H, Zhu X, Gao Q, Qiu S, Shi Y, Wang X et al (2022) 711P Sintilimab combined with bevacizumab biosimilar as a conversion therapy in potentially resectable intermediate stage hepatocellular carcinoma (HCC): a phase II trial. Ann Oncol 33:S867–S868

Sun C, Wang Q, Hou L, Zhang R, Chen Y, Niu L (2023) A contrast-enhanced ultrasound-based nomogram for the prediction of therapeutic efficiency of anti-PD-1 plus anti-VEGF agents in advanced hepatocellular carcinoma patients. Front Immunol 14:1229560

Wang FH, Zhang XT, Li YF, Tang L, Qu XJ, Ying JE et al (2021) The Chinese Society of Clinical Oncology (CSCO): clinical guidelines for the diagnosis and treatment of gastric cancer, 2021. Cancer Commun (Lond) 41(8):747–795

Wang L, Wang H, Cui Y, Liu M, Jin K, Xu D et al (2023) Sintilimab plus Lenvatinib conversion therapy for intermediate/locally advanced hepatocellular carcinoma: a phase 2 study. Front Oncol 13:1115109

Weik MH (2000) Logical product. Comput Sci Commun Dict 924

Wen W, Zhang Y, Zhang H, Chen Y (2023) Clinical outcomes of PD-1/PD-L1 inhibitors in patients with advanced hepatocellular carcinoma: a systematic review and meta-analysis. J Cancer Res Clin Oncol 149(3):969–978

Windisch J (2015) Biosimilars versus original biologics: similarities and differences from development to approval. Z Rheumatol 74:672–681

Xie S, Wang M, Zeng C, Ou Y, Zhao L, Wang D et al (2023) Research progress of targeted therapy combined with immunotherapy for hepatocellular carcinoma. Front Oncol 13:1197698

Yau T, Park JW, Finn RS, Cheng AL, Mathurin P, Edeline J et al (2022) Nivolumab versus sorafenib in advanced hepatocellular carcinoma (CheckMate 459): a randomised, multicentre, open-label, phase 3 trial. Lancet Oncol 23(1):77–90

Zhang L, Sun J, Wang K, Zhao H, Zhang X, Ren Z (2022a) First-and second-line treatments for patients with advanced hepatocellular carcinoma in China: a systematic review. Curr Oncol 29(10):7305–7326

Zhang W, Gong C, Bi X, Xuenan P, Sun Y, Wang Y et al (2022b) Final clinical outcome and biomarker analysis of sintilimab plus a bevacizumab biosimilar (IBI305) for advanced hepatocellular carcinoma: a phase Ib clinical trial. American Society of Clinical Oncology. J Clin Oncol 40:455

Zhang H, Lu W, Fu Y, Li X, Dong W, Zhu N et al (2023) 90P HAIC plus sintilimab and bevacizumab biosimilar as treatment for patients with advanced hepatocellular carcinoma (HCC): a phase II trial. Ann Oncol 20(suppl_1):100535

Biosimilars in Ovarian Cancer

12

Pallavi Manish Lavhale, Payal Kesharwani, Anoop Kumar, and Shiv Kumar Prajapati

Abstract

Ovarian cancer is a deadly malignancy and remains to be a formidable challenge in respect to oncology, that is characterized by its insidious onset, aggressive nature, and limited treatment options. Despite the research in conventional and targeted delivery, the management of ovarian cancer requires a novel therapeutic strategy to improve the quality of life. Biologics have been increasingly used for the management of ovarian cancer, but their high cost and stability add a significant burden among the pharmaceutical companies and patients. Therefore, a promising alternative in reference to biologics is used providing improved efficacy and safety and reduced cost. Leveraging the principles of biotechnology and molecular biology, biosimilars have paved the way for enhanced accessibility to biologic therapies, addressing unmet medical needs across various disease states, including cancer. In response to these challenges, researchers and clinicians are increasingly focusing on advancing drug delivery systems to optimize the efficacy, safety, and accessibility of biosimilar therapies. From nanotechnology-based formulations to targeted delivery strategies and innovative biomaterials, recent developments offer promising avenues for overcoming delivery-related obstacles and enhancing the clinical utility of biosimilars in ovarian cancer treatment. This chapter aims to provide a comprehensive overview of the utilization of biosimilars in ovarian cancer management, elucidating

P. M. Lavhale (✉) · P. Kesharwani
Ram-Eesh Institute of Vocational and Technical Education, Greater Noida, Uttar Pradesh, India

A. Kumar
Delhi Pharmaceutical Sciences and Research University (DPSRU), New Delhi, India

S. K. Prajapati
Institute of Pharmaceutical Research, GLA University, Mathura, Uttar Pradesh, India

S. Bhatt et al. (eds.), *Biosimilars for Cancer Treatment*,
https://doi.org/10.1007/978-981-97-3001-8_12

the challenges faced in their delivery and exploring recent advancements in drug delivery systems. By integrating insights from clinical research, regulatory perspectives, and technological innovations, this discussion seeks to inform and empower stakeholders in navigating the dynamic landscape of biosimilar-based therapies for ovarian cancer.

Keywords

Biosimilars · Biologics · Ovarian cancer · Drug delivery · Targeting

12.1 Ovarian Cancer—An Introduction (Etiology, Occurrence, Prevalence)

Due to the lifestyle spent nowadays, women are suffering from ovarian cancer, that is a life-threatening cancer for the female. From the data collected worldwide, it is the seventh most occurring cancer among women. It is the second most fatal cancer, after breast cancer which can be developed after 40 years in a woman. The high rate of occurrence is found in Western and Northern Europe, and less cases are found in China and Africa (La Vecchia 2017). Epidemiologic research has demonstrated the variation in occurrence of ovarian cancer worldwide, and it was found that Sweden carries the highest lifetime risk for ovarian cancer (1.73%), followed by the USA (1.53%), UK (1.25%), South Europe (1.11%), South America (0.87%), India (0.75%), and Japan (0.47%). These findings imply that ovarian cancer risk may be influenced by ethnicity (Cherif et al. 2021). The survival rate for a woman suffering from ovarian cancer is 47.4%. Despite increased knowledge of ovarian cancer, treatment approaches and survival rates have not altered much, since early diagnosis is still difficult to identify its occurrence. The possibility of diagnosis is at stage III and IV, when the cancer has progressed, therefore is considered to be a silent killer. Although the primary reason for the lack of an early diagnosis is yet unknown, it can generally be assumed that the absence of a conclusive screening method, nebulous symptoms, and indicators may masquerade as other benign illnesses. The only way to diagnose ovarian cancer in women is the physical examination of the presence of pelvic mass in a woman body.

Though it is uncertain about the molecular process leading to growth of ovarian cancer but epidemiologic research suggests that carcinogenesis can result from a variety of factors, including those related to diet, reproduction, endocrine function, viruses, aging, postmenopausal hormone treatment usage, infertility, nulliparity, and environmental factors. Family aggregation was present in around 7% of the affected women, indicating a potential importance of the BRCA1 or BRCA2 genes as a genetic component (Momenimovahed et al. 2019). Pregnancies, using oral contraceptives, breastfeeding, tubal ligation, and hysterectomy are factors linked to a chance of developing ovarian cancer (Yoshikawa et al. 2000). Even an ovulation process in a woman causes a wound each month, which is subsequently healed by an enhanced proliferation process in a surface ovarian epithelial cell, according to the theory of ceaseless ovulation. It has been suggested that the main causes of ovarian carcinomas are either postovulatory inclusion cysts or the ovarian surface epithelium, which develops after follicular rupture and repair, increasing the risk of mutation and harm to DNA (Mara et al. 2020). This hypothesis, however, is at odds with the observation that those with polycystic ovarian syndrome who appear to have fewer ovulatory cycles are more likely to acquire ovarian cancer (Koshiyama et al. 2014). Also, there an greater likelihood of ovarian cancer after using infertility drugs gonadotropin stimulation of the ovarian surface epithelium, increases the risk of ovarian cancer development (Harris and Terry 2016).

Traditionally, a combination of both chemotherapy and surgery has been adopted for treating cancer, which includes tumor debulking surgery, thereafter including chemotherapy once the afflicted tissue has been surgically staged. One of the chemotherapy's drawbacks is that resistant cells might proliferate and eventually form tumors. Biologics have been increasingly used for the management of ovarian cancer, but their high cost and stability add a significant burden among pharmaceutical companies and patients. Therefore, new approaches need to be focused (Coleridge et al. 2021).

New drugs are examined in ongoing clinical studies, which focus on effectivity and primarily target molecular entities and pathways leading to the growth of tumors by enhancing the proliferation of cancerous cells, and their resistance to apoptotic signals and immune surveillance. One such product is biosimilars which is being focused nowadays for the treatment of ovarian cancer (Debela et al. 2021). A biosimilar is a biological product that is highly similar to an FDA-approved reference product currently on the market in terms of potency, purity, and safety. It offers better safety and efficacy at a lower cost than biologics and has no clinically significant differences from them. Follistin alfa and rFSH compounds are the two biosimilars that have been approved and are now in use (Bergandi et al. 2020). The primary concern of physicians about the use of biosimilars is immunogenicity, since even minor structural modifications to the original biologics may result in reduced efficacy and an increased incidence of adverse events. In addition, contaminants, various delivery methods, storage conditions, and patient characteristics can all cause immunogenicity (Konstantinidou et al. 2020).

This chapter aims to provide a comprehensive overview of the utilization of biosimilars in ovarian cancer management, elucidating the challenges faced in their delivery and exploring recent advancements in drug delivery systems. By integrating insights from clinical research, regulatory perspectives, and technological innovations, this discussion seeks to inform and empower stakeholders in navigating the dynamic landscape of biosimilar-based therapies for ovarian cancer.

12.2 Conventional Therapy of Ovarian Cancer and Their Drawbacks

Ovarian cancer once detected can be addressed through surgical intervention aimed at excising the cancerous tissues, taking necessary steps to avert the disease resurgence. The surgical measures exhibit efficacy in context to stage I phase of ovarian cancer, followed by chemotherapy treatment. Given the inadequacy of singular pharmaceutical agents in achieving comprehensive remission of ovarian carcinoma, the imperative lies in the pursuit of combination therapy, wherein an assortment of anti-neoplastic drugs is employed for the protracted management of ovarian malignancy (Peres et al. 2019).

12.2.1 Surgical Treatment

Surgical treatment is a comprehensive management of ovarian cancer and is a cornerstone in the treatment paradigm (Greer et al. 2004). The protocol entails cytoreductive surgical debulking and platinum-based chemotherapy. During surgery, the disease's staging and diagnostic confirmation are carried out first because, in any event, it is important to identify the tumor's histological type, including grading (Rawahi et al. 2013). The Gynaecologic Oncology group (GOG) formerly defined optimum cytoreduction as remaining tumor nodules with a maximum diameter of no more than 1 cm (Elattar et al. 2011). Achieving optimal cytoreduction is the primary goal of primary debulking surgery (PDS) for ovarian cancer. PDS entails removing both the visible metastatic lesions and the main tumor mass. PDS is an essential part of multimodal therapy regimens because studies show that maximum cytoreduction is linked to increased overall survival (Cummings et al. 2022). Following surgery, patients receive intravenous platinum/taxane regimens for six cycles every 21 days (first-line chemotherapy). Higher stages (III/IV) are inaccessible for cytoreduction because of a lesion in the liver hilum and a small bowel mesentery seizure (Cortez et al. 2018). Consequently, after performing three cycles of induction (neoadjuvant) chemotherapy and receiving a favorable response, such an inoperable tumor undergoes interval debulking surgery (IDS). To enhance results, IDS attempts to eliminate the remaining tumor burden following a partial response to chemotherapy by undergoing six cycles (Coleridge et al. 2021). Although surgery is crucial, it also entails certain risks (Tian et al. 2021). There might be complications including bleeding, infection, and harm to nearby structures. Furthermore, particularly in individuals with comorbidities, the quantity of surgery necessary for adequate cytoreduction may present difficulties (Mehta et al. 2016). Surgery for ovarian cancer, especially large-scale debulking operations, may need a protracted recuperation period after the treatment. This might lower the patient's quality of life and delay the initiation of other treatments like chemotherapy (Schorge et al. 2010). Because of the extensive metastases seen in late stages of ovarian cancer, obtaining full cytoreduction may be difficult. In such circumstances, residual illness may exist after maximal surgical effort, requiring the integration of other therapy methods (Shih and Chi 2010). In such cases, patients have to be observed regularly for the presence CA125 acting as an early symptom of reoccurrence (Charkhchi et al. 2020).

In cases when the cancer is limited to the pelvic region, meticulous staging becomes imperative in precisely identifying the level of disease present at the time of diagnosis. Considering possible routes of spread, apart from the removal of all visible tumors, essential surgical steps include peritoneal lavage, peritoneal biopsies that obtain the entire abdominal environment, and a retroperitoneal assessment including the pelvic and para-aortic regions. The overall prognosis for affected individuals might worsen as a result of suboptimal surgical staging, which can also result in erroneous categorization and insufficient postoperative treatment (Kurnit and Frumovitz 2022).

12.2.2 Chemotherapeutic Agents Used in Treatment of Ovarian Cancer

Ovarian cancer is a formidable adversary that often exhibits subtle symptoms, requiring an advanced diagnosis. Because of this, the care requires extensive and vigorous treatment techniques, with chemotherapy being a key component (Board et al. 2023). Therefore, chemotherapy intervention remains the gold standard for restoring patient survival. Cytotoxic drugs including cisplatin (CP), paclitaxel (PTX), doxorubicin (DOX), decitabine, and gemcitabine (GB), as well as their synergistic combinations, are primarily used in the treatment paradigm (Lheureux et al. 2019). A plethora of randomized trials, employing combinations of hexamethyl melamine, cyclophosphamide (CPP), methotrexate (MTX), and fluorouracil, has yielded superior endurance outcomes when contrasted with the administration of single therapeutic agents. Commonly used drugs include platinum-based compounds (e.g., cisplatin, carboplatin) and taxanes (e.g., paclitaxel), which interfere with DNA synthesis and mitosis, inducing apoptosis in cancer cells. This pursuit of optimizing treatment efficacy underscores the relentless commitment to advancing the frontiers of ovarian cancer therapeutics (Pokhriyal et al. 2019).

Chemotherapy is essential for treating advanced or recurrent ovarian cancer, where surgery may not be as practical. Combining platinum-based and taxane chemotherapy regimens has been shown to improve response rates and overall survival in a number of clinical trials, making it a standard treatment approach (Lheureux et al. 2019). Chemotherapy has shown to be an extremely effective treatment for ovarian cancer, particularly in the form of cytoreduction, that focus on lowering tumor load. When given before surgery, neoadjuvant chemotherapy can reduce tumor size, increasing the likelihood of a successful surgical excision. Following surgery, adjuvant chemotherapy is used to target any remaining cancer cells and lower the chance of recurrence (Monk et al. 2019).

The majority of chemotherapeutic drugs used to treat ovarian cancer demonstrate significant therapeutic success, but their use is limited by intrinsic disadvantages. First of all, these drugs frequently cause a variety of negative side effects that jeopardize patients' general health (Liu et al. 2021). Furthermore, their efficacy in treating ovarian cancer recurrences is limited, which poses a barrier to the long-term management of the illness (Desai et al. 2008). Another significant disadvantage is chemo-resistant, a noteworthy issue in which these drugs develop the tendency to gradually build resistance, which reduces their long-term therapeutic effect. Treatment failure may result from cancer cells developing defence mechanisms against the cytotoxic effects of chemotherapy over time. This complex interplay of efficacy and limitations underscores the ongoing challenges in optimizing chemotherapy for ovarian cancer management (Mikuła-Pietrasik et al. 2019). Women receiving chemotherapy treatment suffer from bad quality of sleep which can lead to serious depression problems (Wu et al. 2022). Moreover, chemotherapy involves the administration of cytotoxic drugs to target, where rapidly dividing cancer cells are present. But in case of ovarian cancer, chemotherapy is primarily systemic, aiming to eradicate cancer cells not only within the ovaries but also in surrounding tissues and potential metastatic sites (Tchounwou et al. 2021). This major problem with chemotherapy has been its toxicity to quickly proliferating normal cells, which can result in a variety of adverse effects (Schirrmacher 2019). In addition, many women with ovarian cancer worry about how chemotherapy would affect their ability to conceive. Chemotherapy medications have the potential to impair ovarian function and result in infertility. In cancer care, the preservation of fertility by methods such as cryopreservation of eggs or embryos is a developing field (Harada and Osuga 2019). Hematological toxicity, neurotoxicity, gastrointestinal problems, and exhaustion are also the other possible side effects found in patients. During therapy, controlling these side effects is essential preserving the patient's quality of life (Techata et al. 2022). Ongoing research endeavors persistently aim to discern novel therapeutic entities and enhance the efficacy of established treatment modalities. Furthermore, a continuum of investigations is underway to identify innovative therapeutic agents, augmenting the effectiveness of existing modalities.

12.3 Need of Biosimilars

The clinical findings of biosimilars are expected to be similar to those of the original product, notwithstanding their deviations from it in terms of purity, effectiveness, and safety. Improved access to life-saving drugs, greater treatment alternatives, and lower total treatment costs are all anticipated benefits of the biosimilars. Such advantages rely on the number of available biosimilar drugs, their development, approval, and their availability to the patients. Different types of biologics such as blood and related components, hormones, antiallergy drugs, and vaccines are used. Also, as biologics, antibodies particularly monoclonal antibodies are utilized to treat a variety of diseases, including cancer. These large complex molecules help by targeting the specific pathways of cancer growth and their progression. Nonetheless, the cost of such products is quite high and often affects their reach to the patient. In such cases, biosimilars in cancer treatment address financial obstacles by providing reasonably priced alternatives, whereas maintaining the product's therapeutic potential including efficacy and safety.

The significance of biosimilars in cancer treatment depends on their ability to target the molecular and cellular components of cancer development. By mimicking the action of standard drug/biologics, the biosimilar will help to modulate the immune responses, promote apoptosis, and reduce cell proliferation, restrain angiogenesis in cancer. The biosimilar-based drug product in cancer therapy helps in advancing personalized and precision medicine, this will help clinicians and therapists to plan tailored treatment as per individual patient needs to optimize the therapeutic outcomes. Biosimilars and their reference brand biologics differ in price by an average of 30% within each molecular type.

While individual countries and hospitals eventually decide which biosimilars are included in their local formularies (Alnahar et al. 2017). However, these potentially advantageous biosimilars face several hurdles. Some of the key challenges include immunogenicity concerns, the complex concept of interchangeability, dose, and route of biosimilar administration, limited awareness and acceptance among healthcare professionals and patients, the requirement for extensive pre-approval clinical trials, stringent regulatory frameworks that can limit potential cost savings, and even political obstacles (Farhat et al. 2018; Chopra and Lopes 2017).

12.4 Why Biosimilars in Treatment of Ovarian Cancer?

Conventional treatment modalities including surgery and chemotherapy have traditionally been employed to address ovarian cancer (Lheureux et al. 2019). However, recent advancements in the development of biosimilars—lower-cost alternatives to existing drugs—have provided additional options for treatment (Wirth et al. 2023). Despite these advancements, significant challenges remain in the battle against ovarian cancer, including high recurrence rates and the often late-stage diagnosis of the disease. A product is considered biosimilar if its physical, chemical, and biological characteristics closely reflect those of the reference product. Although the approval pathway for biosimilars differs from that of original drugs, the criteria of quality, safety, and efficacy are followed in the similar manner (Cohen et al. 2023; Manwatkar and Kumar 2023).

It is important to understand the needs and demands of the healthcare industry, particularly in the field of ovarian cancer treatment. One such need arises from the unique challenges posed by this disease, including late-stage diagnoses and high recurrence rates. To address these challenges effectively, innovative and accessible treatment options are necessary (Caeiro et al. 2022).

Biosimilars provide a plethora of potential advantages in the treatment of ovarian cancer. The affordability of these medications allows for wider patient access to high-quality therapies, reducing the economic burden associated with cancer care. This enhanced accessibility is significantly important in reference to ovarian cancer, where late-stage diagnoses often result in more complex and costly treatment regimens. Moreover, biosimilars can offer a viable alternative to expensive biologics, without compromising on efficacy or safety. Patients now have more alternatives for their treatments, and healthcare professionals are empowered to make well-informed decisions that balance patient results with budgetary considerations (Ingrasciotta et al. 2018; Advani et al. 2021).

The demand for biosimilars in ovarian cancer treatment is driven by the need for innovative and accessible treatment options. By providing affordable alternatives to costly biologics, biosimilars have the potential to improve patient accessibility to superior treatments and mitigate the financial strain linked to cancer care. Ovarian cancer is not particularly treated with any biosimilar; instead, existing biosimilars are used in conjunction with other drugs to treat the disease. Thus, it is crucial to recognize the potential benefits of biosimilars in addressing the unique challenges of ovarian cancer treatment and to actively explore opportunities of biosimilars in the development of safe and effective biosimilars.

The field of biosimilars has become a viable means of addressing the demand for ovarian cancer therapy alternatives that are both accessible and effective. This disease is characterized by late-stage diagnoses and high recurrence rates, posing unique challenges to patients and healthcare providers. However, the emergence of biosimilar technology has created a new avenue for the treatment of ovarian cancer.

Enhancing therapeutic accessibility is one of the primary benefits of biosimilars. By offering cost-effective alternatives to expensive biologics, biosimilars allow a wider range of patients to benefit from high-quality therapies (Dutta et al. 2020). When it comes to ovarian cancer, this is especially important because there aren't many treatment choices and early identification is sometimes challenging. Biosimilars offer a viable solution to bridge this gap, ensuring that more patients have access to the necessary treatments they need. Moreover, biosimilars can also alleviate the economic burden associated with ovarian cancer care. With their lower costs compared to the original biologics, biosimilars reduce the financial strain on patients and healthcare systems (Reardon 2020). This not only reduces the burden on individuals and families but also frees up resources for further research and development in the field of ovarian cancer treatment. By leveraging the cost advantage of biosimilars, healthcare providers can optimize resource allocation and potentially invest in other areas of research to advance the field (Moorkens et al. 2021). Furthermore, biosimilars expand the range of treatment options available to healthcare providers, thereby improving the outcomes for patients with ovarian cancer. The complex and diverse nature of this disease demands a personalized approach to treatment. With the availability of biosimilars, healthcare providers have the flexibility to tailor treatment plans to individual patient needs, enhancing the overall efficacy of ovarian cancer management (Kalia 2013).

12.4.1 Potential Benefits of Biosimilars

The advent of biosimilars holds promise for providing a number of advantages in the medical profession. Firstly, biosimilars have the ability to mitigate drug shortages by providing an increased choice of available agents. This means that healthcare providers have a wider range of options when it comes to treating patients, ultimately preventing shortages and ensuring that medication is readily accessible. Additionally, the lower acquisition costs associated with biosimilars can significantly increase patient access to medication. The reduced cost of these alternatives means that more individuals can afford the necessary treatments, ensuring that they receive the care they need for their medical conditions. Not only do biosimilars help patients, but they also bring about improved cost–benefit ratios that free up healthcare budgets. By making use of biosimilars, healthcare systems can allocate resources more efficiently, allowing for the allocation of funds to other vital areas within the healthcare sector. Additionally, the use of biosimilars encourages sustainability and competitiveness in the pharmaceutical sector, creating a more lively and varied market for medical goods.

12.4.2 Potential Barriers to the Use of Biosimilars

Despite the numerous advantages, some barriers hinder the widespread usage of biosimilars. One major obstacle is the limited knowledge about biosimilars among healthcare professionals. This lack of understanding can create doubts and resistance towards the use of these agents, hindering their integration into clinical practice. There is also a lack of trust in the quality and effectiveness of biosimilars, especially when compared to reference medicines and especially in treatments that have the potential to be curative. This doubt can be attributed to the complex manufacturing processes and the unique nature of biologic drugs. The lack of physician engagement in the decision-making process encompassing the availability of biosimilars in hospitals is another important obstacle. This lack of physician input can result in limited access to these alternatives, as decisions are made without considering their perspectives and expertise.

12.4.3 Recent Advances in Biosimilars for the Treatment of Ovarian Cancer

Biosimilars that target important molecules involved in the growth of ovarian cancer have been developed at a remarkable rate in recent years. These biosimilars focus on inhibiting critical molecules implicated in the uncontrolled growth of ovarian cancer cells (Dhritlahre and Saneja 2021). This advancement is significant not only for its potential economic benefits but also for its impact on expanding treatment options and addressing the diverse nature of ovarian cancer which diversifies the treatment option. Different subtypes of ovarian cancer may respond differently to various treatments, and biosimilars targeting specific molecules offer a tailored approach to intervention.

As biosimilars targeting key molecules in ovarian cancer continue to progress, ongoing research, including clinical trials and real-world evidence, contributes valuable insights into their safety and efficacy. This collective effort aims to improve patient outcomes, advance scientific understanding, and create a more sustainable healthcare landscape.

Bevacizumab, also known by its commercial name Avastin, is a humanized monoclonal antibody that is used to block the function of vascular endothelial growth factor (VEGF), a vital protein that promotes the development of new blood vessels. By inhibiting VEGF, bevacizumab aims to inhibit tumor growth and spread (Booth and Pilz 2022). Bevacizumab has a wide range of uses in oncology, proving to be beneficial against several types of cancer. It is approved for use in treating ovarian, breast, cervical, non-small cell lung, and renal cell carcinomas (Petrelli et al. 2009). Biosimilars have been developed for the biologic medicine bevacizumab, these biosimilars include Mvasi, Zirabev, Alymsys, Vegzelma, and Avzivi. Remarkably, some of these biosimilars can be used to treat ovarian cancer, especially when combined with other medications.

MVASI represents a remarkable advancement in the field of biosimilars. Specifically, MVASI is a biosimilar product of bevacizumab, also referred to as ABP215, bevacizumab-awwb, and Amgen. The USFDA has granted approval for this biosimilar product. It is important to note that MVASI is a biosimilar of the US-licensed Avastin, which is frequently utilized in the management of cancer. This approval ensures that patients and healthcare providers have increased access to a more affordable alternative to Avastin while keeping therapeutic effectiveness and safety at the same level. The development and approval of biosimilars like MVASI contribute to the overall advancement of medical treatments, providing more options for patients and improving their overall quality of life (Casak et al. 2018). The European Union (EU) has authorized MVASI for a number of indications, including the core indications already approved in the United States (US). This approval includes using it in conjunction with chemotherapy to treat patients who have metastatic, recurring, or persistent epithelial ovarian cancer. Furthermore, MVASI has been authorized for use in conjunction with erlotinib exclusively for patients with EGFR activating mutations and non-small-cell lung cancer (NSCLC), exclusive to the EU market. The European Union's approval of MVASI signifies its recognition and acceptance as an effective therapeutic option for these specific indications. By granting this approval, the EU acknowledges the potential benefits and efficacy of MVASI combined with chemotherapy for patients with ovarian cancer that is chronic, recurring, or metastatic. It is important to note that these approved indications for MVASI are specific to the European Union market and may differ from the approved indications in other regions, such as the United States. Nevertheless, recognition and authorization of MVASI in the European Union for these specific indications highlight its potential as a valuable treatment option for the people living in the country (Avastin 2018). MVASI is approved for the treatment of adult patients with their first recurrence of primary peritoneal, fallopian tube, or platinum-sensitive epithelial ovarian cancer. It can be used in combination with carboplatin and gemcitabine or with paclitaxel. This indication applies to those patients who have not yet being treated with bevacizumab or any other VEGF inhibitors or VEGF receptor-targeted agents. Such patients should be given a maximum of two cycles of chemotherapy regimens and should not have received previous therapy with bevacizumab or other VEGF inhibitors or VEGF receptor-targeted agents (Thomas et al. 2019).

ZIRABEV is a biosimilar drug of bevacizumab, categorized as a targeted therapy drug within the monoclonal antibody class. On February ninth, 2021, Pfizer granted FDA acceptance to add epithelial ovarian, fallopian tube, and primary peritoneal cancer to the list of indications for ZIRABEV® (bevacizumab-bvzr). Its mode of action includes suppression of the vascular endothelial growth factor (VEGF) protein. While healthy cells produce VEGF, cancer cells tend to over-produce it. By blocking VEGF, Zirabev impedes the formation of new blood vessels that tumors require for growth and nourishment. Consequently, it "starves" cancerous tumors, inhibiting their growth and metastasis (Chawla 2022; Zirabev 2023).

ZIRABEV is a biosimilar that has been authorized for its use in epithelial ovarian cancer. It is indicative for its use either as a combination therapy with carboplatin and paclitaxel, or ZIRABEV alone, for the patients at advanced stage (III or IV) after initial surgery. Furthermore, it is utilized in conjunction with topotecan, pegylated liposomal doxorubicin, or paclitaxel for the treatment of platinum-resistant recurrent diseases in female patients who have had no more than two chemotherapy visits. It may also be used in combination with carboplatin and paclitaxel or with carboplatin and gemcitabine, followed by ZIRABEV alone, in patients with platinum-sensitive recurrent disease (Poveda et al. 2015; Yang et al. 2021; Comment 2022). Regarding fertility, pregnancy, and breastfeeding, it should be noted that ZIRABEV has the potential to cause ovarian dysfunction and affect fertility. Its use during pregnancy is not advised due to harm to the fetus, and also if taken, pregnancy should be avoided for at least 6 months following the last dose of ZIRABEV administration. The risks associated with ZIRABEV during breastfeeding are uncertain, and it is suggested to avoid breastfeeding for women during treatment or till 6 months after the last dose of administration, due to serious adverse reactions (UCIR 2023).

ALYMSYS® (bevacizumab-maly) is an FDA-approved biosimilar of and is vascular endothelial growth factor inhibitor utilized in the United States for treating epithelial ovarian cancer in combination with paclitaxel (Amneal Pharmaceuticals, Inc 2023).

12.4.4 Pharmacological and Clinical Aspects of Biosimilars

As the word biosimilars itself suggests that something is similar to the structure and function of original and innovative biological product (Schellekens 2004). The US FDA give approval to any biosimilars that can pass the different phases of the safety and efficacy including human clinical trials to evaluate the pharmacologically resemblance and adverse reaction between reference drugs and biosimilars. There should also be a comparative efficacy trial between the two. The data obtained from the clinical trial of the biosimilars must be reviewed by the US FDA for their efficacy and safety (Calvo and Zuniga 2012).

Biosimilars have emerged as a promising approach for the treatment of various diseases, including ovarian cancer. As a biosimilar to Avastin, Avzivi is a VEGF inhibitor that can be used to treat a variety of cancers, including primary peritoneal, ovarian, epithelial, fallopian tube, cervical, renal cell, and colorectal cancers. Avzivi functions by recognizing and binding to VEGF molecules, preventing them from connecting with the Flt-1 and KDR VEGF receptors on the surface of endothelial cells. This slows the growth and progression of cancer by limiting the proliferation of endothelial cells and reducing the formation of new blood vessels (FDA 2023).

The pharmacokinetics of biosimilars for ovarian cancer should be comparable to those of the reference biologic drugs, which include their absorption, distribution, metabolism, and elimination from the body. The pharmacodynamics of biosimilars should also be similar to those of the reference biologic drugs, which includes their mechanism of action and therapeutic effect. In addition, the immunogenicity of biosimilars should be evaluated to ensure that they do not result in any clinically significant immune responses, such as the development of anti-drug antibodies (Konstantinidou et al. 2020).

12.5 Biosimilars and Pharmacovigilance

The science and practices pertaining to the identification, evaluation, comprehension, and avoidance of side effects or any other drug-related issues are referred to by the World Health Organization (WHO) as pharmacovigilance (Moore et al. 2019) (Kirchhoff et al. 2017). Biologics are used in a special case of treatment where few patients are exposed to the biologics in clinical trials due to which post-marketing surveillance (pharmacovigilance) is very critical for biosimilars. Due to the smaller number of subjects as well as less studies done in clinical trials, the number of identified adverse effects for biosimilars is also less as compared to the reference drug (Casadevall et al. 2013). When biologics and biosimilars were first approved, very few safety reports and profiles were focused sue to less no. of population involved in clinical studies which was less satisfactory. So, phase 4 (pharmacovigilance) studies are required to identify the novel signal associated with biosimilars (Oza et al. 2019).

Some key considerations in the pharmacovigilance of biosimilars are described below, and the flowchart of the pharmacovigilance of biosimilars is described in Fig. 12.1.

1. Post-Marketing Surveillance:
   (a) After approval and commercialization, continuous monitoring of biosimilars is essential to detect any unexpected adverse events or changes in the safety profile.
   (b) In order to gather and evaluate actual data about the efficacy and safety of biosimilars, strong post-marketing monitoring systems have been implemented.

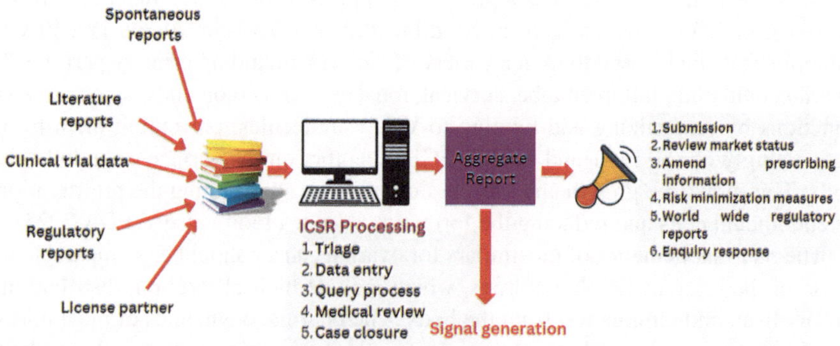

Fig. 12.1 Workflow of the pharmacovigilance process of biosimilars

2. Traceability and Identifiability:
   (a) Clear product identification and traceability are crucial in pharmacovigilance. Proper labeling and tracking systems help link specific adverse events to the correct product.
   (b) Regulatory authorities often require unique identifiers for biosimilars to distinguish them from the reference product.
3. Risk Management Plans:
   (a) Manufacturers of biosimilars are typically required to develop risk management plans (RMPs) outlining how they will monitor and manage the risks associated with their products.
   (b) RMPs may include strategies for risk minimization, additional studies, or post-authorization safety studies (PASS) to further investigate specific safety concerns.
4. Signal Detection and Management:
   (a) Signal detection involves the systematic review of safety data to identify any potential new risks or changes in the known safety profile.
   (b) Once a signal is detected, appropriate actions, such as further investigation or changes to product labeling, may be initiated.
5. Communication and Education:
   (a) Effective communication is essential for pharmacovigilance. This includes communication between healthcare professionals, regulatory agencies, patients, and the pharmaceutical industry.
   (b) Measures to educate patients and healthcare professionals on the appropriate use of biosimilars and how to report adverse occurrences may be taken.
6. Collaboration with Healthcare Providers:
   (a) Collaboration with healthcare providers is crucial for collecting accurate and timely information on adverse events.
   (b) Establishing reporting systems where health care professional can be encouraged and report any suspected adverse reactions contribute to effective pharmacovigilance.
7. Regulatory Oversight:
   (a) Regulatory agencies are essential in monitoring the safety of biosimilars. They review safety data, assess compliance with regulatory requirements, and may take regulatory actions if safety concerns arise (Grampp and Felix 2015).

12.6 Regulatory Challenges in Use of Biosimilars

The clinical behavior and molecular makeup of biosimilars might differ based on the culture conditions under which they are cultivated. This may introduce pollutants, viral contamination, and residues of DNA and cell protein. Thus, due to the complicated biological quality parameters, there are batch-to-batch variations that should be within acceptable ranges. Producers and regulators are aware that the usage of living cells to produce target molecules will result in variations in the process of production, controls, limits, and standards. These products are also less accessible to a larger population due to their high value of patients and is also a factor propelling the biosimilars market. Additionally, there is a severe lack of the appropriate analytical methods to accurately distinguish between the original therapeutic protein and the biosimilar due to the molecular structures of biosimilars being extremely complicated. Although the degree of biological and chemical identity resemblance between molecules can be demonstrated, the capacity to clearly establish that every biosimilar can precisely function as a medication with the same amount of clinical safety and efficacy since the original product is still missing. Each of these elements has the potential to significantly shorten the time needed for approval and production. The regulatory agencies require biosimilar makers to provide sufficient proof of clinical safety, and processing and packaging big molecules might provide hurdles.

12.7 Conclusion and Future Prospective

Biologics are frequently employed as therapeutic agents in cancer therapy and palliative care. However, because of their high cost, global healthcare budget problems, and the expiration of biologics patents, biosimilars have entered the market. Although biosimilars are less expensive than reference biologics, they have the same safety and toxicity profiles and aren't different in any way which makes them clinically significant. Thus, biosimilars facilitate a decrease in healthcare expenses while expanding a patient's access to innovative therapy alternatives. Biosimilars have the potential to improve access to effective and affordable cancer treatments for ovarian cancer. The pharmacological aspects of biosimilars for ovarian cancer should be carefully evaluated to ensure that they are safe and effective for use in patients and that they provide a cost-effective alternative to the reference biologic drugs. Patients and physicians may not be fully aware of the advantages and difficulties associated with using biosimilars, which might be the reason for the underutilization of biosimilars in clinical practice at the moment. Strict regulatory frameworks and diligent post-marketing monitoring of approved biosimilars are necessary to guarantee the safety and efficacy of these new drugs in real-world situations. To guarantee the effective integration of biosimilars into standard cancer treatment, healthcare providers and the general public need to be well informed on the various facets of biosimilars.

References

Advani SH, Doval DC, Koppikar SB, Vijay Anand Reddy P, Prasad SV, Bapna A, Narayanan P, Dwary AD, Krishna V, Dhanushkodi M, Joel A (2021) Use of bevacizumab in advanced ovarian cancer: consensus from an expert panel oncologists. Indian journal of. Gynecol Oncol 19:1–8

Al Rawahi T, Lopes AD, Bristow RE, Bryant A, Elattar A, Chattopadhyay S, Galaal K (2013) Surgical cytoreduction for recurrent epithelial ovarian cancer. Cochrane Database Syst Rev 2:CD008765

Alnahar S, Elliott RA, Smith MD (2017) Biosimilar uptake by British local formularies: a cross sectional study. Int J Clin Pharm 39:1055–1060

Amneal Pharmaceuticals, Inc. Amneal Launches First Biosimilar with ALYMSYS® (bevacizumab-maly) in the United States. (2023). https://investors.amneal.com/news/press-releases/press-release-details/2022/Amneal-Launches-First-Biosimilar-with-ALYMSYS-bevacizumab-maly-in-the-United-States/default.aspx. Accessed 19 Dec 2013

Bergandi L, Canosa S, Carosso AR, Paschero C, Gennarelli G, Silvagno F, Benedetto C, Revelli A (2020) Human recombinant FSH and its biosimilars: clinical efficacy, safety, and cost-effectiveness in controlled ovarian stimulation for in vitro fertilization. Pharmaceuticals 13(7):136

Booth JP, Pilz J (2022) Retrospective indication-matched cohort study of reference product and biosimilar: bevacizumab versus bevacizumab-awwb. Hosp Pharm 57(4):455–461

Caeiro C, Leão I, Oliveira I, Sousa I, André T (2022) Recurrent ovarian cancer with BRCAness phenotype: a treatment challenge. Adv Ther 39(11):5289–5299

Calvo B, Zuniga L (2012) The US approach to biosimilars: the long-awaited FDA approval pathway. BioDrugs 26:357–361

Casadevall N, Edwards IR, Felix T, Graze PR, Litten JB (2013) Strober BE Warnock DG, pharmacovigilance and biosimilars: considerations, needs and challenges. Expert Opin Biol Ther 13(7):1039–1047

Casak SJ, Lemery SJ, Chung J, Fuchs C, Schrieber SJ, Chow EC, Yuan W, Rodriguez L, Gwise T, Rowzee A, Lim S (2018) FDA's approval of the first biosimilar to bevacizumab. Clin Cancer Res 24(18):4365–4370

Charkhchi P, Cybulski C, Gronwald J, Wong FO, Narod SA, Akbari MR (2020) CA125 and ovarian cancer: a comprehensive review. Cancers 12(12):3730

Chawla K (2022) Anti VEFG agents in retinal diseases. Delhi J Ophthalmol 32(3):93–96

Cherif S, Amine A, Thies S, Taube ET, Braicu EI, Sehouli J, Kaufmann AM (2021) Prevalence of human papillomavirus detection in ovarian cancer: a meta-analysis. Eur J Clin Microbiol Infect Dis 40(9):1791–1802

Chopra R, Lopes G (2017) Improving access to cancer treatments: the role of biosimilars. J Global Oncol 3(5):596–610

Cohen HP, Turner M, McCabe D, Woollett GR (2023) Future evolution of biosimilar development by application of current science and available evidence: the developer's perspective. BioDrugs 37(5):583–593

Coleridge SL, Bryant A, Kehoe S, Morrison J (2021) Neoadjuvant chemotherapy before surgery versus surgery followed by chemotherapy for initial treatment in advanced ovarian epithelial cancer. Cochrane Database Syst Rev 7:CD005343

Comment O (2022) Relugolix take at Loading Dose: 360 mg on Day then, 120 mg once dailyis ongoing; approximately the same time each 1; day; If treatment with interrupted for followednot> 7 crush days, by or 120 resume chew. APRN and PA's Complete Guide to Prescribing Drug Therapy: 124

Cortez AJ, Tudrej P, Kujawa KA, Lisowska KM (2018) Advances in ovarian cancer therapy. Cancer Chemother Pharmacol 81:17–38

Cummings M, Nicolais O, Shahin M (2022) Surgery in advanced ovary cancer: primary versus interval cytoreduction. Diagnostics 12(4):988

Debela DT, Muzazu SG, Heraro KD, Ndalama MT, Mesele BW, Haile DC, Kitui SK, Manyazewal T (2021) New approaches and procedures for cancer treatment: current perspectives. SAGE Open Med 9:20503121211034366

Desai AG, Qazi GN, Ganju RK, El-Tamer M, Singh J, Saxena AK, Bedi YS, Taneja SC, Bhat HK (2008) Medicinal plants and cancer chemoprevention. Curr Drug Metab 9(7):581–591

Dhritlahre RK, Saneja A (2021) Recent advances in HER2-targeted delivery for cancer therapy. Drug Discov Today 26(5):1319–1329

Dutta B, Huys I, Vulto AG, Simoens S (2020) Identifying key benefits in European off-patent biologics and biosimilar markets: it is not only about price! BioDrugs 34:159–170

Elattar A, Bryant A, Winter-Roach BA, Hatem M, Naik R (2011) Optimal primary surgical treatment for advanced epithelial ovarian cancer. Cochrane Database Syst Rev 8:CD007565

Farhat F, Torres A, Park W, de Lima LG, Mudad R, Ikpeazu C, Abi AS (2018) The concept of biosimilars: from characterization to evolution—a narrative review. Oncologist 23(3):346–352

FDA (n.d.) https://www.accessdata.fda.gov/drugsatfda_docs/label/2023/761198s000lbl.pdf. Assessed 20 Dec 2023

Genentech, Inc (2018) Avastin (bevacizumab) prescribing information. Genentech, Inc, CA, USA

Grampp G, Felix T (2015) Pharmacovigilance considerations for biosimilars in the USA. BioDrugs 29:309–321

Greer BE, Swensen RE, Gray HJ (2004) Surgery for ovarian cancer: rationale and guidelines. J Natl Compr Cancer Netw 2(6):561–568

Harada M, Osuga Y (2019) Fertility preservation for female cancer patients. Int J Clin Oncol 24:28–33

Harris HR, Terry KL (2016) Polycystic ovary syndrome and risk of endometrial, ovarian, and breast cancer: a systematic review. Fertil Res Pract 2(1):1–9

Ingrasciotta Y, Cutroneo PM, Marcianò I, Giezen T, Atzeni F, Trifirò G (2018) Safety of biologics, including biosimilars: perspectives on current status and future direction. Drug Saf 41:1013–1022

Kalia M (2013) Personalized oncology: recent advances and future challenges. Metabolism 62:S11–S14

Kirchhoff CF, Wang XZM, Conlon HD, Anderson S, Ryan AM, Bose A (2017) Biosimilars: key regulatory considerations and similarity assessment tools. Biotechnol Bioeng 114(12):2696–2705

Konstantinidou S, Papaspiliou A, Kokkotou E (2020) Current and future roles of biosimilars in oncology practice. Oncol Lett 19(1):45–51

Koshiyama M, Matsumura N, Konishi I (2014) Recent concepts of ovarian carcinogenesis: type I and type II. Biomed Res Int 2014:934261

Kurnit KC, Frumovitz M (2022) Primary mucinous ovarian cancer: options for surgery and chemotherapy. International journal of gynecologic. Cancer 32(11):ijgc-2022-003806

La Vecchia C (2017) Ovarian cancer. Eur J Cancer Prev 26(1):55–62

Lheureux S, Braunstein M, Oza AM (2019) Epithelial ovarian cancer: evolution of management in the era of precision medicine. CA Cancer J Clin 69(4):280–304

Liu YQ, Wang XL, He DH, Cheng YX (2021) Protection against chemotherapy-and radiotherapy-induced side effects: a review based on the mechanisms and therapeutic opportunities of phytochemicals. Phytomedicine 80:153402

Manwatkar S, Kumar B (2023) Biosimilars: promising and rapidly emerging biotherapeutics. In: Novel Technologies in Biosystems, Biomedical & Drug Delivery. Springer Nature Singapore, Singapore, pp 45–67

Mara JN, Zhou LT, Larmore M, Johnson B, Ayiku R, Amargant F, Pritchard MT, Duncan FE (2020) Ovulation and ovarian wound healing are impaired with advanced reproductive age. Aging (Albany NY) 12(10):9686

Mehta SS, Gelli M, Agarwal D, Goéré D (2016) Complications of cytoreductive surgery and HIPEC in the treatment of peritoneal metastases. Indian J Surg Oncol 7:225–229

Mikuła-Pietrasik J, Witucka A, Pakuła M, Uruski P, Begier-Krasińska B, Niklas A, Tykarski A, Książek K (2019) Comprehensive review on how platinum-and taxane-based chemotherapy of ovarian cancer affects biology of normal cells. Cell Mol Life Sci 76:681–697

Momenimovahed Z, Tiznobaik A, Taheri S, Salehiniya H (2019) Ovarian cancer in the world: epidemiology and risk factors. Int J Women's Health 30:287–299

Monk BJ, Randall LM, Grisham RN (2019) The evolving landscape of chemotherapy in newly diagnosed advanced epithelial ovarian cancer. Am Soc Clin Oncol Educ Book 39:e141–e151

Moore N, Berdai D, Blin P, Droz C (2019) Pharmacovigilance–the next chapter. Therapies 74(6):557–567

Moorkens E, Vulto AG, Huys I (2021) Biosimilars in Belgium: a proposal for a more competitive market. Acta Clin Belg 76(6):441–452

Oza B, Radhakrishna S, Pipalava P, Jose V (2019) Pharmacovigilance of biosimilars–why is it different from generics and innovator biologics? J Postgrad Med 65(4):227
PDQ Adult Treatment Editorial Board (2023) Ovarian epithelial, fallopian tube, and primary peritoneal cancer treatment (PDQ®). In: PDQ cancer information summaries. National Cancer Institute (US)
Peres LC, Cushing-Haugen KL, Köbel M, Harris HR, Berchuck A, Rossing MA, Schildkraut JM, Doherty JA (2019) Invasive epithelial ovarian cancer survival by histotype and disease stage. JNCI J Natl Cancer Inst 111(1):60–68
Petrelli NJ, Winer EP, Brahmer J, Dubey S, Smith S, Thomas C, Vahdat LT, Obel J, Vogelzang N, Markman M, Sweetenham JW (2009) Clinical cancer advances 2009: major research advances in cancer treatment, prevention, and screening-a report from the American Society of Clinical Oncology. J Clin Oncol 27(35):6052
Pokhriyal R, Hariprasad R, Kumar L, Hariprasad G (2019) Chemotherapy resistance in advanced ovarian cancer patients. Biomarkers in cancer 11:1179299X19860815
Poveda AM, Selle F, Hilpert F, Reuss A, Savarese A, Vergote I, Witteveen P, Bamias A, Scotto N, Mitchell L, Pujade-Lauraine E (2015) Bevacizumab combined with weekly paclitaxel, pegylated liposomal doxorubicin, or topotecan in platinum-resistant recurrent ovarian cancer: analysis by chemotherapy cohort of the randomized phase III AURELIA trial. J Clin Oncol 33(32):3836–3838
Reardon G (2020) Pharmacoeconomics of biologic medicines and biosimilars. In: Biologics, biosimilars, and biobetters: an introduction for pharmacists, physicians, and other health practitioners, pp 195–212
Schellekens H (2004) How similar do 'biosimilars' need to be? Nat Biotechnol 22(11):1357–1359
Schirrmacher V (2019) From chemotherapy to biological therapy: a review of novel concepts to reduce the side effects of systemic cancer treatment. Int J Oncol 54(2):407–419
Schorge JO, McCann C, Del Carmen MG (2010) Surgical debulking of ovarian cancer: what difference does it make? Rev Obstet Gynecol 3(3):111
Shih KK, Chi DS (2010) Maximal cytoreductive effort in epithelial ovarian cancer surgery. J Gynecol Oncol 21(2):75–80
Tchounwou PB, Dasari S, Noubissi FK, Ray P, Kumar S (2021) Advances in our understanding of the molecular mechanisms of action of cisplatin in cancer therapy. J Exp Pharmacol 13:303–328
Techata A, Muangmool T, Wongpakaran N, Charoenkwan K (2022) Effect of cancer stage on health-related quality of life of patients with epithelial ovarian cancer. J Obstet Gynaecol 42(1):139–145
Thomas M, Thatcher N, Goldschmidt J, Ohe Y, McBride HJ, Hanes V (2019) Totality of evidence in the development of ABP 215, an approved bevacizumab biosimilar. Immunotherapy 11(15):1337–1351
Tian H, Ma D, Tan X, Yan W, Wu X, He C, Zhong L, Zhang Y, Yu B, Zhang Y, Qi X (2021) Platinum and taxane based adjuvant and neoadjuvant chemotherapy in early triple-negative breast cancer: a narrative review. Front Pharmacol 12:770663
UCIR (2023) Bevacizumab-BVZR (Zirabev) Drug information. https://www.ucir.org/immunotherapy-drugs/bevacizumab-bvzr. Accessed 15 Dec 2023
Wirth K, Boes S, Näpflin M, Huber C, Blozik E (2023) Initial prescriptions and medication switches of biological products: an analysis of prescription pathways and determinants in the Swiss healthcare setting. BMJ Open 13(11):e077454
Wu HJ, Chuang CM, Chien CH, Wang TJ, Liang SY (2022) Changes in depression and sleep quality and associated factors in women receiving chemotherapy for ovarian cancer: an observational study. Cancer Nurs 45(4):271–279
Yang J, Liu R, Ektare V, Stephens J, Shelbaya A (2021) Does biosimilar bevacizumab offer affordable treatment options for cancer patients in the USA? A budget impact analysis from US commercial and medicare payer perspectives. Appl Health Econ Health Policy 19:605–618

Yoshikawa H, Jimbo H, Okada S, Matsumoto K, Onda T, Yasugi T, Taketani Y (2000) Prevalence of endometriosis in ovarian cancer. Gynecol Obstet Investig 50(Suppl. 1):11–17

Zirabev (2023)Taking ZIRABEV™ (bevacizumab-bvzr). https://www.zirabev.com/takingzirabev, Accessed 15 Dec 2023

Biosimilars in Blood Cancer

13

Gauri Pathak, Aryaa Nigade, Dishank Purandare,
Vaishnavi Thorat, and Shvetank Bhatt

Abstract

Biosimilars are the medicines that are similar in structure and functions to biologics. They are very much similar to their reference products—this means the product. The biosimilars have shown great efficacy in the treatment of cancers including blood cancer. These biosimilars are more effective and preferred for the treatment of blood cancer over traditional chemotherapeutic agents due to their effectiveness and less side effects. Trastuzumab, rituximab, filgrastim and pegfilgrastim are the biosimilars that are approved recently for the treatment of various types of blood cancer. This chapter discussess various FDA-approved biosimilars in blood cancers and regulatory considerations and future directions of biosimilars for the treatment of blood cancers.

Keywords

Biosimilars · FDA · Rituximab · Lymphoma · Tratuzumab

13.1 Introduction

In 2024, cancer is one of the leading causes of death all around the word. It is a disease that is characterized by uncontrolled division of cells and their multiplication. It involves a myriad of gene mutations, which results in the emergence of tumour and hence development of cancer. A diverse variety of treatment options have been planned for the treatment of the same, but none of it has been proved to cure it 100%. There are always conditions where if a patient has been cleared of

G. Pathak · A. Nigade · D. Purandare · V. Thorat · S. Bhatt (✉)
School of Health Sciences and Technology, Dr. Vishwanath Karad MIT World Peace University, Pune, Maharashtra, India

S. Bhatt et al. (eds.), *Biosimilars for Cancer Treatment*,
https://doi.org/10.1007/978-981-97-3001-8_13

cancer, many years later it may still relapse. Cancer cells are such kind of cells that hijack the nutrients of normal cells. They are capable of escaping from the immune system of the body (Upadhyay 2021). Cancer is a deadly disease and is considered as the main causes of mortality throughout the world. There are more than a hundred types of cancer (Takabe and Benesch 2022), of which blood cancer is a type of cancer in which liquid tumours form in the body fluids depending upon its progression in the body, while other types of cancer have solid tumours in the affected organ (Jiang et al. 2022). Blood cancer is a type of cancer that affects the production and function of blood cells, e.g. by preventing them from fighting infection or severe bleeding, through uncontrolled growth of abnormal blood cell types. Available treatments include chemotherapy, immunotherapy, radiotherapy and blood transfusion or transplantation (Gill and June 2015), but the mortality rate in this type of cancer is high due to the aggressive course, bone metastasis (invasion of primary tumours into the bones) and other factors, which include high doses or frequent administrations of chemotherapeutic agents to achieve the desired therapeutic effect, which can lead to severe side effects (Jiang et al. 2014). The likelihood of relapse is a higher risk due to the low efficacy of the cancer drugs administered and the resulting low survival rate. In the majority of cases, the treatment plan includes surgery, chemotherapy, radiation, hormonal therapy, immunotherapy, blood transfusion or transplantation, which is solely dependent on the severity and type of the tumour either benign or malignant.

Cancer can exist as carcinoma, leukaemia, sarcoma and lymphoma, while three main types of blood cancer include: leukaemia, lymphoma and myeloma. Leukaemia is a cancer that occurs in the bone marrow and blood and leads to increased production of abnormal white blood cells, affecting their ability to fight off infection (Mullighan 2012); lymphoma affects the lymphatic system and leads to the formation of lymphomas (abnormal lymphocytes) that rapidly divide and accumulate in lymph nodes and other tissues; and finally, multiple myeloma, a cancer of the plasma cells, i.e. the white blood cells. This is a cancer of the plasma cells, the white blood cells responsible for producing antibodies to fight infection and disease in the body. In this disease, the plasma cells prevent the production of antibodies, resulting in a weakened immune system and increased susceptibility to infections (Palumbo and Anderson 2011). Leukaemia is the cancer of blood. It occurs when healthy normal blood cells begin to multiply rapidly (Zhang et al. 2023). It is subclassified into 4 types: acute lymphoblastic leukaemia, chronic lymphocytic leukaemia, acute myeloid leukaemia and chronic myeloid leukaemia. Acute myeloid leukaemia is the most predominant type of blood cancer in individuals. Symptoms are generally immensely vague and may seem like as that of normal infections such as fever, fatigue, easy bruising, profuse bleeding and loss in weight (Davis et al. 2014). The incidence of lymphomas and leukaemia is more prevalent in developed regions such as that in the United States. Leukemia is more common in males than in females, but the gender gap decreases after a particular age. In 2018, it was the 15th commonly diagnosed cancer and the 11th most common cause of mortality due to cancer. 29% of all childhood cancers are interpreted as leukaemia (Bispo et al. 2020). The occurrence of leukemia increases with age with approximately an average of 1.3 individuals per

100,000 individuals who are below 65 years of age and an average of 12.2 per 100,000 individuals above 65 years of age. Thus, haematological malignancies contribute to global tumour burden as one of the leading causes of death. Genetics factors play a predominant role in the etiology of cancer. Various factors including the Down syndrome, Klinefelter syndrome, Neurofibromatosis, Fanconi anaemia, Bloom Syndrome may be concluded as the risk factors of cancer. Viral infections like Epstein Barr, human T lymphotropic virus also play a role in the development of cancer. Other risk factors involve exposure to radiation for a prolonged period, nuclear bombing, persistent smoking and exposure to benzene (Chennamadhavuni et al. 2024). Treatment of leukaemia involves the use of monoclonal antibodies and stem cell transplantation. Proven evidence suggests that avoiding smoking and increasing the number of fruits and vegetables in the daily diet can reduce the rate of occurrence of cancer (Ames et al. 1995). Biosimilars are a new wave of therapeutics that have revolutionized in the treatment of various diseases, including blood cancer. Biological pharmaceuticals, or biologics, are a novel class of medicines that are derived (Adami et al. 2019) from living things or their components and provide specialized treatments for a wide range of illnesses. In the present book chapter, we are going to discuss the potential of biosimilars in the treatment of blood cancer.

13.2 Introduction to Biologics and Biosimilars

Biosmilars are biological products that are very similar to an existing FDA-approved reference product and have no clinically meaningful differences from it (Kaida-Yip et al. 2018). Biosimilars for several important biologic drugs used in the treatment of blood cancers are now available. The use of biosimilars in the treatment of blood cancer has several potential advantages. They offer a more cost-effective treatment option and make them more accessible to patients. The rigorous approval process for biosimilars ensures that their safety and efficacy are comparable to that of the reference biological medicine (Bachu et al. 2022). Biosimilars are synthetic versions of biological medicines, which are complex compounds produced on a huge scale inside living cells, as opposed to traditional medications. Since it is difficult to duplicate biologics (Kaida-Yip et al. 2018) precisely due to their intrinsic complexity, biosimilars are designed to closely resemble the original product. Thorough testing is used to confirm their resemblance and show that they are safe and effective in comparison to the reference biologic. Biosimilars, in contrast to generic medications (van de Vooren et al. 2015), require express professional permission before being substituted for the original biologic.

However, the use of biosimilars is also associated with challenges. There are concerns regarding the interchangeability of biosimilars and their biological reference medicinal products as well as possible immunogenicity. Biosimilars represent a significant advance in the treatment of blood cancer. They offer the potential for more affordable and accessible treatment options without compromising safety or efficacy. As our understanding and use of biosimilars evolves, they will play an increasingly important role in the field of oncology.

This broad range of therapeutic agents includes gene treatments, monoclonal antibodies (Pham-Huy et al. 2021), vaccinations (Adams 2009) and other bioproducts with biological origins. Biologics are distinct from their traditional small-molecule equivalents because of their complex chemical structures and exact mechanisms of action and depend on live biological systems for manufacture. For example, biosimilar versions are now available for rituximab, a monoclonal antibody used to treat non-Hodgkin's lymphoma and chronic lymphocytic leukaemia (Brown et al. 2018). Trastuzumab biosimilar is used for the treatment of HER2-positive breast cancer and gastric cancer.

Biologics represent a revolutionary class of drugs made from life forms or their components, signalling highly focused and creative treatment approaches for a wide range of illnesses (Martins et al. 2022). Due to their structural complexity, accuracy and reliance on biological systems for manufacture, these diverse therapeutic agents—which include monoclonal antibodies, vaccines, gene treatments and other biological derivatives—stand in sharp contrast to traditional small molecule medications. The development of biologics has transformed medical treatments, providing effective treatment for ailments that were previously difficult to treat. These therapies frequently target particular molecules or physiological pathways, allowing for a more individualized and accurate approach to the treatment of illness. As a result, biologics have shown great promise in the treatment of several autoimmune diseases (Drosos et al. 2021), cancer, rheumatoid arthritis (Curtis and Singh 2011) and chronic illnesses (Diaz-Borjon et al. 2006).

Like biological replicas, biosimilars precisely replicate licensed original biologics, also referred to as reference products (García et al. 2020); (Kaida-Yip et al. 2018). Strict procedures are followed during its development to guarantee safety and efficacy comparable to the reference biologic. The emergence of biosimilars has signalled a paradigm shift in the pharmaceutical industry by providing patients with more affordable treatment options (Mysler et al. 2021) and boosting competition in the biologics market. Because biological molecules are complex, biosimilars entail a more involved replication process than their small-molecule counterparts. Before approving a biosimilar, regulatory agencies thoroughly inspect it to make sure that certain guidelines are followed.

The emergence of biosimilars has the potential to increase competition in the biopharmaceutical market, which will promote affordability and increase patient access to innovative treatments. Although biosimilars mimic the main mechanisms and therapeutic advantages of reference biologics, they provide medical professionals and patients with an extended therapeutic repertoire and may result in significant cost savings in the healthcare environment. This development has the potential to completely alter therapy paradigms by providing a combination of affordability, accessibility and efficacy in the field of advanced pharmaceuticals.

In the world of medicine, biologics and biosimilars are two different groups, each having special traits and regulations (Chan and Chan 2017). Bioengineering techniques such as recombinant DNA technology are used to manufacture complex molecules called biologics, which are derived from live creatures or cells. Their huge size and structural complexity allow them to include a wide range of therapeutic substances, such as proteins, antibodies, hormones and nucleic acids. Biosimilars, on the other hand, are made to closely mimic biologics that have previously received approval—also referred to as reference products—in terms of composition, effectiveness, safety and quality. Although they aim for resemblance, biosimilars cannot be exact replicas of their reference counterparts due to intrinsic biological variability.

Comprehensive clinical trials to prove safety and efficacy are required for regulatory clearance of biologics (Rahalkar et al. 2021), and they are supervised by strict regulatory agencies such as the FDA and EMA. To verify the product's safety and efficacy before market authorization, these trials include preclinical investigations as well as several stages of clinical testing. A comprehensive comparative analytical study, encompassing structural and functional characterization, pharmacokinetic and pharmacodynamic assessments and clinical trials where required, is the regulatory pathway that biosimilars must take to prove their similarity to the reference product (Wang and Chow 2012). Regulatory bodies provide both biologics and biosimilars non-proprietary names; however, biosimilars frequently include a distinctive suffix to denote that they are biosimilar goods. When it comes to competition and market access, biologics usually have a window of exclusivity before biosimilars may join the market, usually when patents or other exclusivity rights expire (Moorkens et al. 2016). The gradual release of biosimilars may encourage competition, which might lower costs and broaden access to these treatment choices.

As of now, biosimilars have shown promise as effective and safe alternatives to reference biologics for the treatment of blood cancers, with the added benefit of perhaps lower treatment costs. Biosimilars, such as Zarxio and Nivestim, which increase the formation of white blood cells and aid in the management of chemotherapy-induced neutropenia in patients with leukaemia or lymphoma, are examples of filgrastim biosimilars used in this context (Joshi et al. 2022). In contrast to the reference medication Neulasta, pegfilgrastim biosimilars like Fulphila and Udenyca decrease the risk of febrile neutropenia in patients receiving myelosuppressive chemotherapy for blood cancers. Truxima and Ruxience are two examples of rituximab biosimilars that target B-cell malignancies, equivalent in effectiveness and safety to the reference product and include non-Hodgkin lymphoma and chronic lymphocytic leukaemia (Humphreys et al. 2022). Darbepoetin alfa biosimilars, such as Nesp, also promote the synthesis of red blood cells, which helps cure anaemia brought on by chemotherapy and blood malignancies such as myelodysplastic syndromes. All things considered, biosimilars are vital to the expansion of therapy choices for blood malignancies since they guarantee patients' access to necessary medications while upholding strict safety and effectiveness criteria (Grossi et al. 2007).

13.3 FDA Approved Biosimilars in Blood Cancer

The emergence of FDA-approved biosimilars represents a significant advancement in the treatment landscape for blood cancer (Nahleh et al. 2022). Biosimilars are biologic drugs that are highly similar to their reference products, offering comparable efficacy and safety profiles (Hung et al. 2017). In recent years, several biosimilars have gained FDA approval for various blood cancer indications, including leukaemia, lymphoma and multiple myeloma. These biosimilars provide healthcare providers and patients with more treatment options, fostering competition in the market and potentially lowering treatment costs (Farhat et al. 2018). Notably, biosimilars undergo rigorous evaluation by the FDA to ensure they meet stringent standards for similarity, quality, purity and potency compared to their reference products (Kaida-Yip et al. 2018). This thorough evaluation process assures healthcare professionals and patients of the safety and efficacy of biosimilars. Moreover, the availability of FDA-approved biosimilars enhances patient access to essential therapies, particularly in cases where the reference biologics may be costly or in short supply. By expanding treatment choices and improving affordability, biosimilars contribute to advancing patient care and outcomes in the management of blood cancer. Additionally, the introduction of biosimilars fosters innovation and competition within the pharmaceutical industry, potentially driving further advancements in treatment modalities and healthcare delivery. As biosimilars continue to gain acceptance and utilization in clinical practice, ongoing research and real-world evidence will further elucidate their role in optimizing treatment strategies and improving patient outcomes in the ever-evolving landscape of blood cancer management (Ngo and Chen 2021). In conclusion, FDA-approved biosimilars represent a pivotal development in the treatment armamentarium for blood cancer, offering comparable efficacy, safety and affordability compared to their reference biologics. Through stringent regulatory evaluation and increased market competition, biosimilars enhance patient access to essential therapies and drive innovation in the field, ultimately improving the quality of care and outcomes for individuals affected by blood cancer.

The approval of biosimilars by the U.S. Food and Drug Administration (FDA) has transformed the landscape of blood cancer treatment, providing patients with more accessible and cost-effective options while upholding stringent standards of safety and efficacy (Sochacka-Ćwikła et al. 2021). Biosimilars are biologic products highly similar to an already FDA-approved biologic, known as the reference product, with no clinically meaningful differences in terms of safety, purity and potency. In the realm of blood cancer, several FDA-approved biosimilars have emerged, targeting critical therapeutic areas such as leukaemia, lymphoma and multiple myeloma. The evolvement of biosimilars was based on the only main reason; reduction of price. The US Food and Drug administration (FDA) has to approve these biological products prior to the marketing of the same. Blood cancer, stomach cancer, breast cancer and all other types of cancers can be treated with biosimilars. The following biosimilars are approved in oncology for the treatment of cancer and supportive care namely; Rituximab, Trastuzumab & Bevacizumab, and supportive agents including filgrastim, pegfilgrastim, epoetin α & epoetin.

The first biosimilar, Somatotropin, was approved in 2006 by the European Medical Agency (EMA). (Mascarenhas-Melo et al. 2024). As of now, the US-FDA has approved 29 biosimilars; out of which 17 of them are related to oncology. (Nahleh et al. 2022). Out of these 17, only 7 have been marketed so far (Zhai et al. 2019). The clinical trials performed for the following FDA approved biosimilars showcased absolutely no notable difference in their safety, efficacy and their absorption, distribution, metabolism, excretion when compared with their reference products. A study conducted in the United States specified that almost 70% of the participants did not have any prior knowledge regarding biosimilars (Nahleh et al.

2022) Erythropoietin, a hormone released by the kidneys, promotes the formation of red blood cells and if it fails to do so, then laboratory-based erythropoietin can be utilized. Epoetin alpha-epbx; a biosimilar for the similar condition, is marketed in the United States. Roferon-A, a biosimilar of interferon, is approved for its use in the United States for the treatment of renal carcinoma, T-cell lymphoma, hairy cell leukemia and other melanomas and lymphomas (Joshi et al. 2022).

One notable FDA-approved biosimilar is rituximab, a monoclonal antibody used in the treatment of various B-cell malignancies, including non-Hodgkin lymphoma (NHL) and chronic lymphocytic leukaemia (CLL). Rituximab biosimilars have undergone rigorous evaluation to demonstrate their similarity to the reference product, ensuring comparable efficacy and safety profiles. These biosimilars offer patients and healthcare providers a cost-effective alternative to the reference product, potentially expanding access to life-saving therapies for those affected by blood cancer.

Trastuzumab (McKeage and Perry 2002) is another critical biologic used in the treatment of HER2-positive breast cancer, which can also metastasize to the blood and bone marrow. FDA-approved trastuzumab biosimilars provide patients with HER2-positive blood cancers, such as leukaemia and lymphoma, access to a more affordable treatment option without compromising effectiveness or safety. These biosimilars have undergone extensive clinical trials to establish their similarity to the reference product, ensuring confidence among healthcare professionals and patients alike.

Filgrastim (Rastogi et al. 2021), a granulocyte colony-stimulating factor (G-CSF) used to stimulate the production of white blood cells, is essential in the management of patients undergoing chemotherapy for various blood cancers. FDA-approved filgrastim biosimilars offer a cost-effective solution for managing chemotherapy-induced neutropenia, reducing the risk of infections and treatment delays in patients with blood cancer. Through robust clinical evaluation, these biosimilars have demonstrated equivalence to the reference product, providing reassurance regarding their safety and efficacy. Epoetins are used for the treatment of chemotherapy-induced anemia (CIA). Treatment with epoetins reduce the need for transfusion of blood and also improve the quality of life of patient. Bevacizumab (Avastin®) is a recombinant humanized monoclonal antibody, that acts on vascular endothelial growth factor (VEGF-A) and hampers angiogenesis and thereby the growth of new tumors.

The approval process for biosimilars by the FDA involves comprehensive comparative analytical studies, preclinical evaluations and clinical trials to establish similarity to the reference product. Additionally, post-marketing surveillance ensures ongoing monitoring of safety and effectiveness, further enhancing confidence in the use of biosimilars in clinical practice. Healthcare professionals play a crucial role in evaluating the suitability of biosimilars for individual patients, considering factors such as disease status, treatment history and cost considerations (Jacobs et al. 2017).

The availability of FDA-approved biosimilars for the treatment of blood cancer represents a significant advancement in oncology care, offering patients increased access to essential therapies while potentially reducing healthcare costs. By fostering competition in the biologics market, biosimilars drive innovation and affordability, ultimately benefiting patients, healthcare providers and the healthcare system as a whole. As research and development continue to expand the repertoire of FDA-approved biosimilars, the future holds promise for further improving outcomes and quality of life for individuals affected by blood cancer.

13.4 Regulatory Considerations

Regulatory considerations are basically a set of rules or standards applicable in the respective country you are operating in, including the exchange of data with govern security practices. There can be additional rules and regulations depending upon the country. There exist regulatory authorities of countries such as Food and Drug Administration (FDA),European Medicines Agency (EMA), and World Health Organization (WHO), which are responsible to set the standard rules according to which similarities should exist between the biosimilars and the reference products (Joshi et al. 2022). It is their job to perform comparative studies, which help us analyse the quality, efficacy and safety of the biosimilar in oncology.

The first regulatory authority that provided standards and rules for biosimilars was the EMA in 2003, and the approval for usage of these products was granted in 2006. In 2009, FDA gave specific guidelines for the approval of biosimilars, and since 2012, many other countries such as India, South Korea, Japan, Canada and Australia have formatted their own guidelines which are correlating with EMA guidelines (Bennett et al. 2014).

For Europe, the EMA states that based on the preclinical and clinical data which compared safety and efficacy with respect to reference biological products, there should exist a biological product under EMA for 10 years at least, and on the patent expiry, biosimilars can be approved. Their guidelines are classified depending upon the different classes of biosimilars in oncology such as epoetin and filgrastim. It addresses toxicology, pharmacology, pharmacodynamics, pharmacokinetics, manufacturing, etc. Continuous observations are conducted by post-marketing and risk management studies. Similar to EMA, WHO has also developed a regulatory framework for approving similar biotherapeutic products, which evaluates study clinical and non-clinical measures as compared to the original biopharmaceutic. This comparison is done with respect to their dosage terms, route of exposure and a license and must be granted based on complete documentation.

The United States sets up standards by Biologics Price Competition Act (BPCIA) for biosimilars approval, for which the reference product must be in market for at least 12 years (Cutroneo et al. 2014). It performs evaluations based on preclinical and clinical trial data, and these guidelines also states requirements for evaluations and analysis of its safety, efficacy and immunogenicity in human clinical trials. Scientific data can also be discussed in collaborative meetings of EMA and FDA, which proves to be beneficial for FDA for developing guidelines with better accuracy (Renwick et al. 2016).

After approval of around 20 biosimilars, India developed its own regulatory guidelines for approving further biosimilars in 2012, Guidelines on Similar Biologics that stated the requirements for authorization of marketing biosimilars in oncology (Malhotra 2011). Similar to other countries, their guidelines also focus on preclinical and clinical study data that ensures safety and efficacy of the biosimilars.

South Korea was the second country to develop its regulatory guidelines after Europe. The patent period is for 6 years after which licensing of biosimilars can be approved. Japan has Pharmaceutical and Medical Devices Agency (PDMA) to deal with the approval process of biosimilars. Unlike other countries, Japan does not require clinical trials as a compulsion for approving biosimilars.

Australia also adopted the EMA guidelines and formed the Australian Therapeutic Goods Administration (TGA). Some of its contents were inspired from the ICH as well and they provide only 5 years of data protection for novel biologics (Ahmed et al. 2012). Health Canada is the regulatory authority for Canada, and it approves biosimilars solely on the basis of similarity between biosimilars and the reference used. They also adopted few points from the ICH. Many other countries have their own developed regulatory guidelines and are constantly working on improving them for more accurate and effective biosimilars in the market.

13.5 Future Directions

Biosimilars are widely developing so that they can be used instead of biologics. Since biologics exhibit high cost for their use in treatment of various diseases, biosimilars emerge as an alternative option with the main objective of having reduced cost of treatment. The US FDA has approved 29 such biosimilars. Most of the biologics having their use in multiple diseases do not have their existing counterpart which only increases the prevalence of biosimilars in the near future.(Kaida-Yip et al. 2018). Use of these products due to their variability can find its purpose in a wide range of diseases ranging from acute to chronic conditions.

The main reason of developing biosimilars is the nearing expiration of patents on biologics that compels pharmaceutical industries to focus more on the research and development of such biological medicines (Roger 2010). A survey conducted in the United States came to a conclusion where out of 150 physicians, a majority of them would be in favor of prescribing biosimilars if only they were cheaper by 25% (Park et al. 2020).

A change in the manufacturing sites and the processes and the raw materials that may occur while manufacturing biosimilars have to be kept in check so that there is no overruling of the guidelines set by the International Council for Harmonization (ICH), i.e. the ICH guideline Q5E. Future advances in the field of analytical development will allow the extensive and thorough study of many complex biosimilars. Innovations in biosimilars can also give rise to new novel methods of manufacturing and processing. Scientists and developers should make sure that the biosimilars developed should not lack in any way where it may be in their efficacy, safety and quality (Kim et al. 2020). Since the term biosimilars is still a novel and not a well-known one by many; the US FDA did not have a specific set of regulatory guidelines for the same which resulted in the delay for their approval and in turn a diminished percentage of their application in the United States compared to that of Europe as a whole (Bachu et al. 2022). The interchangeability between the biosimilars and reference drugs has shown consistent positive results regarding their safety and efficacy which paves a way for the use of these products clinically (Uhlig and Goll 2017).

A leap in the development of biosimilar monoclonal antibodies has also been observed hence enhancing the technological advances. Thus, educating the population on such biological agents and their use for increasing their confidence in the therapeutic applications of biosimilars is extensively important considering it as the future of medicine.

13.6 Conclusion

Biosimilar covers major market for the effective treatment of various blood cancer subtypes. The use of biosimilars are evolving very fast in last decade. A number of biosimilars are approved in recent years, and so many are in clinical trials. The acceptance of these biosimilar drugs are more as compared to generic drugs because of their safety and effectiveness. These biosimilars increase the overall survival of patients with blood cancer as well as improve the quality of life of these patients.

References

Adami G et al (2019) Balancing benefits and risks in the era of biologics. Ther Adv Musculoskelet Dis 11:1759720X19883973. https://doi.org/10.1177/1759720X19883973

Adams KT (2009) Biologic vaccines: a panacea for infectious diseases? Biotechnol Healthc 6(1):12–20

Ahmed I, Kaspar B, Sharma U (2012) Biosimilars: impact of biologic product life cycle and European experience on the regulatory trajectory in the United States. Clin Ther 34(2):400–419. https://doi.org/10.1016/j.clinthera.2011.12.005

Ames BN, Gold LS, Willett WC (1995) The causes and prevention of cancer. Proc Natl Acad Sci USA 92(12):5258–5265. https://doi.org/10.1073/pnas.92.12.5258

Bachu RD et al (2022) Oncology biosimilars: New developments and future directions. *s* 5(11):e1720. https://doi.org/10.1002/cnr2.1720

Bennett CL et al (2014) Regulatory and clinical considerations for biosimilar oncology drugs. *Lancet Oncol* 15(13):e594–e605. https://doi.org/10.1016/S1470-2045(14)70365-1

Bispo JAB, Pinheiro PS, Kobetz EK (2020) Epidemiology and etiology of leukemia and lymphoma. Cold Spring Harb Perspect Med 10(6):a034819. https://doi.org/10.1101/cshperspect.a034819

Brown JR et al (2018) The role of rituximab in chronic lymphocytic leukemia treatment and the potential utility of Biosimilars. Oncologist 23(3):288–296. https://doi.org/10.1634/theoncologist.2017-0150

Chan JCN, Chan ATC (2017) Biologics and biosimilars: what, why and how? ESMO Open 2(1):e000180. https://doi.org/10.1136/esmoopen-2017-000180

Chennamadhavuni A et al (2024) Leukemia, Treasure Island (FL)

Curtis JR, Singh JA (2011) Use of biologics in rheumatoid arthritis: current and emerging paradigms of care. Clin Ther 33(6):679–707. https://doi.org/10.1016/j.clinthera.2011.05.044

Cutroneo PM et al (2014) Safety profile of biological medicines as compared with non-biologicals: an analysis of the italian spontaneous reporting system database. Drug Saf 37(11):961–970. https://doi.org/10.1007/s40264-014-0224-1

Davis AS, Viera AJ, Mead MD (2014) Leukemia: an overview for primary care. Am Fam Physician 89(9):731–738

Diaz-Borjon A, Weyand CM, Goronzy JJ (2006) Treatment of chronic inflammatory diseases with biologic agents: opportunities and risks for the elderly. Exp Gerontol 41(12):1250–1255. https://doi.org/10.1016/j.exger.2006.10.010

Drosos AA et al (2021) Biologic therapies and autoimmune phenomena. Mediterr J Rheumatol 32(2):96–103. https://doi.org/10.31138/mjr.32.2.96

Farhat F et al (2018) The concept of Biosimilars: from characterization to evolution-a narrative review. Oncologist 23(3):346–352. https://doi.org/10.1634/theoncologist.2017-0126

García JJ, Raez LE, Rosas D (2020) A narrative review of biosimilars: a continued journey from the scientific evidence to practice implementation. Transl Lung Cancer Res 9(5):2113–2119. https://doi.org/10.21037/tlcr-20-601

Gill S, June CH (2015) Going viral: chimeric antigen receptor T-cell therapy for hematological malignancies. Immunol Rev 263(1):68–89. https://doi.org/10.1111/imr.12243

Grossi A, Balestri F, Santini S (2007) Darbepoetin alpha in the treatment of cancer chemotherapy-induced anemia. Ther Clin Risk Manag 3(2):269–275. https://doi.org/10.2147/tcrm.2007.3.2.269

Humphreys SZ, Geller RB, Walden P (2022) Pegfilgrastim Biosimilars in US supportive oncology: a narrative review of administration options and economic considerations to maximize patient benefit. Oncol Therapy 10(2):351–361. https://doi.org/10.1007/s40487-022-00207-2

Hung A, Vu Q, Mostovoy L (2017) A systematic review of U.S. biosimilar approvals: what evidence does the FDA require and how are manufacturers responding? J Manag Care Spec Pharm 23(12):1234–1244. https://doi.org/10.18553/jmcp.2017.23.12.1234

Jacobs I et al (2017) Biosimilars for the Treatment of Cancer: A Systematic Review of Published Evidence. *Bio Drugs* 31(1):1–36. https://doi.org/10.1007/s40259-016-0207-0

Jiang T et al (2014) Poly aspartic acid peptide-linked PLGA based nanoscale particles: potential for bone-targeting drug delivery applications. Int J Pharm 475(1–2):547–557. https://doi.org/10.1016/j.ijpharm.2014.08.067

Jiang Y, Lin W, Zhu L (2022) Targeted drug delivery for the treatment of blood cancers. *Molecules* 27(4):1310. https://doi.org/10.3390/molecules27041310

Joshi D et al (2022) Biosimilars in oncology: latest trends and regulatory status. Pharmaceutics 14(12):2721. https://doi.org/10.3390/pharmaceutics14122721

Kaida-Yip F et al (2018) Biosimilars: Review of current applications, obstacles, and their future in medicine. World J Clin Cases 6:161–166. https://doi.org/10.12998/wjcc.v6.i8.161

Kim H et al (2020) The future of Biosimilars: maximizing benefits across immune-mediated inflammatory diseases. Drugs 80(2):99–113. https://doi.org/10.1007/s40265-020-01256-5

Malhotra H (2011) Biosimilars and non-innovator biotherapeutics in India: an overview of the current situation. Biologicals 39(5):321–324. https://doi.org/10.1016/j.biologicals.2011.06.018

Martins AC et al (2022) Trends and perspectives of biological drug approvals by the FDA: a review from 2015 to 2021. Biomedicines 10(9):2325. https://doi.org/10.3390/biomedicines10092325

Mascarenhas-Melo F et al (2024) An overview of Biosimilars—development, quality, regulatory issues, and Management in Healthcare. Pharmaceuticals 17(2):235. https://doi.org/10.3390/ph17020235

McKeage K, Perry CM (2002) Trastuzumab: a review of its use in the treatment of metastatic breast cancer overexpressing HER2. Drugs 62(1):209–243. https://doi.org/10.2165/00003495-200262010-00008

Moorkens E et al (2016) Overcoming barriers to the market access of Biosimilars in the European Union: the case of biosimilar monoclonal antibodies. Front Pharmacol 7:193. https://doi.org/10.3389/fphar.2016.00193

Mullighan CG (2012) Molecular genetics of B-precursor acute lymphoblastic leukemia. J Clin Invest 122(10):3407–3415. https://doi.org/10.1172/JCI61203

Mysler E et al (2021) Biosimilar-to-biosimilar switching: what is the rationale and current experience? Drugs 81(16):1859–1879. https://doi.org/10.1007/s40265-021-01610-1

Nahleh Z et al (2022) Use of biosimilar medications in oncology. JCO Oncol Pract 18(3):177–186. https://doi.org/10.1200/OP.21.00771
Ngo D, Chen J (2021) A clinical review of Biosimilars approved in oncology. Ann Pharmacother 55(3):362–377. https://doi.org/10.1177/1060028020944596
Palumbo A, Anderson K (2011) Multiple myeloma. N Engl J Med 364(11):1046–1060. https://doi.org/10.1056/NEJMra1011442
Park SH et al (2020) Biosimilars: concept, current status, and future perspectives in inflammatory bowel diseases. Intest Res 18(1):34–44. https://doi.org/10.5217/ir.2019.09147
Pham-Huy A et al (2021) The use and impact of monoclonal antibody biologics during pregnancy. CMAJ 193(29):E1129–E1136. https://doi.org/10.1503/cmaj.202391
Rahalkar H et al (2021) Current regulatory requirements for Biosimilars in six member countries of BRICS-TM: Challenges and Opportunities. *Front Med* 8:726660. https://doi.org/10.3389/fmed.2021.726660
Rastogi S et al (2021) Efficacy and safety of Filgrastim and its Biosimilars to prevent febrile neutropenia in cancer patients: a prospective study and meta-analysis. Biology 10(10):1069. https://doi.org/10.3390/biology10101069
Renwick MJ et al (2016) Postmarket policy considerations for biosimilar oncology drugs. Lancet Oncol 17(1):e31–e38. https://doi.org/10.1016/S1470-2045(15)00381-2
Roger SD (2010) Biosimilars: current status and future directions. Expert Opin Biol Ther 10(7):1011–1018. https://doi.org/10.1517/14712591003796553
Sochacka-Ćwikła A, Mączyński M, Regiec A (2021) FDA-approved drugs for hematological malignancies-the last decade review. Cancers 14(1):87. https://doi.org/10.3390/cancers14010087
Takabe K, Benesch MGK (2022) Types of Cancer and Research Covered in *World Journal of Oncology*. *World J Oncol* 13:325–328. https://doi.org/10.14740/wjon1558
Uhlig T, Goll GL (2017) Reviewing the evidence for biosimilars: key insights, lessons learned and future horizons. Rheumatology 56(suppl_4):iv49–iv62. https://doi.org/10.1093/rheumatology/kex276
Upadhyay A (2021) Cancer: an unknown territory; rethinking before going ahead. Genes Dis 8(5):655–661. https://doi.org/10.1016/j.gendis.2020.09.002
van de Vooren K, Curto A, Garattini L (2015) Biosimilar versus generic drugs: same but different? Appl Health Econ Health Policy 13:125–127. https://doi.org/10.1007/s40258-015-0154-9
Wang J, Chow S-C (2012) On the regulatory approval pathway of biosimilar products. *Pharmaceuticals* 5(4):353–368. https://doi.org/10.3390/ph5040353
Zhai MZ, Sarpatwari A, Kesselheim AS (2019) Why are Biosimilars not living up to their promise in the US? AMA J Ethics 21(8):E668–E678. https://doi.org/10.1001/amajethics.2019.668
Zhang N et al (2023) Global burden of hematologic malignancies and evolution patterns over the past 30 years. Blood Cancer J 13(1):82. https://doi.org/10.1038/s41408-023-00853-3

Biosimilars in Clinical Trials

14

Pallavi M. Chaudhari, Pranav Shah, Harita Desai,
Neha Raghuvanshi, and Bhagwat Patil

Abstract

Clinical trials impose validation of safety and efficacy, and biosimilars have contributed large share to the pipeline of clinical trials. These biosimilars are the therapeutic alternatives to the reference biologic drugs, developed through rigorous regulatory requirements. When a commercial biologic loses patent protection, there are augmented opportunities for launch of biosimilars. On the basis of clinical trials, a biosimilar can meet the requirements of approval. Once the safety and efficacy are proved, FDA approves the biosimilar via an abbreviated licensure pathway implemented under the Biologics Price Competition and Innovation Act (BPCIA) and licensed under section 351(k) of the Public Health Service (PHS) Act. As of now, most of the biosimilars are used for treatment of cancers, diabetes, autoimmune diseases, etc. Eventually, more of the biosimilar drugs are approved by means of clinical studies that result in cost-effective drugs for patients. Thus, due to the biosimilars treatments are more tailored, with novel therapies that can be developed, but those seems to be challenging. So, the sponsors/investigators of biosimilars need to understand these challenges and refer the FDA guidelines to facilitate and expedite approval. Thus, biosimilars present a potential for new plays in the industry that can act as a game changer for essential and critical drugs. The presented chapter acknowledges about biosimilar

P. M. Chaudhari (✉) · B. Patil
Department of Pharmaceutics, Dr. D. Y. Patil College of Pharmacy, Akurdi, Pune, India
e-mail: pallavichaudhari@dyppharmaakurdi.ac.in

P. Shah
Maliba Pharmacy College, UkaTarsadia University, Surat, Gujarat, India
e-mail: pranav.shah@utu.ac.in

H. Desai · N. Raghuvanshi
Department of Pharmaceutics, Bombay College of Pharmacy, Santacruz East, Mumbai, India
e-mail: harita.desai@bcp.edu.in; neha.raghuvanshi@bcp.edu.in

S. Bhatt et al. (eds.), *Biosimilars for Cancer Treatment*,
https://doi.org/10.1007/978-981-97-3001-8_14

development, studies involved in clinical trials for biosimilars, considerations for clinical trials of biosimilars, regulatory complexities and challenges of biosimilars, challenges, and licensing of biosimilars with potential of biosimilars.

Keywords

Biosimilars · Clinical trials · Study designs · Regulatory considerationsl · Licensing and potential of biosimilars.

14.1 Biologics

The term biologics or moreover known as natural items are medications that are made from living beings (plant, creature cell, or microorganism), by the utilization of complex fabricating forms. These are basically used for the treatment of severe and life-threatening diseases. Examples of biologics include—vaccines, gene therapies, stem cell therapy, growth factors, tissue transplant, monoclonal antibodies, and recombinant proteins (Kaida-Yip et al. 2018).

The difference between biologics and traditional drugs is in terms of their source, complex structures, and manufacturing processes and regulatory considerations. Thus, biologics can impose higher price that in turn deny access to some patients.

14.2 Biosimilars

As the title states, biosimilars are fundamentally the biologic that is comparative and has no clinical contrast from an existing biologic medicine (moreover known as reference thing) that is approved by the US Food and Drug Administration (FDA). Thus, these biosimilars improve access for the patients and those who need treatment (García et al. 2020).

As per U.S. Food and Drug Administration (FDA), biosimilars are defined as as "a biological product that is highly similar to and has no clinically meaningful differences from an existing FDA-approved reference product" when considering safety and potency. In 2009, Biologics Price Competition and Innovation (BPCI) Act was enacted for regulation toward approval of biosimilars. The purpose of this act was to access the medicines, increase treatment options, and promote competition for low healthcare costs (Koyfman 2013).

14.3 Biosimilar Development

14.3.1 Scientific Research and Target Identification

During the initial phase, researchers conduct in-depth scientific research to identify a suitable target biologic for biosimilar development. This involves a comprehensive review of literature to understand the therapeutic landscape and identify biologics with significant clinical relevance. Advanced analytical techniques, such as mass spectrometry, nuclear magnetic resonance (NMR) spectroscopy, and X-ray crystallography, are employed to characterize the structure, composition, and functional attributes of the target biologic. This detailed characterization helps in elucidating the critical quality attributes (CQAs) of the reference product.

Computational modeling and bioinformatics tools may also be utilized to predict and analyze the structural and functional properties of the target biologic. These tools facilitate the identification of potential differences between the biosimilar candidate and the reference product, guiding subsequent development strategies (Cohen 2019).

14.3.2 Analytical Characterization and Comparative Assessment

Analytical characterization involves a comprehensive assessment of the biosimilar candidate to establish its similarity to the reference product. This includes evaluating physicochemical properties, such as molecular weight, size, charge, and glycosylation patterns, using sophisticated analytical techniques.

Comparative studies are conducted to compare the biosimilar candidate directly with the reference product. These studies assess similarities and differences in structural attributes, such as primary sequence, higher order structure, posttranslational modifications, and biological activity. Bioassays and functional assays are performed to compare the biological activity, potency, and efficacy of the biosimilar candidate with the reference product. These assays provide critical data on the similarity in biological function and therapeutic effect (Mirasol 2023).

14.3.3 Preclinical Testing

Preclinical studies are manufactured to evaluate the pharmacokinetics, pharmacodynamics, and safety profile of the biosimilar candidate in relevant animal models. These studies aim to identify potential toxicities, immunogenicity, and target-mediated effects. Toxicology studies assess the safety of the biosimilar candidate by investigating its effects on various organ systems, potential for immunogenicity, and overall toxicity profile. These studies follow regulatory guidelines and help establish safe starting doses for clinical trials.

Pharmacokinetic (PK) and pharmacodynamic (PD) studies evaluate the absorption, distribution, metabolism, and excretion (ADME) of the biosimilar candidate and its pharmacological effects in animal models. These studies provide valuable data for dose selection, regimen optimization, and prediction of human exposure (Bui et al. 2015).

14.3.4 Clinical Trials

Clinical development of biosimilars involves a stepwise approach, starting with Phase 1 trials in healthy volunteers and progressing to Phase 2 and Phase 3 trials in patients with the target disease or condition.

Phase 1 trials focus on assessing the safety, tolerability, PK, and PD profiles of the biosimilar candidate. These studies provide initial insights into dose ranging, formulation optimization, and potential adverse effects.

Phase 2 trials further assess the security and viability of the biosimilar candidate in patients, typically using a randomized, controlled design. These studies aim to establish preliminary evidence of efficacy and refine dosing regimens for subsequent Phase 3 trials.

Phase 3 trials are pivotal studies designed to demonstrate the comparative efficacy, safety, and immunogenicity of the biosimilar candidate compared to the reference product. These large-scale trials are rigorously designed to meet regulatory requirements and provide robust evidence of biosimilarity (Alten and Cronstein 2015).

14.3.5 Regulatory Approval

Regulatory approval of biosimilars involves submission of a comprehensive regulatory dossier to health authorities, such as the FDA or EMA. The dossier includes detailed data on the quality, safety, and efficacy of the biosimilar candidate, along with comparative analytical and clinical data demonstrating similarity to the reference product.

Regulatory agencies conduct a thorough review of the submitted data to assess the biosimilar candidate's quality, safety, and efficacy. This review involves evaluating the totality of evidence, including analytical, preclinical, and clinical data, to determine whether the biosimilar meets the required standards for approval.

Regulatory approval may be granted based on an assessment of demonstrating the biosimilarity. A biosimilar is similar to a generic product; there is no difference in safety, efficacy, or immunogenicity.

14.3.6 Market Entry and Post-Marketing Surveillance

Upon regulatory approval, the biosimilar can be launched into the market, providing patients with access to a more affordable treatment option. Post-marketing surveillance activities, including pharmacovigilance and risk management, are essential to monitor the safety and effectiveness of the biosimilar in real-world clinical practice.

Pharmacovigilance activities involve ongoing monitoring of adverse events, product complaints, and safety-related information associated with the biosimilar. These activities aim to detect and evaluate any potential safety signals or emerging risks associated with the biosimilar.

Risk management plans may be implemented to mitigate identified risks and ensure the safe use of the biosimilar. These plans include strategies for monitoring and managing known or potential risks, as well as communication and education initiatives for healthcare professionals and patients (Liese et al. 2022).

14.4 Studies Involved in Clinical Trials for Biosimilars (Huneycutt et al. 2015)

- Human PK/PD studies
- Clinical immunogenicity studies

14.4.1 Human PK/PD Studies

1. *Pharmacokinetics (PK) Study*: Pharmacokinetic studies are pivotal in elucidating the intricate journey of biosimilars within the human body, encompassing a multidimensional exploration of absorption, distribution, metabolism, and excretion (ADME) kinetics. These studies are meticulously designed, incorporating considerations such as drug formulation, physicochemical properties, and biological variability to ensure robust data collection. The advent of advanced analytical methodologies, including high-resolution mass spectrometry, liquid chromatography-tandem mass spectrometry (LC—MS/MS), and hybrid immunoaffinity techniques, empowers researchers to quantify minute drug concentrations with unparalleled sensitivity and specificity. Moreover, the integration of novel sampling strategies, such as microdialysis and microsampling techniques, enables real-time monitoring of drug disposition kinetics at localized tissue sites, shedding light on the spatial and temporal dynamics of drug distribution and elimination. Furthermore, the application of physiologically-based pharmacokinetic (PBPK) modeling, coupled with in vitro–in vivo extrapolation (IVIVE) approaches, allows for the prediction of complex drug behaviors in diverse patient populations, guiding dose optimization strategies and enhancing therapeutic outcomes. Through the synergy of cutting-edge technologies and innovative methodologies, pharmacokinetic studies serve as the bedrock of biosimilar development, facilitating the translation of preclinical insights into clinical practice and paving the way for the delivery of safe, efficacious, and patient-centric therapeutic interventions.
2. *Pharmacodynamics (PD) Study*: Pharmacodynamic studies represent the pinnacle of biosimilar characterization, offering a profound insight into the intricate interplay between drug molecules and biological systems. These studies transcend traditional paradigms, delving deep into the molecular mechanisms underpinning drug action, cellular signaling cascades, and physiological responses elicited by biosimilars. Biomarkers, the molecular beacons of biological processes, emerge as indispensable tools in pharmacodynamic investigations, providing real-time insights into drug–target interactions, downstream signaling

events, and therapeutic efficacy. Furthermore, the advent of cutting-edge technologies, including omics-based approaches such as genomics, proteomics, and metabolomics, revolutionizes the landscape of pharmacodynamic research, enabling the comprehensive profiling of drug-induced changes in cellular pathways and biological networks. Through the integration of pharmacodynamic data with systems biology approaches and computational modeling, the intricate tapestry of drug action is meticulously deciphered, unlocking novel insights into disease pathogenesis, therapeutic mechanisms, and drug response variability. Moreover, pharmacodynamic studies play a pivotal role in driving precision medicine initiatives, facilitating the development of personalized therapeutic strategies tailored to individual patient needs, and ushering in a new era of transformative breakthroughs in patient care and clinical practice.

3. *PK/PD Modeling and Simulation*: PK/PD modeling and simulation emerge as the linchpin of biosimilar development, heralding a paradigm shift in drug discovery and development paradigms. These computational methodologies harness the power of mathematical modeling, systems biology, and machine learning algorithms to unravel the hidden complexities of drug action, predict clinical outcomes, and optimize therapeutic interventions with unprecedented precision and efficiency. Complex pharmacokinetic-pharmacodynamic models, ranging from mechanistic frameworks based on physiological principles to empirical models derived from clinical data, offer a comprehensive platform for understanding the dynamics of drug exposure and response across diverse patient populations. Moreover, the integration of quantitative systems pharmacology (QSP) approaches with mechanistic modeling techniques enables the elucidation of complex drug–disease interactions, guiding the identification of novel therapeutic targets and predictive biomarkers of treatment response. Furthermore, the advent of in silico modeling and simulation techniques enables researchers to conduct virtual experiments, explore hypothetical scenarios, and accelerate the drug development process with unparalleled speed and cost-effectiveness. Through the seamless integration of PK/PD modeling and simulation with experimental data and clinical observations, biosimilar developers gain unprecedented insights into drug behavior, therapeutic efficacy, and patient outcomes, driving innovation, and excellence in biosimilar development and advancing the frontiers of modern medicine.
4. *Regulatory Considerations*: Regulatory considerations represent a cornerstone in development and approval of biosimilar drugs, with regulatory agencies serving as custodians of patient safety, efficacy, and quality. These agencies provide meticulous guidelines and stringent requirements for conducting PK/PD studies, ensuring the comparability of biosimilars to their reference products in terms of pharmacokinetic and pharmacodynamic profiles. Bridging studies, the linchpin of biosimilar development, play a pivotal role in establishing the similarity between the biosimilar and the reference product through rigorous comparative assessments of their pharmacokinetic and pharmacodynamic parameters. Moreover, regulatory agencies leverage advanced statistical methodologies,

bioinformatics tools, and data-driven approaches to scrutinize the robustness, reliability, and reproducibility of PK/PD data, safeguarding the integrity of the regulatory process and upholding the highest standards of scientific rigor and transparency. Through collaborative interactions, regulatory agencies, biosimilar developers, and academic researchers foster a culture of innovation, continuous improvement, and excellence in biosimilar development, ensuring the timely providing safe, effective, and efficient healthcare to patients around the world.

14.4.2 Clinical Immunogenicity Studies

14.4.2.1 Purpose and Importance

Clinical immunogenicity studies are fundamental for evaluating how the human immune system responds to pharmaceutical biosimilars. The primary purpose is to assess the potential of the biosimilar to induce immune reactions, such as the production of ADAs and NAbs, in patients. These studies are vital because they provide critical insights into the biosimilar's safety, efficacy, and overall therapeutic profile. Immunogenicity assessments help identify and quantify immune responses that could impact the biosimilar's effectiveness or lead to adverse reactions. Understanding the immunogenicity profile of the biosimilar is crucial for assessing its clinical risk–benefit profile and ensuring patient safety during treatment.

14.4.2.2 Study Design

Designing clinical immunogenicity studies requires meticulous planning to ensure the validity and reliability of the data collected. Several key considerations include selecting an appropriate patient population representative of the intended use of the biosimilar, determining the sampling schedule to capture potential immune responses over time, and choosing sensitive and specific assays for detecting ADAs and NAbs. The study design may incorporate various factors such as treatment duration, dosing regimen, and concomitant medications. Additionally, it is essential to consider the potential impact of patient-related factors, such as age, gender, ethnicity, and underlying diseases, on immunogenicity.

14.4.2.3 Data Analysis and Interpretation

Analyzing data from clinical immunogenicity studies involves robust statistical methods and careful interpretation to derive meaningful conclusions. Quantitative analysis focuses on determining the incidence, magnitude, and persistence of ADAs and NAbs in treated patients. Statistical techniques such as regression analysis, survival analysis, or mixed-effects modeling may be employed to explore relationships between immunogenicity and clinical outcomes. Furthermore, factors influencing immunogenicity, such as patient demographics, treatment regimen, and disease characteristics, are evaluated to identify potential risk factors. Longitudinal data analysis allows for the assessment of immunogenicity trends over time and their impact on treatment efficacy and safety.

14.5 Considerations for Clinical Trials of Biosimilars

- Endpoints
- Study population
- Biological product characteristics
- Sample size
- Duration of study
- Study design
- Statistical analysis
- *As per definition of Biosimilar given by FDA* "The biologic product is highly similar to the reference product not withstanding minor differences in clinically inactive components and that there are no clinically meaningful differences between the biologic product and the reference product in terms of safety, purity, and potency of the product," (Huneycutt et al. 2015). As interpreted from this definition, biosimilars are not exactly identical to the reference product, and any differences in potency and purity between the biosimilar and reference product must not have a clinically significant impact (Alten and Cronstein 2015).

14.5.1 What Distinguishes Biosimilars from the Reference Product?

- Owing to a high level of complexity of biologicals and biosimilars as compared to small-molecule pharmaceuticals, biologics and biosimilars require a more difficult manufacturing procedure (FDA n.d.). Since there is a close bond between the manufacturers manufacturing procedures usually maintain proprietary information (FDA n.d.; Scheinberg and Kay 2012; Mellstedt et al. 2008a) in line with the biologic turing process and the features of the end product (Scheinberg and Kay 2012). Producing an identical replica of a reference product is therefore not feasible with today's technology. Even slight changes to the source materials always result in the biological molecule's chemical structure, as well as its biological actions and breakdown products (FDA n.d.).

14.5.2 Clinical Development of Biosimilars

- To develop a biosimilar, a variety of data that compares the biosimilar *candidate* to the reference product is required (Lee et al. 2012). A series of structural analyses and functional assays comparative in nature are required along with determination of antibody-dependent cellular cytotoxicity (ADCC). A step-by-step hierarchical process is generated that starts with a thorough characterization of the reference product and the suggested biosimilar. This study forms the basis for demonstrating biosimilarity. Depending upon the stringency of evaluation, the demonstration of a similar structure and function is made, which makes it necessary for conducting highly precise study involving nonclinical in vivo testing and clinical research. Based on all available data from all phases of research, biosimilarity is determined (Stebbing et al. 2020; European Medicines Agency 2014a; US Food and Drug Administration 2015; WHO 2009). During the biosimilar development process, diverse data that compares the candidate biosimilar and the reference product is needed. This is produced through the aforementioned step-by-step hierarchical process (WHO 2017; Gabioline n.d.; Serra-Matencio et al. 2016; Stebbing et al. 2020; European Medicines Agency 2014b).

- The studies on nonclinical in vivo testing and clinical research are required. The quality of this first stage lays the groundwork for a biosimilarity demonstration. The tougher this evaluation is in demonstrating comparable structure and function. Every piece of evidence from every step of development is considered for determining biosimilarity (WHO 2017; Gabioline n.d.; Serra-Matencio et al. 2016; Stebbing et al. 2020; European Medicines Agency 2014b).

14.5.3 Comparative Clinical Trials for the Creation of Biosimilars

14.5.3.1 Study Population

- There is no requirement to establish the therapeutic benefit of reference product through the clinical efficacy study, as it is independently verified. The objective of the study is to demonstrate that biosimilar and reference products are clinically equivalent based on prespecified data. For this, there is a need that the study population and end point are sensitive enough to demonstrate product-related differences without being influenced by factors dependent on patient or diseases. A study population with a demonstrated robust therapeutic effect of the reference product is a sensitive study population (US Food and Drug Administration 2015; Stebbing et al. 2020; European Medicines Agency 2012, 2014c). In such cases, it would be essential to undertake a study with clinical efficacious end points quantifying pharmacological activity in a consort of homogenous patient population.

14.5.3.2 End Points Selection

- Together with an assessment of immunogenicity, these investigations ought also contain a thorough safety analysis. The chosen end point may not match the one used in pivotal studies to show the reference product's efficacy. For *demonstration* of clinical benefits of registration trials taken for novel anticancer therapeutics, there is a requirement for the use of end points such as survival without disease, survival with no progression of disease, complete survival, surrogate end points that are short term, and specifically determined overall response rate. A study design demonstrating equivalency is generally recommended so that statistical assessment can be made indicating a comparison between biosimilar and reference product with respect to effectivity (European Medicines Agency 2014b; US Food and Drug Administration 2015; Stebbing et al. 2020).
- As per regulatory guidelines, a comparative program involving a clinical study involving one therapeutic indication can establish biosimilarity, which further can be useful for approval of additional indications of a biosimilar (European

Medicines Agency 2014b; US Food and Drug Administration 2015; Stebbing et al. 2020). From an ethical, financial, and scientific standpoint, research on biosimilars should not aim to duplicate the reference product's safety and effectiveness information for every indication (Serra-Matencio et al. 2016). Extrapolation involves consideration of the clinical, nonclinical, and analytical data supporting biosimilarity, and it must also be supported by science (European Medicines Agency 2014b; US Food and Drug Administration 2015; Stebbing et al. 2020). For instance, if the active substance's mechanism of action (MOA) contains multiple receptors or binding sites, the contribution of each receptor or binding site may make extrapolation difficult.

- It is crucial that prescribers and doctors understand the unique role of clinical studies *which* are comparative in nature and the role they play in the development of biosimilar paradigm because they are accustomed to assessing new medications based on clinical studies (Stebbing et al. 2020).

14.5.4 Comparative Clinical Effectiveness Trials of Adalimumab, A Biosimilar

- There is a requirement to conduct phase I research with healthy volunteers to prove that the pharmacokinetic (PK) characteristics of the product that is branded and the biosimilar product are comparable. In order to further establish the biosimilar nature of the biosimilar and branded products in terms of safety and effectiveness, a clinical efficacy trial is frequently necessary (Krendyukov and Schiestl 2018; European Medicines Agency 2012; Curigliano et al. 2016).
- While the design of effectiveness trials for biosimilars seems reasonably easy and uncomplicated at first, there are several design elements that need to be carefully considered below typical objectives. Regulators have stated that the trial goals is to establish comparable clinical efficacy between biosimilar and original medicine rather than establishing the clinical effectiveness of biosimilar. This is based on the supposition that the branded product has clearly shown safe nature and effectiveness in a prior development program.
- Effect sizes that are consistent across research are typically used for this. Comparative end point study uses both branded and biosimilar products, that makes it essential for randomised, parallel group double blind comparative clinical trials by use of effective end points in non-inferiority design (European Medicines Agency 2014c).
- We illustrate design elements using adalimumab as an illustration. Some of the elements include the therapeutic indications, transition design (transition from the originator to the biosimilar product), background therapy, blinding, target patient population stratification, and alternative statistical considerations.
- Herewith we discuss design of phase III clinical efficacious trials that have achieved compliance from European Medicines Agency (EMA) and/or Food and Drug Administration (FDA) biosimilar guidelines (European Medicines Agency 2014c).

- Prescription medicine sales of adalimumab (Humira) are the highest worldwide. For the treatment of various inflammatory diseases, including psoriatic arthritis (PsA), plaque psoriasis (PsO), rheumatoid arthritis (RA), ankylosing spondylitis (AS), Crohn's disease (CD), and ulcerative colitis (UC), this biological tumor necrosis factor (TNF) inhibitor has been approved for sale in over 87 countries (Christl et al. 2017; Lai and La Noce 2016).
- Many pharmaceutical companies are creating biosimilar copies of adalimumab because it is likely for the patent to expire in the coming years.
- Recently, the first biosimilar adalimumab was authorized for marketing in India under the brand Exemptia.6. Nevertheless, neither the EU nor the USA have yet authorized Exemptia. In order to receive guidelines for obtaining clearance in both the EU and the United States, Zydus Cadila, the developer of Exemptia, has arranged talks with European and US authorities for 2015 (Food and Drug Administration 2015).
- Currently, seven ADL biosimilars have received approval in the EU and/or the USA, namely PF-06410293, ABP 501, BI 695501, SB5, GP2017, MSB11022, FKB327, and PF-06410293. These biosimilars have all been shown to be safe and effective alternatives to the licensed reference product (RP) (Curigliano et al. 2016).
- The FDA and EMA authorized ABP 501 as the first ADL biosimilar in 2016 and 2017, respectively. The function and PK of ABP 501 and ADL were shown to be comparable in the preclinical investigation and phase I clinical trial. The structural similarity, which encompassed general characteristics, carbohydrate structure, primary and higher-order structure, purity and impurities, isoelectric profile, and thermally-forced degradation profiles, was also validated (European Medicines Agency 2014c; Castañeda-Hernández et al. 2014; FDA 2008).The safety, effectiveness, and immunogenicity equivalence between ABP 501 and ADL were assessed in 526 patients with moderate-to-severe rheumatoid arthritis who were receiving methotrexate (MTX) in a subsequent randomized, double-blind, phase III equivalency study (EMA 2015).

14.5.4.1 Cyltezo, BI 695501

- In 2017, the FDA and EMA approved BI 695501. Its structural, functional, and PK similarities to ADL have been validated by earlier studies (Lee 2014; Lu et al. 2021).*Cohen* et al. carried out a 58-week, randomized, double-blind, parallel-arm study to show that BI 695501 and ADL were clinically equivalent. All 635 participants in this trial with moderate-to-severe RA were receiving stable MTX (15–25 mg/week baseline therapy for at least 12 weeks prior to enrollment; 10–14 mg/week was also allowed if patients were intolerant to higher dose). Patients were randomized 1:1 to receive 40 mg subcutaneously of either ADL (n = 321) or BI 695501 (n = 324) once every two weeks for a total of twenty-four weeks. Patients were rerandomized at week 24 to either remain on ADL or switch to BI 695501 (ADL/ADL, BI695501/BI695501, ADL/ADL, and ADL/BI695501). The ACR20 responses at week 12 (as requested by the FDA) and week 24 (as requested by the EMA), based on FAS, were co-primary efficacy endpoints

14.5.4.2 Hadlima/Imraldi SB5

- After being shown to have a structure, function, and PK similar to ADL, SB5 was approved by the FDA and EMA in 2017 and 2019, respectively (Landewe and Van Der Heijde 2012; Velayudhan et al. 2016). The phase III, randomized, double-blind, parallel group research by Weinblatt et al. compared the effectiveness, safety, and immunogenicity of SB5 with ADL in 544 patients with moderate to severe RA who were receiving MTX therapy.

14.5.4.3 GP2017 (Halimatoz, Hefya, and Hyrimoz)

- In 2018, the FDA and EMA approved GP2017. Prior research has shown that GP2017 and Humira share the same amino acid sequence, are functionally and *pharmacologically* equivalent to the reference medication, have identical secondary and tertiary structures, and have the same degree of posttranslational changes (Kang et al. 2021; Mellstedt et al. 2008b). A phase III trial led by Wiland et al. involved 353 patients with moderate-to-severe RA who were not responding well to disease-modifying anti-rheumatic medications. Patients were receiving biologics and MTX at a steady dose. Patients in the ADL group were moved to get GP 2017 up until week 46 at week 24 (Liu et al. 2016; Kaur et al. 2017).

14.5.4.4 MSB11022 (Idacio)

- In 2019, the EMA authorized MSB11022. The same amino acid sequence, N- and C-terminal modifications, the relative distribution of the intact and glycated forms of both the light chain and heavy chain, and the C-terminal lysine truncation *of* a heavy chain level of MBS11022 are among the structural and functional similarities between the drug and the reference drug that preclinical studies have demonstrated. MSB11022 displayed lower levels of oxidation than ADL. Additionally, there was no variation in the high-order structure (Cohen et al. 2017; Cohen et al. 2019a; Cohen et al. 2019b; Grozdanova et al. 2016).

14.5.4.5 Hulio's FKB327

- The FDA and EMA approved FKB327 in 2020 and 2018, respectively. Its structural and functional similarities to ADL have been established through a preclinical experiment. Similarities in PK and safety have been shown in a follow-up phase I trial using healthy participants (Batran et al. 2022). In an open-label extension of their phase III, randomized, double-blind study, Genovese et al. enrolled 730 active patients with rheumatoid arthritis who were at least 18 years old and taking MTX at a stable dose. Patients in this trial were randomized 1:1 to receive either FKB327 or RP for 24 weeks in period.

14.5.4.6 PF-06410293 (Amsparity, Abrillada)

- The FDA and EMA approved PF-06410293 in 2019 and 2020, respectively. The amino acid sequences of PF-06410293 and ADL were shown to be similar by peptide *mapping* analysis. Previous trials have demonstrated that the efficacy, safety, tolerance, and immunogenicity profiles of PF-06410293 are comparable to those of ADL. Fleischmann et al. carried out a double-blind, randomized research to compare the immunogenicity, PK, pharmacodynamics, safety, and effectiveness of PF-06410293597 against ADL.

14.5.4.7 Study Design

- Trials such as the superiority trials are used to show the effectiveness of one treatment on another (placebo), but such trials are not recommended for biosimilar development. Biosimilar evaluation requires nonsuperiority trials. Non*superiority* trials are classified as equivalence or noninferiority designs. Such trials are included in the 2012 FDA draft guidance. Practically equivalence design is used that signifies the equality of biosimilar with reference product (Batran et al. 2022).
- An equivalence trial uses two-sided test. A specified range is determined. Null hypothesis *selected* is biosimilar and is inferior to the reference product or superior to reference product. The selection of equivalence margins is done such that it helps to detect differences with clinically significance in effectiveness between biosimilar and reference product. A 95% confidence interval is maintained. The study is planned such that the null hypothesis is rejected for nonequivalence, and the alternative hypothesis is accepted indicating that the biosimilar and reference product are equivalent.
- For a smaller sample size as compared to equivalence design, a one-sided noninferiority design is preferred. The design is best fitting for thc establishment that *reference* products use is at or near maximal level of clinical effect. A noninferiority design is sufficient for evaluating immunogenicity and other safety outcomes considering the lower immunogenicity and safety events do not have efficacy implications (Landewe and Van Der Heijde 2012; Velayudhan et al. 2016).

14.6 Regulatory Complexities and Challenges of Biosimilars

The European Union has set up a legitimate framework, and the Europe, Center East, and Africa locale has too set up a legitimate framework for the endorsement of biosimilars. Since there are critical contrasts between biopharmaceutical items, the endorsement handle may change by item. In specific, therapeutic data may depend on the distinction between last execution and the accessibility of the recognized score. In common, endorsement of biosimilars will be based on show that the unused item is comparable and secure (i.e., "comparable") within the important persistent populace. The EMEA Order will permit the dispersal of data from one clinical zone to another, where suitable, and the utilize of biosimilars in signs that have not however been considered. Since biopharmaceuticals are in some cases, but not more often than not, related with unfavorable occasions, the EMEA mandate moreover requires clinical trials and pharmacovigilance to screen the viability and security of biosimilar items after endorsement. The as of late endorsed biosimilar development hormone omnitrope highlights the significance of immunogenicity testing and pharmacovigilance. Amid improvement, the generation of products is moved from one manufacturing plant to another. Whereas quality testing did not reveal significant contrasts between the ultimate products from these sources, the distinction was found within the immunogenicity indicated by the producer prior to endorsement. There is right now no administrative preparation for endorsing biosimilars within the Joined together States, and so the FDA has not, however, created direction with respect to such items (Kang et al. 2021).

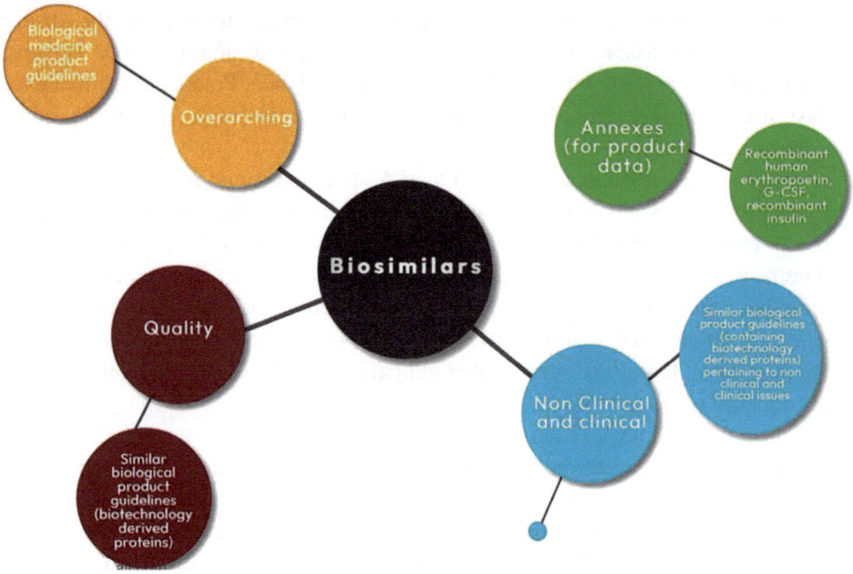

14.6.1 Regulatory Landscape

1. *Development and Approval Process*
   (a) Biosimilars undergo a rigorous development and approval process, including extensive analytical and clinical comparability studies to demonstrate similarity to the reference biologic. Regulatory agencies, such as the U.S. Food and Drug Administration (FDA) and the European Medicines Agency (EMA), establish a unique process for the development and approval of biosimilars.
2. *Reference* Product Data Exclusivity
   (a) Reference biologics are granted a period of data exclusivity during which biosimilar developers cannot rely on the reference product's data for their regulatory submissions. This period varies by region and can impact the timing of biosimilar market entry.
3. *Extrapolation of Indications.*
   (a) Regulatory agencies may allow biosimilar approval for certain indications without requiring separate clinical trials if the biosimilar demonstrates sufficient similarity in other approved indications. Extrapolation is a key concept that can expedite the approval process but requires a robust scientific rationale.

14.7 Challenges of Biosimilars (Mellstedt et al. 2008b)

1. Demonstrating Biosimilarity:
   (a) Complexity of Molecules: Biologics are large, complex molecules with unique structures and functions. Not at all like small-molecule drugs, which are chemically synthesized and have well-defined structures, biologics are delivered in living beings (such as microscopic organisms, yeast, or mammalian cells) and exhibit structural heterogeneity due to posttranslational modifications (PTMs) and protein folding. This complexity poses challenges in replicating the exact structure and function of the reference biologic.
   (b) Variability: Biologics can exhibit inherent variability due to differences in manufacturing processes, raw materials, cell lines, and environmental factors. Even small changes in these parameters can affect the structure, stability, and biological activity of the biologic. Biosimilar developers must identify and control sources of variability to ensure consistent quality and similarity to the reference product.
2. Analytical Techniques:
   (a) Sophisticated Analytical Methods: Biosimilar developers utilize a wide range of advanced expository procedures to characterize the physicochemical and organic properties of their items. These techniques include mass spectrometry, chromatography (e.g., high-performance liquid chromatography, size-exclusion chromatography), spectroscopy (e.g., nuclear magnetic resonance, infrared spectroscopy), and immunoassays. These methods allow for in-depth analysis of critical quality attributes (CQAs) such as primary and higher-order protein structure, glycosylation pattern, charge variants, impurities, and biological activity.
   (b) Comparability Studies: Comparability studies are comprehensive assessments conducted to demonstrate similarity between the biosimilar and the reference product. These studies involve side-by-side comparisons of multiple lots of the biosimilar and the reference product using a variety of analytical techniques. By demonstrating analytical similarity in CQAs, biosimilar developers can provide evidence of biosimilarity to regulatory agencies.
3. Clinical Trial Design:
   (a) Study Design Challenges: Designing clinical trials for biosimilars requires careful consideration of various factors, including patient selection, study duration, sample size, statistical methodology, and clinical endpoints. The goal is to design studies that can reliably demonstrate equivalence or noninferiority of the biosimilar compared to the reference product while minimizing patient risk and ethical concerns. This may involve innovative trial designs such as adaptive trials, which allow for modifications based on interim data analysis, and Bayesian statistical methods, which incorporate prior knowledge into the analysis.
   (b) Endpoints Selection: Selecting appropriate clinical endpoints is critical for demonstrating biosimilarity. Clinical endpoints should be sensitive, clinically relevant, and reflective of the mechanism of action of the biologic.

Common endpoints include disease activity scores, patient-reported outcomes, biomarker measurements, and adverse event assessments. The choice of endpoints may vary depending on the therapeutic indication and regulatory requirements.

4. Immunogenicity:
    (a) Risk Assessment: Immunogenicity refers to the ability of a biologic to induce immune responses in patients, leading to the production of anti-drug antibodies (ADAs) and potential adverse reactions. Biosimilar developers conduct comprehensive risk assessments to evaluate the potential immunogenicity of their products. Factors such as structural similarity to the reference product, impurities, formulation, route of administration, and patient population characteristics are considered in the risk assessment.
    (b) Immunogenicity Testing: Immunogenicity testing involves assessing the immunogenic potential of the biosimilar in preclinical models and clinical trials. These tests aim to detect the presence of ADAs and neutralizing antibodies (NAbs) and evaluate their impact on safety, efficacy, and pharmacokinetics. Immunogenicity assays should be sensitive, specific, and validated to reliably detect immune responses.
5. Extrapolation of Indications:
    (a) Data Requirements: Extrapolation involves using data from one indication to support approval in extra signs for which the reference item is endorsed. Regulators assess the totality of evidence, including pharmacokinetic/pharmacodynamic (PK/PD) data, clinical efficacy, safety, and mechanism of action. The goal is to determine whether the biosimilar behaves similarly across different disease states, patient populations, and treatment settings.
    (b) Regulatory Considerations: Extrapolation decisions are based on a thorough understanding of the underlying biology and clinical data available for the biosimilar and the reference product. Regulators evaluate factors such as the similarity of the disease pathophysiology, patient populations, and treatment responses across indications. Extrapolation may be supported by robust scientific rationale, comparative PK/PD data, and clinical experience with the reference product.
6. Switching and Interchangeability:
    (a) Switching Studies: Switching studies assess the safety, efficacy, and immunogenicity of transitioning patients from the reference product to the biosimilar or vice versa. These studies typically involve multiple switches between the two products and may include clinical, immunological, and pharmacokinetic assessments. The goal is to demonstrate that switching between products does not result in clinically meaningful differences in outcomes or safety.
    (b) Interchangeability Criteria: Interchangeability refers to the ability of a biosimilar to replace the product without the need for physician intervention. Regulators establish criteria for biosimilars to be considered interchangeable, such as demonstrating comparable clinical outcomes, safety, and immunogenicity following multiple switches. Biosimilar developers must

provide sufficient evidence to meet these criteria, which may include data from switching studies and additional comparative analyses.

7. Regulatory Requirements:
   (a) Divergent Regulatory Pathways: Distinctive administrative organizations have created particular pathways and rules for the endorsement of biosimilars. These pathways may vary in terms of data requirements, scientific principles, and regulatory expectations. For example, the European Medicines Agency (EMA) has established a centralized approval process for biosimilars, while the US Food and Drug Administration (FDA) follows an abbreviated pathway under the Biologics Price Competition and Innovation Act (BPCIA). Biosimilar developers must navigate these diverse regulatory landscapes to obtain approval in multiple jurisdictions.
   (b) Biosimilar Development Guidelines: Regulatory agencies publish guidelines outlining the scientific and regulatory principles for demonstrating biosimilarity. These guidelines provide recommendations on analytical characterization, nonclinical studies, clinical trials, and post-approval requirements. Biosimilar developers rely on these guidelines to design and conduct their development programs and to interact with regulatory authorities throughout the approval process.
8. Market Access and Reimbursement:
   (a) Pricing Strategies: Biosimilar pricing strategies involve setting the price of the biosimilar relative to the reference product and other competing products in the market. Pricing decisions consider factors such as production costs, market competition, reference product pricing, healthcare system preferences, and payer reimbursement policies. Biosimilar developers may adopt pricing strategies aimed at maximizing market penetration and access while ensuring profitability.
   (b) Health Economics and Outcomes Research (HEOR): HEOR studies assess the economic value, cost-effectiveness, and patient outcomes associated with biosimilars compared to the reference product. These studies provide evidence to support reimbursement decisions by payers and healthcare decision-makers. HEOR analyses may include cost-minimization, cost-effectiveness, budget impact, and quality-of-life assessments, among others.
9. Education and Acceptance:
   (a) Healthcare Provider Training: Education initiatives aim to increase awareness and understanding of biosimilars among healthcare providers, including physicians, nurses, and pharmacists.

14.8 Licensing of Biosimilars (Liu et al. 2016)

14.8.1 Preclinical Development

- *Comparability Studies*:

  Manufacturers conduct comparative studies to confirm the similarity between the biosimilar and the product before clinical trials. These studies encompass a range of analytical tests employing advanced techniques such as mass spectrometry, chromatography, and spectroscopy. The aim is to assess physicochemical properties, structural attributes, and the product studies biosimilars to ensure consistency with the product.

- *Toxicology Studies*:

  Preclinical toxicology studies are pivotal in evaluating the biosimilars safety profile prior to human trials. These studies, conducted in accordance with regulatory guidelines, involve testing in animal models to assess acute and chronic toxicity, immunogenicity, and potential adverse effects on various organ systems. Emphasis is placed on identifying potential immune responses, which could significantly impact safety and efficacy. The findings from these studies guide dosing strategies for subsequent clinical trials and provide crucial safety data for regulatory submissions.

14.8.2 Clinical Development

- *Phase 1 Clinical Trials*:

  Phase 1 trials mark the initial evaluation of the biosimilar's safety and pharmacokinetic profile in humans. These trials typically involve a small cohort of healthy volunteers and aim to determine the maximum tolerated dose, pharmacokinetics, and pharmacodynamics of the biosimilar. Data from Phase 1 trials provide critical insights into the biosimilar's safety profile and help establish initial dose ranges for further clinical investigation.

- *Phase 2 Clinical Trials*:

  Phase 2 trials expand the evaluation to include patients with the target disease or condition. These studies are designed to assess preliminary efficacy, further characterize safety, and optimize dosing regimens. Employing randomized, controlled designs, Phase 2 trials compare the biosimilar to the reference product or placebo, utilizing clinical endpoints such as disease response rates and progression-free survival to evaluate therapeutic efficacy.

- *Phase 3 Clinical Trials*:

  Phase 3 trials represent pivotal studies conducted in larger patient populations to confirm the safety and efficacy of the biosimilar. Typically randomized and controlled, these trials may involve hundreds to thousands of patients. Utilizing non-inferiority or equivalence designs, developers aim to demonstrate comparable efficacy to the reference product. Primary endpoints, such as overall survival or disease progression, are clinically meaningful measures used to assess efficacy and safety in Phase 3 trials.

14.8.3 Regulatory Submission

- *Marketing Authorization Application (MAA).*

  The MAA is a comprehensive submission to regulatory agencies, providing detailed data on the biosimilar's quality, safety, and efficacy. It includes comprehensive information on preclinical and clinical studies, manufacturing processes, quality control measures, and pharmacovigilance plans. Supported by a thorough review of scientific literature and regulatory guidelines, the MAA undergoes rigorous assessment to ensure compliance with regulatory standards and suitability for approval.

- *Scientific Advice*:

  Biosimilar developers often seek scientific advice from regulatory agencies before submitting the MAA. These interactions provide opportunities to address specific questions or concerns regarding study design, endpoints, or data interpretation. Feedback obtained during scientific advice meetings is invaluable in optimizing the development program and increasing the likelihood of regulatory approval.

14.8.4 Regulatory Review

- *Assessment of Similarity*:

  Regulatory agencies conduct a meticulous review of the biosimilar's data package to assess its similarity to the reference product. This assessment involves evaluating analytical, preclinical, and clinical data to determine whether the biosimilar meets the criteria for similarity in quality, safety, and efficacy. The primary objective is to ensure that the biosimilar demonstrates high similarity to the reference product, with no clinically meaningful differences that could impact safety or efficacy.
- *Risk–Benefit Assessment*:

  The risk–benefit assessment involves evaluating the potential benefits of the biosimilar against its risks, including safety concerns and uncertainties. Factors considered include the clinical pertinence of any watched contrasts from the reference item, the strength of the clinical data supporting efficacy, and the potential impact on patient outcomes. The goal is to ensure that the benefits of the biosimilar outweigh any potential risks and that it provides a therapeutic alternative with comparable safety and efficacy.
- *Labeling and Product Information*:

  Upon approval, regulatory agencies provide guidance on the biosimilar's labeling and product information. This includes indications, dosing recommendations, contraindications, warnings, precautions, and adverse reactions associated with the biosimilar. The labeling should accurately reflect the available data and inform appropriate use of the biosimilar.

14.8.5 Post-Marketing Surveillance

- *Pharmacovigilance Monitoring*:

  Biosimilar manufacturers are mandated to establish pharmacovigilance systems to monitor the safety of their products in real-world clinical practice. These systems involve continuous surveillance of adverse events, product complaints, and other safety-related information. Data collected through pharmacovigilance activities are regularly reviewed and analyzed to detect potential safety signals and inform risk management decisions.

- *Post-Authorization Studies*:

  Regulatory agencies may require biosimilar manufacturers to conduct post-authorization studies to address specific safety or efficacy concerns identified during the regulatory review process. These studies may include long-term safety assessments, additional efficacy evaluations, or pharmacoeconomic analyses. The results of post-authorization studies contribute to the ongoing evaluation of the biosimilar's benefit–risk profile and may inform future regulatory decisions.

14.9 Potential of Biosimilars (Kaur et al. 2017)

Biosimilars have significant potential in various aspects of healthcare, offering opportunities for improved patient access to effective and potentially more affordable biologic therapies.

1. Increased Access to Biologic Therapies:

   Biosimilars provide alternatives to expensive originator biologics, potentially expanding patient access to critical treatments for various diseases, including cancer, autoimmune disorders, and chronic conditions.
2. Treatment Options and Therapeutic Choice:

   Biosimilars provide physicians and patients with additional treatment options, potentially leading to better therapeutic choices based on factors such as cost, accessibility, and individual patient preferences.
3. Biosimilar Interchangeability:

   Some biosimilars may achieve interchangeability status, allowing them to be substituted for the reference product without intervention from the healthcare provider. This can enhance convenience and potentially lead to broader adoption.
4. Patient Affordability:

   The availability of lower-cost biosimilars can make biologic therapies more affordable for patients, reducing financial barriers to accessing essential treatments.
5. Increased Patient Access to Treatment:

   Biosimilars offer the potential to expand patient access to essential biologic therapies by providing alternative treatment options. Biologic drugs are often associated with high costs, making them inaccessible to many patients, particularly those in developing countries or with limited healthcare coverage. Biosimilars, which are typically priced lower than their reference biologics, can help address this access barrier by providing more affordable alternatives. Increased access to treatment can lead to improved health outcomes and quality of life for patients who would otherwise be unable to afford biologic therapies.

6. Affordability and Cost Savings:

   One of the primary advantages of biosimilars is their potential to reduce healthcare costs. By offering lower-priced alternatives to reference biologics, biosimilars can generate significant cost savings for patients, healthcare systems, and payers. These cost savings can be particularly beneficial in managing the financial burden associated with chronic diseases or conditions that require long-term treatment with biologic therapies. Additionally, biosimilars can help alleviate budget constraints for healthcare providers and governments, allowing them to allocate resources more efficiently to other areas of healthcare.

7. Competition and Market Dynamics:

   The introduction of biosimilars into the market fosters competition, which can have several positive effects on healthcare. Increased competition among biosimilar manufacturers can lead to price reductions, as companies compete for market share. This competitive pressure can incentivize innovation and drive improvements in manufacturing processes, product quality, and patient access programs. Moreover, competition can promote transparency and information-sharing, enabling healthcare providers and patients to make more informed treatment decisions based on the availability of biosimilar options.

8. Innovation and Product Development:

   Biosimilar development requires significant investment in research, development, and manufacturing capabilities. As a result, biosimilar manufacturers often leverage innovative technologies and methodologies to demonstrate biosimilarity and ensure product quality. This investment in innovation can drive advancements in analytical techniques, manufacturing processes, and regulatory science, benefiting both biosimilar development and the broader biopharmaceutical industry. Furthermore, biosimilar development can stimulate research into new therapeutic targets and treatment modalities, leading to the discovery of novel biologic therapies and improved patient outcomes.

9. Global Health Impact:

   Biosimilars have the potential to have a profound impact on global health by increasing access to life-saving treatments, particularly in low- and middle-income countries. These regions often face significant challenges in accessing expensive biologic therapies due to cost constraints and limited healthcare infrastructure. Biosimilars offer a cost-effective solution for addressing unmet medical needs and improving health outcomes in these populations. By providing affordable alternatives to reference biologics, biosimilars can help reduce disparities in healthcare access and promote health equity on a global scale.

10. Supporting Healthcare Sustainability:

    With rising healthcare costs and budgetary constraints, biosimilars offer a sustainable solution for improving access to biologic therapies while managing healthcare expenditures. By offering cost-effective alternatives to reference biologics, biosimilars help mitigate the financial burden on healthcare systems and payers, ensuring that patients receive the treatments they need without imposing undue strain on healthcare budgets. This sustainability is essential for ensuring the long-term viability and resilience of healthcare systems worldwide, particularly in the face of evolving healthcare challenges and economic uncertainties.

References

Alten R, Cronstein BN (2015) Clinical trial development for biosimilars. Semin Arthritis Rheum 44(6):S2–S8

Batran RA, Elmoshneb M, Hussein AS, Hussien OM, Adel F, Elgarhy R, Morsi MI (2022) Biosimilars: science, implications, and potential outlooks in the Middle East and Africa. Biologics 16:161–171

Bui LA, Hurst S, Finch GL, Ingram B, Jacobs IA, Kirchhoff CF, Ng CK, Ryan AM (2015 May) Key considerations in the preclinical development of biosimilars. Drug Discov Today 20(Suppl 1):3–15

Castañeda-Hernández G, Szekanecz Z, Mysler E et al (2014) Biopharmaceuticals for rheumatic diseases in Latin America, Europe, Russia, and India: innovators, biosimilars, and intended copies. Joint Bone Spine 81:471–477

Christl LA, Woodcock J, Kozlowski S (2017) Biosimilars: the US regulatory framework. Annu Rev Med 68:243–254

Cohen H (2019) Biosimilar development: approval of biosimilar medicines through totality of the evidence. Drug Development Delivery 19:40–45

Cohen S, Genovese MC, Choy E, Perez-Ruiz F, Matsumoto A, Pavelka K et al (2017) Efficacy and safety of the biosimilar ABP 501 compared with adalimumab in patients with moderate to severe rheumatoid arthritis: a randomised, double- blind, phase III equivalence study. Ann Rheum Dis 76:1679–1687. https://doi.org/10.1136/annrheumdis-2016-210459

Cohen S, Pablos JL, Pavelka K, Müller GA, Matsumoto A, Kivitz A et al (2019a) An open-label extension study to demonstrate long-term safety and efficacy of ABP 501 in patients with rheumatoid arthritis. Arthritis Res Ther 21:1–10. https://doi.org/10.1186/s13075-019-1857-3

Cohen SB, Czeloth N, Lee E, Klimiuk PA, Peter N, Jayadeva G (2019b) Long- term safety, efficacy, and immunogenicity of adalimumab biosimilar BI 695501 and adalimumab reference product in patients with moderately- to-severely active rheumatoid arthritis: results from a phase 3b extension study (VOLTAIRE-RAext). Expert Opin Biol Ther 19:1097–1105. https://doi.org/10.1080/14712598.2019.1645114

Curigliano G, O'Connor DP, Rosenberg JA et al (2016) Biosimilars: extrapolation for oncology. Crit Rev Oncol Hematol 104:131–137

EMA 2015 Humira drug label (summary of product characteristics) approved by the EMA.http://www.ema.europa.eu/docs/en_GB/document_library/EPAR_-_Product_Information/human/000481/WC500050870.pdf. Accessed 14 May 2015

European Medicines Agency. Guideline on Similar Biological Medicinal Products Containing Monoclonal Antibodies—Non-Clinical and Clinical Issues, 2012. http://www.ema.europa.eu/docs/en_GB/document_library/Scientific_guideline/2012/06/WC500128686.pdf

European Medicines Agency. Guideline on similar biological medicinal products. 2014a. https://www.ema.europa.eu/documents/scientific-guideline/guidelinesimilar-biological-medicinal-products-rev1_en.pdf

European Medicines Agency. Guideline on similar Biological Medicinal Products Containing Biotechnology-Derived Proteins as Active Substance: Non-Clinical and Clinical Issues, 2014b. https://www.ema.europa.eu/en/documents/scientific-guideline/guideline-similar-biological-medicinal-products-containingbiotechnology-derived-proteins-active_en-2.pdf

European Medicines Agency. Guideline on similar biological medicinal products containing biotechnology-derived proteins as active substance: non-clinical and clinical issues on similar biological medicinal products. 2014c. http://www.ema.europa.eu/docs/en_GB/document_library/Scientific_guideline/2015/01/WC500180219.pdf. Accessed 26 Mar 2015

FDA Humira drug label approved by the FDA. 2008. http://www.accessdata.fda.gov/drugsatfda_docs/label/2008/125057s0110lbl.pdf. Accessed 26 Mar 2015
FDA (n.d.). https://www.fda.gov/regulatory-information/search-fda-guidance-documents/scientific-considerations-demonstrating-biosimilarity-reference-product
Food and Drug Administration. Guidance for industry. Scientific considerations in demonstrating biosimilarity to a reference product. 2015. http://www.fda.gov/downloads/drugs/guidancecompliance regulatoryinformation/guidances/ucm291128.pdf. Accessed 26 Mar 2015
Gabioline (n.d.). https://gabionline.net/biosimilars/research/Biologicals-characteristics-that-make-them-unique-and-special
García JJ, Raez LE, Rosas D (2020 Oct) A narrative review of biosimilars: a continued journey from the scientific evidence to practice implementation. Transl Lung Cancer Res 9(5):2113–2119
Grozdanova A, Netkovska KA, Sterjev Z, Naumovska Z, Zarevski R, Dimovski A, Suturkova L (2016) Biosimilar medical products - licensing, pharmacovigilance and interchangeability. Pril (Makedon Akad Nauk Umet Odd Med Nauki) 37(1):27–36
Huneycutt B, Gillespie E, Woollett G (2015) Clinical trials in the development of biosimilars: future considerations. Biosimilars 5:49–63
Kaida-Yip F, Deshpande K, Saran T, Vyas D (2018) Biosimilars: review of current applications, obstacles, and their future in medicine. World J Clin Cases 6(8):161–166
Kang HN, Thorpe R, Knezevic I, Casas Levano M, Chilufya MB, Chirachanakul P, Chua HM, Dalili D, Foo F, Gao K, Habahbeh S, Hamel H, Kim GH, Perez Rodriguez V, Putri DE, Rodgers J, Savkina M, Semeniuk O, Srivastava S, Tavares Neto J, Wadhwa M, Yamaguchi T (2021) Regulatory challenges with biosimilars: an update from 20 countries. Ann N Y Acad Sci 1491(1):42–59
Kaur P, Chow V, Zhang N, Moxness M, Kaliyaperumal A, Markus R (2017) A randomised, single-blind, single-dose, three-arm, parallel-group study in healthy subjects to demonstrate pharmacokinetic equivalence of ABP 501 and adalimumab. Ann Rheum Dis 76:526–533. https://doi.org/10.1136/annrheumdis-2015-208914
Koyfman H (2013 Aug) Biosimilarity and interchangeability in the biologics Price competition and innovation act of 2009 and FDA's 2012 draft guidance for industry. Biotechnol Law Rep 32(4):238–251
Krendyukov A, Schiestl M (2018) Extrapolation concept at work with biosimilar: a decade of experience in oncology. ESMO Open 3:e000319
Lai Z, La Noce A (2016) Key design considerations on comparative clinical efficacy studies for biosimilars: adalimumab as an example. RMD open. 2(1):e000154
Landewe RBM, Van Der Heijde DMFM (2012) Clinical trial design and analysis. In: Firestein GS, Budd RC, Gabriel SE, O'Dell JR, IB MI (eds) Kelley's Textbook of Rheumatology, 9th edn. Elsevier, Philadelphia, PA, pp 453–461
Lee H (2014) Is extrapolation of the safety and efficacy data in one indication to another appropriate for biosimilars? AAPS J 16:22–26
Lee JF, Litten B, Grampp G (2012) Comparability and biosimilarity:considerations for the health-careprovider. Curr Med Res Opin 28:1053–1058
Liese B, Allary M, Steven S, Paul D, Vulto AG, Isabelle H (2022) Regulatory information and guidance on Biosimilars and their use across Europe: a call for strengthened one voice messaging. Front Med 9:1–16
Liu J, Eris T, Li C, Cao S, Kuhns S (2016) Assessing analytical similarity of proposed amgen biosimilar ABP 501 to adalimumab. BioDrugs 30:321–338. https://doi.org/10.1007/s40259-016-0184-3
Lu X, Hu R, Peng L, Liu M, Sun Z (2021) Efficacy and safety of Adalimumab Biosimilars: current critical clinical data in rheumatoid arthritis. Front Immunol 12:638444
Mellstedt H, Niederwieser D, Ludwig H (2008a) The challenge of biosimilars. Ann Oncol 19:411–419
Mellstedt H, Niederwieser D, Ludwig H (2008b) The challenge of biosimilars. Ann Oncol 19(3):411–419

Mirasol F (2023) Assessing biosimilarity with analytical advancements. BioPharm International 36(6):30–31

Scheinberg MA, Kay J (2012) The advent of biosimilar therapies in rheumatology—"O brave new world". s 8:430–436

Serra-Matencio JM, Ramirez-Herraiz E, Morell-Baladron A, Castañeda S (2016) Biosimilars: a new challenge in the current pharmacology. Hospital de La Princesa, IIS-IP, Madrid, Universidad Autónoma, Madrid, Spain

Stebbing J, Mainwaring PN, Curigliano G, Pegram M, Latymer M, Bair AH, Rugo HS (2020) Understanding the Role of Comparative Clinical Studies in the Development of Oncology Biosimilars. J Clin Oncol. 38(10):1070–1080. https://doi.org/10.1200/JCO.19.02953

US Food and Drug Administration: Scientific considerations in demonstrating biosimilarity to a reference product: Guidance for industry, 2015. https://www.fda.gov/media/82647/download

Velayudhan J, Chen Y, Rohrbach A, Pastula C, Maher G, Thomas H et al (2016) Demonstration of functional similarity of proposed biosimilar ABP 501 to adalimumab. BioDrugs. 30:339–351. https://doi.org/10.1007/s40259-016-0185-2

WHO: Guidelines on Evaluation of Similar Biotherapeutic Products (SBPs), 2009. http://www.who.int/biologicals/areas/biological_therapeutics/ BIOTHERAPEUTICS_FOR_WEB_22APRIL2010.pdf.

WHO: Guidelines on evaluation of monoclonal antibodies as similar biotherapeutic products (SBPs), 2017. https://www.who.int/biologicals/biotherapeutics/WHO_TRS_1004_web_Annex_2.pdf?ua=111.

Regulatory Considerations of Biosimilars in Cancer 15

Lokesh Nagar, Annu Saini, Nisha Gulati, Neeta Solanki, and Harish Dureja

Abstract

In the fast-evolving field of cancer treatment, the introduction of biosimilars has brought about significant changes in the available therapeutic options. Biosimilars, which are identical to the original biologic drugs, offer a promising opportunity to increase patient access to effective and more affordable cancer treatments. However, the rules and regulations governing biosimilars in oncology play a crucial role in ensuring their safety, effectiveness, and successful integration into clinical practice. Effectively managing the existing regulations is essential to establish a robust environment for innovation, prioritize patient welfare, and ensure fair access to quality cancer care. Biosimilar regulations are designed to guarantee that medicines similar to existing ones are both safe and effective. A well-established regulatory environment builds confidence in the safety and efficacy of biosimilars, forming the basis for advancements in cancer treatment. These regulations also help maintain quality standards for affordable alternatives to biological drugs. This chapter emphasizes the regulatory considerations linked with biosimilars in cancer treatment, shedding light on the complexities and implications for healthcare professionals, regulatory authorities, and, most importantly, patients.

Keywords

Biosimilar · Food and Drug Administration · Interchangeability · Patient · Regulatory compliance

L. Nagar · A. Saini · N. Gulati · N. Solanki · H. Dureja (✉)
Department of Pharmaceutical Sciences, Maharshi Dayanand University, Rohtak, India
e-mail: harishdureja@mdurohtak.ac.in

S. Bhatt et al. (eds.), *Biosimilars for Cancer Treatment*,
https://doi.org/10.1007/978-981-97-3001-8_15

15.1 Introduction

In the recent years, there has been significant progress in the field of biopharmaceuticals, particularly with the emergence of biosimilars. These are biological products highly similar to existing biologics, ensuring no meaningful differences in safety, efficacy, and quality. This development opens up new possibilities in healthcare, providing patients with increased access to life-saving treatments and contributing to cost containment. However, the successful development and commercialization of biosimilars demand a robust regulatory framework to guarantee patient safety and uphold quality standards (Kirchhoff et al. 2017).

The product lifecycle management of an oncology biosimilar involves several key stages from development to post-marketing, all under careful regulatory oversight. In the early stages, rigorous preclinical studies are conducted to establish biosimilarity with the reference product, adhering to Good Laboratory Practice standards (Makurvet 2020). The subsequent clinical development phase focuses on designing and executing trials to demonstrate safety and efficacy, following guidelines set by regulatory agencies such as the FDA, EMA NMPA, CDSCO, and others. A comprehensive regulatory submission, including data from preclinical and clinical studies, is then prepared and submitted to gain regulatory approval (Kang and Knezevic 2018).

Upon regulatory approval, the biosimilar is launched and commercialized, with a keen focus on compliance with labelling, distribution, and safety monitoring requirements. Post-marketing surveillance becomes a continuous priority, involving the monitoring of the product's safety and efficacy after its launch. Strategies for product improvement, including potential formulation changes, and seeking additional indications through regulatory submissions, contribute to the ongoing lifecycle management. Compliance with evolving regulatory standards, effective communication with regulatory authorities, and proactive engagement with stakeholders are fundamental components throughout the product's lifecycle (Anand et al. 2022).

A reference product is one that has already received FDA approval and is used as a benchmark for evaluating a proposed biosimilar product (Schreiber et al. 2022). Biosimilars are utilized in the diagnosis, treatment, and prevention of illnesses and ailments (Jha et al. 2015). Biosimilars are a broad class of products that are typically big, intricate compounds (Triplitt et al. 2017). These products can be made via biotechnology in a biological system, such as a bacterium, plant cell, or animal cell; however, they are typically more difficult to characterize than small molecule drugs (Morrow and Felcone 2004). Biosimilars have garnered a lot of attention globally in recent years as viable substitutes for biologics. Cancer is undeniably a serious and potentially fatal disease (Chopra and Lopes 2017). Since 1989, the use of biologics in supportive care and primary cancer treatment has been authorized. It has been demonstrated that these biologics enhance haematological and clinical results as well as quality of life connected to health (Curtis and Singh 2011). It is not just primary care that uses biologics; supportive oncologic treatment has also benefited from their development. Biologics provide one of the most cutting-edge treatments available among the different therapy alternatives (Anand et al. 2022). Biologics may be more expensive, less readily available, or require insurance coverage before patients may receive them (Makurvet 2020). Patient access has grown in importance as the patent protection periods for several biologics are coming to an end, weighing the expenses of high-quality care against it (Inotai et al. 2018). Biosimilars are being developed as potential lower cost and more accessible alternatives within the framework of this balancing (Ventola 2013). Biosimilars have the potential to increase patient access to biologic medications while simultaneously lowering the cost of cancer therapy. Biosimilars incur lower development and production expenses in comparison to their reference biologics (Rak Tkaczuk and Jacobs 2014). The price difference between biosimilars and the reference biologic may be as much as 30%. By bringing more competition to the market and lowering prices, biosimilars are able to also lower the cost of biologics. The global impact of biosimilars on the cancer burden is significant, addressing health disparities and promoting a more equitable distribution of cancer therapies. With increased competition, biosimilars emphasize a patient-centric approach to cancer treatment, providing individuals with more choices and fostering collaborative decision-making between healthcare providers and patients. Furthermore, the evolving regulatory processes for biosimilars in oncology ensure that these products meet stringent standards, instilling confidence in healthcare professionals and patients alike (Kvien et al. 2022).

15.2 Overview of Biosimilars

Although it is noteworthy that regulatory bodies worldwide have not established a singular, universally accepted definition for the term "biosimilar", an examination of these varied definitions underscores the presence of significant similarities are mentioned below (Mysler et al. 2021):

- Central Drugs Standard Control Organization (CDSCO): *"A similar biologic product is that which is similar in terms of quality, safety, and efficacy to an approved reference biological product based on comparability"* (CDSCO 2016a, 2016b).
- *Food and Drug Administration (FDA)*: *"A biosimilar has no clinically meaningful differences from an existing FDA-approved biologic, called a reference product"* (FDA 2023).
- *The European Medicines Agency (EMA)*: *A biosimilar is a biological medicine highly similar to another already approved biological medicine (the 'reference medicine')"* (EMA 2006).
- *The World Health Organization (WHO)*: *"A biotherapeutic product which is similar in terms of quality, safety and efficacy to an already licensed reference biotherapeutic product"* (WHO 2018).

Two main themes arise from the shared definitions of biosimilars provided by various regulatory agencies. These key concepts represent fundamental principles commonly acknowledged among these authorities.

- *Biosimilars are duplicates of previously approved products that have demonstrated similarities to the original product—also referred to as reference products—and are not generics and.*
- *A biologic product that has been approved must be comparable to a biosimilar in terms of quality, safety, and effectiveness* (Sylvester et al. 2013).

While structure, safety, effectiveness, and immunogenicity are all quite comparable between biosimilars and reference products, they differ slightly from one another (Kurki et al. 2021). In essence, the reference product and the biosimilar are two distinct formulations of the same active ingredient. Therefore, the analytical similarity exercise to prove biosimilarity can also be based on the same scientific ideas that support the comparability exercise for modifications in the originator product's manufacturing method (Kumar and Singh 2014). The basic concepts apply to both activities, but because of changes in cell lines and manufacturing procedures, among other factors, further steps are typically required in the development of biosimilars to justify similarity, either analytically or clinically (Vulto and Jaquez 2017). Ultimately, it is important for sponsors—including the product's creators and regulators—to comprehend how structural variations may impact the product's functionality. In order to guarantee the product's efficacy and safety, they must also make sure that these distinctions are carefully considered (American Cancer Society 2018a, b).

15.3 Health Economics and Market Access of Biosimilars in Oncology

Cancer is one of the leading causes of death worldwide. Approximately one in six fatalities worldwide is caused by cancer, which is higher than the combined death rates from HIV, TB, and malaria (World Health Organization 2020). 10.3 million deaths worldwide were attributed to cancer in 2020, while 19.3 million new cases were reported. One-fourth of these deaths take place in low- and middle-income nations, where there are insufficient resources and healthcare systems to handle the burden of disease (Sung et al. 2021). Based on population growth and ageing, it is projected that by 2040, there will be 16.3 million cancer-related deaths worldwide and an estimated 27.5 million diagnoses (American Cancer Society 2018a). According to the National Cancer Institute, direct medical expenses associated with cancer treatment in the United States were estimated to be $183 billion in 2015 and are projected to rise by 34% to $246 billion by 2030 (Mariotto et al. 2020). However, this growth is probably underestimated due to inflation and personalized treatment developments. The introduction of biosimilars has increased market competition, which can benefit cancer patients by making treatments more accessible and lowering their financial burden (Blackstone and Joseph 2013).

15.4 Regulatory Agencies and Guidelines

In an effort to achieve global harmonization, biosimilar laws and regulations are becoming more favourable worldwide, especially in the United States and Europe (Joshi et al. 2022). The EMA, FDA, and WHO are some of the primary regulatory bodies that have established guidelines for figuring out how similar reference products and biosimilars are to one another (Gherghescu and Delgado-Charro 2020). The first regulatory agency that set up a framework for biosimilars was the European Medicines Agency (EMA) in 2003; nevertheless, the first approval for these types of drugs was not given until 2006 (Schiestl et al. 2017). In 2009, the FDA established clear rules for the approval of biosimilars. Furthermore, many other nations—including Canada, South Korea, Australia, Japan, and India—have developed their own standards that are comparable to the EMA tenets since 2008 and 2012 (Kabir et al. 2019).

15.4.1 United States (US)

In the United States, biosimilars are regulated under the BPCI Act, which provides the framework for the FDA's approval of biosimilars through the 351(k) pathway (Generics and Biosimilars Initiative 2023). Biological products (also referred to as reference products or biosimilar products) are subject to rigorous testing to ensure patient safety, effectiveness, and quality (Pfizer 2022). As part of its Quality Considerations in Demonstrating Biosimilarity to a Reference Protein Product initiative, the USFDA released guidance documents. As a general matter, these guidelines address clinical and nonclinical issues related to quality (Research 2020b). A set of guidelines for proving biosimilar interchangeability with a reference product was released by the USFDA in May 2019 (Niazi 2023).

The "draft guidance for industry" on biosimilars lays forth specifications for assessing and analyzing the immunogenicity, safety, and effectiveness of biosimilars in human studies. In order for a biosimilar product to be approved by the FDA, its manufacturer must show that the product is extremely similar to the reference product and does not differ clinically from it, rather than producing the same complete profile of nonclinical and clinical outcome data as the FDA-approved product (FDA 2020a). The FDA evaluates each proposed biosimilar on an individual basis and advises producers on the kind and extent of testing needed to prove biosimilarity. Compared to the 351(a) pathway for biologic approval, the 351(k) pathway is shorter, meaning that biosimilars require less time and money to develop than biologics (Center for Drug Evaluation and Research 2022).

351(A) & 351(K) refer to sections of the Public Health Service (PHS) Act in the United States, specifically related to the approval of biological products by the Food and Drug Administration (FDA). They represent two different pathways for licensure as shown in Table 15.1.

In 2022, the FDA initiated the biosimilar regulatory science program intending to promote the development of interchangeable biological products and biosimilar products (Sheridan et al. 2024). A list of approved biosimilars in oncology in the United States is given in Table 15.2.

Table 15.1 Difference between 351 (A) and 351 (K) (Hung et al. 2017)

Feature	351(A)	351(K)
Purpose	Approve innovator biologics	Approve biosimilars
Application type	Stand-alone	Relies on reference product
Data requirements	Extensive	Less comprehensive, focused on biosimilarity
Development time	Longer	Shorter
Cost	Higher	Lower

Table 15.2 "FDA" approved biosimilars in oncology (Nahleh et al. 2022)

Biologics and their biosimilars (US/ FDA)			
Active ingredients	Reference brand	Biosimilar brand (approval year)	Sponsor
Filgrastim	Neupogen®	Zarxio (2015)	Sandoz
		Nyvestym (2018)	Kashiv biosciences
		Releuko (2022)	Pfizer
Pegfilgrastim	Neulasta®	Fulphila (2018)	Biocon/Mylan
		Udenyca (2018)	Coherus biosciences
		Ziextenzo (2019)	Sandoz
		Nyvepria (2020)	Pfizer
		Fylnetra (2022)	Kashiv biosciences
		Stimufend (2022)	Fresenius Kabi
Epoetin alfa	Epogen/ Procrit	Retacrit (2018)	Pfizer
Rituximab	Rituxan®	Truxima (2018)	Teva Pharmaceutical
		Ruxience (2018)	Pfizer
		Riabni (2020)	Amgen
Bevacizumab	Avastin®	Mvasi (2017)	Amgen
		Zirabev (2019)	Pfizer
		Alymsys (2022)	Amneal pharmaceuticals
		Vegzelma (2022)	Celltrion
		Avzivi (2023)	Bio-Thera solutions
Trastuzumab	Herceptin®	Herzuma (2018)	Teva Pharmaceutical
		Ontruzant (2018)	Merck
		Trazimera (2019)	Pfizer
		Kanjinti (2019)	Amgen

15.4.2 Europe

Europe is a leader in the regulation of biosimilars, established a robust regulatory framework for marketing authorization in 2004 and having approved the first biosimilar (Omnitrope®, a somatropin biosimilar) in 2006 (Barbier et al. 2022). The European Union has approved more biosimilars than any other country in the world during the last 10 years, gaining significant knowledge on their application and safety (Medicines for Europe 2022). The Committee for Medicinal Products for Human Use (CHMP), released detailed guidelines to address every facet of the development, manufacturing, testing, and regulation of biosimilar medications after the legal framework for biosimilars was established (Wolff-Holz et al. 2019). All EU recognized biological products, such as biosimilars, are subject to the centralized procedure for marketing authorization grant, as specified Regulation (EC) No 726/2004. This indicates that applications for marketing authorization are assessed scientifically by the EMA and approved by the European Commission (EC) (Kohler 2011).The approval of a biosimilar is contingent upon proving its biosimilarity—that is, its high degree of quality, safety, and efficacy in comparison to the reference product. This is accomplished by carrying out in-depth comparison studies with the reference product. The biosimilar guidelines contain scientific recommendations to facilitate the development of biosimilars through comparability programs with a reference medicinal product. The EMA offers more systematic and comprehensive guidelines that are divided into sections for immunogenicity testing, quality concerns, clinical and nonclinical testing, and pharmacokinetics (Niazi 2022a, 2022b). A list of approved biosimilars in oncology in Europe is given in Table 15.3.

15.4.3 India

India has an established regulatory system that enables biosimilar drugs to be marketed. India has become a major player in the biopharmaceuticals industry, particularly in the production of biosimilars, since it is one of the top producers of affordable, effective, and safe generic drugs (Meher et al. 2019). The first biosimilar, "Biovac-B", was authorized for sale in India in 2000 to treat hepatitis B; however, at that time, no formal guidelines had been set (Esteban et al. 2019). Similar biologics were approved by CDSCO and the Review Committee on Genetic Manipulation (RCGM) on an individual basis up until July 2012, using an abbreviated process for new drugs. In 2012, new guidelines for similar biologics were drafted by the CDSCO and the Department of Biotechnology (DBT), these guidelines were later amended in 2015 (Jois et al. 2020). The Drugs and Cosmetics Act of 1940 and Rules of 1945 for the production, use, import, export, and storage of genetically modified organisms, hazardous microorganisms, and cells as reported under the Environment Protection Act 1986 and Guidelines on Similar Biologics regulates similar biologics in India (Singhania and Tripathi 2012). The development and approval procedure for biosimilars was intended to be addressed and improved

Table 15.3 "EMA" approved biosimilars in oncology (Bachu et al. 2022)

Biologics and their biosimilars (EU/EMA)			
Active ingredients	Reference brand	Biosimilar brand (approval year)	Sponsor
Filgrastim	Neupogen®	Ratiograstin (2008)	Ratiopharm
		Tevagrastin (2008)	Teva generics
		Filgrastim Hexal (2009)	Hexal
		Zarzio (2009)	Sandoz
		Nivestim (2010)	Pfizer
		Grastofil (2013)	Apotex
		Accofil (2014)	Accord healthcare
Pegfilgrastim	Neulasta®	Ziextenzo (2018)	Sandoz
		Pelgraz (2018)	Accord healthcare
		Pelmeg (2018)	Cinfa biotech
		Fulphila (2018)	Mylan
		Grasustek (2019)	Juta pharma
		Cegfila (2019)	Mundipharma Biologics
		Stimufend (2022)	Fresenius Kabi
Epoetin alfa	Eprex®/Erypo®	Abseamed (2007)	Medice Arzneimittel Pütter
		Binocrit (2007)	Sandoz
		Epoetin Alfa Hexal (2007)	Hexal
Epoetin zeta	Eprex®	Retacrit (2007)	Pfizer
		Silapo (2007	Stada Arzneimittel
Rituximab	MabThera®	Truxima (2017)	Celltrion healthcare
		Blitzima (2017)	Celltrion
		Riximyo (2017)	Sandoz
		Rixathon (2017)	Sandoz
		Ruxience (2019)	Pfizer
		Zercepac (2020)	Accord healthcare
Bevacizumab	Avastin®	Mvasi (2018)	Amgen
		Aybintio (2020)	Samsung bioepis
		Onbevzi (2021)	Samsung bioepis
		Oyavas (2021)	Stada Arzneimittel
		Abevmy (2021)	Viatris
		Alymsys (2021)	mAbxience research
Trastuzumab	Herceptin®	Ontruzant (2017)	Samsung bioepis
		Trazimera (2018)	Pfizer
		Ogivri (2018)	Biocon
		Herzuma (2018)	Celltrion healthcare
		Kanjinti (2018)	Amgen
		Zercepac (2020)	Accord healthcare

by these guidelines. A similar biologic's application for market authorization may include a limited preclinical and clinical set of data, contingent on the comparative assessment demonstrating similarity (CDSCO 2016a). A list of approved biosimilars in oncology in the India is given in Table 15.4.

Table 15.4 "CDSCO" approved biosimilars in oncology (Priyarega and Natarajan 2022)

Biologics and their biosimilars (India/CDSCO)			
Active ingredients	Reference brand	Biosimilar brand (approval year)	Sponsor
Filgrastim	Neupogen®	Neukine (2004)	Intas biopharmaceuticals
		Religrast (2008)	Reliance Lifesciences
		Emgrast (2010)	Gennova biopharmaceuticals
		Colstim (2013)	Cadila pharmaceuticals
		Fegrast	Claris Lifesciences
		Filgrastim (2013)	Cadila pharmaceutical
		Grafeel (2001)	Dr. Reddy laboratories
Pegfilgrastim	Neulasta®	Neupeg (2007)	Intas biopharmaceuticals
		Pegex (2010)	Gennova biopharmaceuticals
		Peg-grafeel (2011)	Dr. Reddy laboratories
		Lupifil (2013)	Lupin
Epoetin alfa	Eprex®	Ceriton	Ranbaxy
		Epotin	Claris Lifesciences
		Erykine (2005)	Intas biopharmaceuticals
		Wepox (2001)	Wockhardt
Darbepoetin alfa	Aranesp®	Cresp (2010)	Dr. Reddy laboratories
		Darbatitor (2014)	Torrent pharmaceuticals
Rituximab	Rituxan®	Reditux (2007)	Dr. Reddy laboratories
		MabTas (2013)	Intas biopharmaceuticals
		Rituximab (2013)	Zenotech laboratories
		RituxiRel (2015)	Reliance life sciences
		Maball (2015)	Hetero drugs
		Acellbia (2017)	Biocad
Bevacizumab	Avastin®	Bevacirel (2016)	Reliance life sciences
		Cizumab (2016)	Hetero drugs
		Krabeva (2017)	Biocon
		Bevaro (2020)	Cadila pharmaceuticals
Trastuzumab	Herceptin®	CanMab (2013)	Biocon

15.4.4 Japan

The Pharmaceutical and Medical Devices Agency (PMDA) is the regulatory body in charge of assessing and approving, as well as is in charge of managing the registration and regulation of biologics and biosimilars in Japan (Consulting 2023). Biosimilar guidelines specific to a class are not published by the PMDA. However, the guidelines for ensuring the quality, safety, and efficacy of follow-on biologics were released in 2009 and updated in 2020 (Klein et al. 2022). The updated guidelines make it very clear that Japanese participants must be enrolled in at least one of the biosimilar's comparative pharmacokinetic and clinical efficacy studies. Furthermore, there are no laws governing interchangeability, and it is legal in Japan to move between reference products and biosimilars (Arato 2016). Hospitals administering oncology biosimilars in Japan have the authority to choose whether to utilize the reference product or the biosimilar. Professional oncology societies publish guidelines regarding biosimilars in oncology (Tan et al. 2020). The clinical practice guidelines for breast cancer published by the Japanese Breast Cancer Society (2018), for instance, indicate that biosimilars can save costs and have been demonstrated to be safe, effective, and of high quality (Iwata et al. 2019). Additionally, since biosimilar filgrastim has comparable efficacy and safety to the reference medication, it is advised for use in stimulating neutrophil production after chemotherapy by the Japan Society of Clinical Oncology (Hara et al. 2019).

15.4.5 China

In China, 40% of the biologics market is made up of biosimilars. Biosimilars are being used more frequently in oncology since cancer is now the second most common cause of death in the nation (Luo et al. 2023). The guidelines for biopharmaceuticals regulation used in oncology were set in the late 1990s, and any biological product must be marketed for 6 years before it may be utilized. The China Food and Drug Administration (CFDA) currently known as National Medical Product Administration (NMPA) published regulatory guidelines for biosimilars in 2015 (Liu et al. 2022). Biosimilars are approved in terms of the biological products. There have been fewer submissions for new biosimilars as a result of some criteria requiring enormous volumes of data regarding preclinical and clinical trials (Lemery et al. 2016). It mostly depends on the circumstances and the features of the reference product whether a biosimilar medicine can be eligible for fast-track prescription registration. At the moment, the NMPA is working on creating a set of regulations for specific biosimilars (Iskit 2021). A list of approved biosimilars in oncology in China is given in table 15.5.

15.4.6 Australia

The Therapeutic Goods Administration (TGA) oversees the regulation of biologics and biosimilars in Australia. TGA is in charge of evaluating and controlling the

Table 15.5 'NMPA' approved biosimilars in oncology (AJMC 2023)

Biologics and their Biosimilars (China/NMPA)			
Active Ingredients	Reference brand	Biosimilar Brand (Approval year)	Sponsor
Rituximab	MabThera®	Hanlikon (HLX01) (2019)	Shanghai Fuhong Hanlin bio-pharmaceutical
		Halpryza (2020)	Innovent Biologics/ Eli Lilly
Bevacizumab	Avastin®	Ankada (2019)	Qilu pharmaceutical
		Byvasda (2020)	Innovent Biologics
Trastuzumab	Herceptin®	Zercepac (HLX02) (2020)	Shanghai Henlius biotech

Table 15.6 "TGA" approved biosimilars in oncology

Biologics and their biosimilars (Australia/TGA)			
Active ingredients	Reference brand	Biosimilar brand (approval year)	Sponsor
Filgrastim	Neupogen®	Nivestim® (2010)	Sandoz
		Tevagrastim® (2011)	Pfizer
		Zarzio® (2013)	Sandoz
Pegfilgrastim	Neulasta®	Fulphila® (2018)	Alphapharm
		Lapelga® (2019)	Apotex
		Pelgraz® (2019)	Accord healthcare
		Neutropeg® (2019)	Accord healthcare
		Ziextenzo® (2019)	Sandoz
		Filpelga® (2022)	Cipla
Epoetin lambda	Eprex®	Grandicrit (2010)	Sandoz
		Aczicrit (2010)	Sandoz
		Novicrit (2010)	Novartis
Rituximab	MabThera®	Riximyo (2017)	Sandoz
		Truxima (2018)	Celltrion
		Ruxience (2021)	Pfizer
Bevacizumab	Avastin®	Zirabev (2019)	Pfizer
		Mvasi (2020)	Amgen
		Abevmy (2021)	Alphapharm
		Bevaciptin® (2021)	Cipla
Trastuzumab	Herceptin®	Herzuma (2018)	Celltrion
		Ogivri (2018)	Alphapharm
		Kanjinti (2019)	Samsung bioepis
		Ontruzant (2019)	Amgen
		Trazimera (2019)	Pfizer
		Trastucip (2022)	Pfizer

therapeutic product's quality, safety, and efficacy, including biologics and biosimilars (Patel et al. 2022). Australia's biosimilar regulations are mostly based on those of Europe, where in 2005, the European Medicine Agency (EMA) created regulations for biosimilar approval (Mounho-Zamora 2013). The Australian TGA embraced the EMA recommendations for biosimilar regulation in 2008, and it continues to do so by implementing many EMA criteria that are unique to product classes while evaluating biosimilar medications (Kang et al. 2021). The TGA released its own regulations on the assessment of biosimilars in 2013, and it was revised in 2015. Biosimilar medications are assessed via the standard medicine registration procedure, and applications must adhere to the same standards and regulations (Power 2013). A list of approved biosimilars in oncology in the Australia is given in table 15.6.

15.4.7 Canada

The Food and Drugs Act's Schedule D lists biologics in Canada, and Health Canada's Biologics and Genetic Therapies Directorate are in charge of overseeing

Table 15.7 "Health Canada" approved biosimilars in oncology

Biologics and their biosimilars (Canada/Health Canada)			
Active ingredients	Reference brand	Biosimilar brand (approval year)	Sponsor
Filgrastim	Nerpogen®	Grastofil (2015)	Apotex
		Nivestym (2020)	Pfizer
		Nypozi (2021)	Tanvex biopharma
Pegfilgrastim	Neulasta®	Fulphila (2018)	BGP pharma
		Lapelga (2018)	Apotex
		Nyvepria (2020)	Pfizer
		Ziextenzo (2020)	Sandoz
Rituximab	Rituxan®	Truxima (2019)	Celltrion healthcare
		Riximyo (2020)	Sandoz
		Ruxience (2020)	Pfizer
		Riabni (2021)	Amgen
Bevacizumab	Avastin®	Mvasi (2018)	Amgen
		Abevmy (2021)	BGP pharma
		Aybinto (2021)	Samsung bioepis
		Bambevi (2021)	Apotex
		Zirabev (2021)	Pfizer
		Vegzelma (2023)	Celltrion healthcare
Trastuzumab	Herceptin ®	Herzuma (2019)	Celltrion healthcare
		Trazimera (2019)	Pfizer
		Ogivri (2019)	BGP pharma
		Kanjinti (2020)	Amgen
		Ontruzant (2022)	Samsung bioepis

their approval and evaluation. According to Canadian regulations, a biosimilar is a biologic medication that has "demonstrated to be highly similar to a brand name drug already authorised for sale"; this latter medication is sometimes referred to as the "reference" biologic medication. In order to authorize as a biosimilar, the medicine producer must submit data to Health Canada demonstrating the high degree of similarity between the biosimilar and the reference biologic medication, as well as the absence of clinically significant deviations in terms of safety and efficacy. Health Canada evaluates all of the data provided by the manufacturer before deciding whether to approve a biosimilar for sale based on a benefit/risk analysis. Health Canada's approval of a biosimilar does not imply that it is bioequivalent to the reference or originator biologic medication. The determination of product interchangeability must be made by each province and territory in accordance with its own laws and regulations. Biosimilars are not considered interchangeable in Ontario, neither with other biosimilar products nor with the biologic originator/reference. To enable a secure and effective dispensing procedure, a prescription must unambiguously state the product's brand (originator or biosimilar) and generic name. A list of approved biosimilars in oncology in the Canada is given in table 15.7.

15.5 Case Studies and Examples

With time regulatory agencies like the FDA, EMA, NMPA, and PMDA have approved over number of biosimilars for various cancer indications, including bevacizumab, trastuzumab, rituximab, filgrastim, peg-filgrastim, and epoetin alfa. The development and approval of biosimilars are expected to continue at a rapid pace in the coming years. (Konstantinidou et al. 2019).

- *Filgrastim and Pegfilgrastim:* These are the two most widely used biosimilars for stimulating granulocyte-colony stimulating factors (G-CSFs). It is also used to shorten the length of fever and the time it takes for neutrophil recovery in patients with acute myeloid leukemia after induction or consolidation chemotherapy (Rastogi et al. 2021). Pegfilgrastim is used to stop chemotherapy-induced neutropenia, which is characterized by a decrease in certain white blood cells. In the European Union in 2009 and the United States in 2015, the filgrastim biosimilar Zarxio (Sandoz) received approval (Trautman et al. 2019). In 2001, regulatory approval was granted to Grafeel (Dr Reddy's Laboratories), another filgrastim biosimilar (Wynne et al. 2018). Other filgrastim biosimilars that have been approved in various countries include Filgrastim Hexal (EU), Filgrastim (India), Nivestim (Australia), Ratiograstim (EU), and Tevagrastim (Australia). Pelgraz, Udenyca, Fulphila, Grasustek, Cegfila, Stimufend, Pelmeg, and Ziextenzo are among the approved pegfilgrastim biosimilars in Europe; in contrast, Ziextenzo, Nyvepria, Fylnetra, and Stimufend are among the approved pegfilgrastim biosimilars in the USA (Cornes et al. 2020).
- *Bevacizumab*: Recombinant humanized monoclonal antibody bevacizumab suppresses the growth of new cancers and the production of new blood vessels (angiogenesis) by targeting vascular endothelial growth factor (VEGF-A) (Gerriets and Kasi 2019). In 2004, bevacizumab was licenced by the FDA to treat metastatic colorectal cancer as a first-line treatment when combined with fluorouracil-based chemotherapy (Strickler and Hurwitz 2012). The first biosimilar to Avastin (bevacizumab-awwb), Mvasi (Amgen), was authorized by the FDA in 2017 and the EMA in 2018. Additional bevacizumab biosimilars that have received approval in several nations include Zirabev (Australia), Bevacirel, Krabeva (India), Alymsys, Vegzelma, Avzivi (US), (Hexal), Abevmy, and Onbevzi (EU), etc. (Booth and Pilz 2021).
- *Rituximab*: It is one the oldest biosimilars being used in the nation to treat rheumatoid arthritis and non-Hodgkin lymphoma (Jurczak et al. 2019). Rituximab is a human monoclonal antibody that has been genetically modified to target the CD20 antigen, which is present on the surface of B cells. Rituximab and its biosimilars bind to the CD20 antigen, increasing the levels of IL-10 and B-cell lymphoma-2 (Bcl-2) and causing cellular death (Smith 2003). The first monoclonal antibody biosimilar to receive an Indian license was Reditux (Dr Reddy's Laboratories), a rituximab biosimilar. This approval was granted in 2007 (Roy et al. 2013).

- *Trastuzumab*: A humanized monoclonal antibody called trastuzumab binds specifically to the extracellular domain of the protein known as the human epidermal growth factor receptor 2 (HER2) (Gajria and Chandarlapaty 2011). The EMA approved Ontruzant, a drug manufactured by Samsung Bioepis, as the first biosimilar of trastuzumab Herceptin in 2017. Ontruzant can be used either alone or in combination with hormone therapy or chemotherapy for metastatic breast cancer. Ogivri® (Mylan), the first trastuzumab-dkst biosimilar in the United States, was authorized by the FDA in 2017. Ogivri has been licensed in combination with cisplatin and either capecitabine or 5-fluorouracil for the treatment of early breast cancer, metastatic breast cancer, and metastatic gastric cancer (Triantafyllidi and Triantafillidis 2022).
- *Epoetin:* An artificial form of natural erythropoietin known as an erythropoiesis-stimulating agent (ESA) can be used if body is unable to produce enough EPO to produce red blood cells. The kidneys secrete erythropoietin, a hormone that stimulates the formation of red blood cells in the bone marrow (Schoener and Borger 2022). In Europe, the following five epoetin biosimilars were approved in 2007: Abseamed, Binocrit, Ritacrit, Epoetin-α Hexal, and Silapo. However, in the United States, only Retacrit has received clearance. Several non-originator biologicals and biosimilars to epoetin alfa are available in India under the trade name "Similar biologic", such as Epofer, Erypro, and Wepox (Bieber 2001).

15.6 Biosimilar Interchangeability Regulatory Concept

The term "interchangeability" describes a product's capacity to be switched out for another without significantly increasing the chance of a negative health consequence (Alvarez et al. 2020). Interchangeability is regarded as a higher standard than biosimilarity under the US Biologics Price Competition and Innovation Act of 2009 (a part of the Affordable Care Act) since it permits the product to be replaced for the reference product without the assistance of a healthcare practitioner (Niazi 2022a). The FDA also proposed in its draft advice that the biosimilar product should be anticipated to have the same impact as the reference biologic product in any given patient to meet this higher standard of interchangeability. The United States biosimilar guidance defines interchangeability as a legally established route (Reinisch and Smolen 2015). The EMA in Europe does not oversee substitution and switching policies in Europe; instead, member states determine their own approaches based on the principle of interchangeability. Forced switching and substitution has been a practice in various European nations for many years, with no negative consequences (Messner et al. 2023). In order to demonstrate that their proposed interchangeable biosimilar may be switched with the reference product, the makers might conduct a switching study wherein a patient switches many times over a predetermined length of time between the reference product and the interchangeable product (Heinemann et al. 2015). Referring to the long-acting insulin Lantus, the FDA authorized Semglee (insulin glargine-yfgn) by Viatris as the first interchangeable biosimilar in the United States in July 2021. However, there aren't any FDA-approved equivalent products available for oncology at the moment (Matli et al. 2021). To our knowledge, to date, only five interchangeable biosimilar products have been approved in the United States. Biosimilars are not automatically interchangeable with reference items in China; there is no interchangeable designation for them. It is legal in Japan to move between reference products and biosimilars, and there are no laws governing interchangeability (Rumore and Randy Vogenberg 2016).

15.7 Regulatory Landscape Evolution

After the European Medicines Agency (EMEA, presently EMA) established the regulatory approval process for biosimilar medicines in 2005, the first biosimilar, Omnitrope, was approved by the agency in 2006 (Ahmed et al. 2012). The Biologics Price Competition and Innovation Act (BPCI), created by the US Congress in 2009, started the US pathway and made it possible for the Food and Drug Administration (FDA) to approve Zarxio, the country's first biosimilar, in 2015 (Ghergheseu and Delgado-Charro 2020). A number of biological products, including follow-on medicines, have received FDA approval as New Drug Applications prior to the creation of the FDA biosimilar program. Some of these products have been added to the FDA's "Purple Book: Lists of Licensed Biological Products with Reference Product Exclusivity and Biosimilarity or Interchangeability Evaluations" and removed from the FDA's Orange Book as of late (March 2020) (White and Goldman 2018). These products were previously considered to be Biologic License Applications (BLAs). A new EU guideline went into force in 2015, and the FDA created the Biosimilars Action Plan in 2018 to help in the development, approval, and US commercialization of biosimilars (Niazi 2023).

15.8 Future Trends and Conclusion

The regulatory landscape for oncology biosimilars is evolving to support their vital role in cancer treatment. Efforts to simplify approval processes by reducing extensive clinical trials and relying more on real-world evidence are underway. Global collaboration on regulatory standards is crucial for faster biosimilar development and ensuring global patient access. Addressing concerns about immunogenicity involves standardized assays and long-term studies to understand potential risks. Ensuring competition and access includes preventing anti-competitive practices and improving affordability through strategies like price transparency. As biosimilars for complex biologics emerge, regulations need to adapt to address unique challenges, including evolving manufacturing technologies.

Realizing interchangeability poses regulatory challenges in oncology biosimilars. Regulatory uncertainty from diverse guidelines across regions requires standardization. Extensive and costly clinical trials hinder timely development, requiring alternative approaches such as pharmacodynamic studies and real-world data analysis. Extrapolating approval to additional indications demands rigorous scientific justification and potential confirmatory trials. The complexity of cancer treatment regimens and individual patient factors complicates establishing interchangeability. Building trust and providing clear information are essential. International regulatory collaboration is vital for harmonized standards. Innovations like adaptive trials and Bayesian approaches offer opportunities to optimize data collection and reduce costs. Leveraging real-world evidence post-approval can bolster confidence. Evidence-based guidelines, education, and communication strategies are crucial to overcoming hesitancy and maximizing biosimilars' benefits in cancer treatment. As a need of the hour, the future regulatory landscape for oncology biosimilars is focused on balancing patient safety, affordability, and innovation to ensure these vital drugs reach those who need them. The development and approval of biosimilars are expected to continue at a rapid pace in the coming years. This will further expand treatment options for cancer patients and make them more affordable, ultimately improving patient outcomes and healthcare sustainability.

References

Ahmed I, Kaspar B, Sharma U (2012) Biosimilars: impact of biologic product life cycle and European experience on the regulatory trajectory in the United States. Clin Ther 34(2):400–419. https://doi.org/10.1016/j.clinthera.2011.12.005

AJMC (2023) Part 1: oncology Biosimilars offer comparable benefits to originators at lower prices. Center for Biosimilars. https://www.centerforbiosimilars.com/view/part-1-oncology-biosimilars-offer-comparable-benefits-to-originators-at-lower-prices. Accessed19 Jan. 2024

Alvarez DF, Wolbink G, Cronenberger C, Orazem J, Kay J (2020) Interchangeability of Biosimilars: what level of clinical evidence is needed to support the interchangeability designation in the United States? BioDrugs 34(6):723–732. https://doi.org/10.1007/s40259-020-00446-7

American Cancer Society (2018a). *Estimated Number of New Cancer Cases by World Area, 2018**. https://www.cancer.org/content/dam/cancer-org/research/cancer-facts-and-statistics/global-cancer-facts-and-figures/global-cancer-facts-and-figures-4th-edition.pdf

American Cancer Society (2018b) Understanding biologic and biosimilar drugs. American Cancer Society Cancer Action Network. https://www.fightcancer.org/policy-resources/understanding-biologic-and-biosimilar-drugs. Accessed 28 Jun 2022

Anand U, Dey A, Chandel AKS, Sanyal R, Mishra A, Pandey DK, De Falco V, Upadhyay A, Kandimalla R, Chaudhary A, Dhanjal JK, Dewanjee S, Vallamkondu J, Pérez de la Lastra JM (2022) Cancer chemotherapy and beyond: current status, drug candidates, associated risks and progress in targeted therapeutics. Genes Dis 10(4):1367–1401. https://doi.org/10.1016/j.gendis.2022.02.007

Arato T (2016) Japanese regulation of biosimilar products: past experience and current challenges. Br J Clin Pharmacol 82(1):30–40. https://doi.org/10.1111/bcp.12931

Bachu RD, Abou-Dahech M, Balaji S, Boddu SHS, Amos S, Singh V, Babu RJ, Tiwari AK (2022) Oncology biosimilars: new developments and future directions. Cancer Rep 5(11):e1720. https://doi.org/10.1002/cnr2.1720

Barbier L, Mbuaki A, Simoens S, Declerck P, Vulto AG, Huys I (2022) Regulatory information and guidance on Biosimilars and their use across Europe: a call for strengthened one voice messaging. Front Med 9:820755. https://doi.org/10.3389/fmed.2022.820755

Bieber E (2001) Erythropoietin, the biology of erythropoiesis and epoetin alfa. An overview. *The*. J Reprod Med 46(5 Suppl):521–530

Blackstone EA, Joseph PF (2013) The economics of Biosimilars. Am Health Drug Benefits 6(8):469–478. https://www.ncbi.nlm.nih.gov/pmc/articles/PMC4031732/

Booth JP, Pilz J (2021) Retrospective indication-matched cohort study of reference product and biosimilar: bevacizumab versus bevacizumab-Awwb. Hosp Pharm 57(4):455–461. https://doi.org/10.1177/00185787211046865

CDSCO (2016a) Guidelines on similar biologics regulatory Requirements for Marketing Authorization in India 2016 Central Drugs Standard Control Organiza on Ministry of Health & Family Welfare. https://birac.nic.in/webcontent/Guidelines_on_Similar_Biologics_06_10_2017.pdf. Accessed 25 May 2021

CDSCO (2016b). Guidelines on similar biologics: Regulatory Requirements for Marketing Authorization in India Central Drugs Standard Control Organization Ministry of Health & Family Welfare Government of India Document Name: guidelines on similar biologics Effective From Year: 15th August, 2016 Validity: Till Further Revision 2 Guidelines on Similar Biologics: Regulatory Requirements for Marketing Authorization in India Content. https://cdsco.gov.in/opencms/resources/UploadCDSCOWeb/2018/Upload Alerts Files/BiosimilarGuideline2016.pdf

Chopra R, Lopes G (2017) Improving access to cancer treatments: the role of Biosimilars. J Glob Oncol 3(5):596–610. https://doi.org/10.1200/jgo.2016.008607

Consulting A (2023) Biologics and biosimilar regulation and registration in Japan (PMDA). Artixio. https://www.artixio.com/post/biologics-and-biosimilar-regulation-and-registration-in-japan-pmda. Accessed 19 Jan 2024

Cornes P, Gascon P, Vulto AG, Aapro M (2020) Biosimilar Pegfilgrastim: improving access and Optimising practice to supportive care that enables cure. BioDrugs 34(3):255–263. https://doi.org/10.1007/s40259-020-00411-4

Curtis JR, Singh JA (2011) Use of Biologics in rheumatoid arthritis: current and emerging paradigms of care. Clin Ther 33(6):679–707. https://doi.org/10.1016/j.clinthera.2011.05.044

EMA (2006). Biosimilar medicines: overview|European medicines agency. www.ema.europa.eu. https://www.ema.europa.eu/en/human-regulatory-overview/biosimilar-medicines-overview#:~:text=in%20the%20EU-

Esteban E, Bustos R-H, García J-C, Jáuregui E (2019) Biosimilars: an approach to some current worldwide regulation frameworks. Curr Clin Pharmacol 14(1):16–40. https://doi.org/10.2174/1574884713666181025142928

FDA (2020a). FDA/FTC workshop on a competitive marketplace for biosimilars a matter of record (301) 890–4188 min-U-script® with word index. https://www.fda.gov/media/136791/download. Accessed 19 Jan. 2024

FDA (2020b) Quality considerations in demonstrating biosimilarity of a therapeutic protein product to a reference product guidance for industry. U.S. Food and Drug Administration. https://www.fda.gov/regulatory-information/search-fda-guidance-documents/quality-considerations-demonstrating-biosimilarity-therapeutic-protein-product-reference-product. Accessed 17 Mar. 2023

FDA (2022) Center for drug evaluation and research. Review and approval. https://www.fda.gov/drugs/biosimilars/review-and-approval

FDA (2023) Overview of biosimilar products. https://www.fda.gov/media/151058/download#:~:text=A%20biosimilar%20has%20no%20clinically

Gajria D, Chandarlapaty S (2011) HER2-amplified breast cancer: mechanisms of Trastuzumab resistance and novel targeted therapies. Expert Rev Anticancer Ther 11(2):263–275. https://doi.org/10.1586/era.10.226

Generics and biosimilars initiative (2023). Biosimilars approved in the US.. www.gabionline.net. https://www.gabionline.net/biosimilars/general/biosimilars-approved-in-the-us#:~:text=In%20the%20US%2C%20a%20legal. Accessed 19 Jan. 2024

Gerriets, V. and Kasi, A. (2019). Bevacizumab. Nih.gov. https://www.ncbi.nlm.nih.gov/books/NBK482126/

Gherghescu I, Delgado-Charro MB (2020) The biosimilar landscape: an overview of regulatory approvals by the EMA and FDA. Pharmaceutics 13(1):48. https://doi.org/10.3390/pharmaceutics13010048

Hara F, Tajima K, Tanabe K (2019) Current situation and challenges regarding biosimilars in Japan: an example of trastuzumab biosimilars for breast cancer. Future Oncol 15(12):1353–1361. https://doi.org/10.2217/fon-2018-0957

Heinemann L, Khatami H, McKinnon R, Home P (2015) An overview of current regulatory requirements for approval of biosimilar insulins. Diabetes Technol Ther 17(7):510–526. https://doi.org/10.1089/dia.2014.0362

Hung A, Vu Q, Mostovoy L (2017) A systematic review of U.S. biosimilar approvals: what evidence does the FDA require and how are manufacturers responding? J Manag Care Spec Pharm 23(12):1234–1244. https://doi.org/10.18553/jmcp.2017.23.12.1234

Inotai A, Csanadi M, Petrova G, Dimitrova M, Bochenek T, Tesar T, York K, Fuksa L, Kostyuk A, Lorenzovici L, Omelyanovskiy V, Egyed K, Kalo Z (2018) Patient access, unmet medical need, expected benefits, and concerns related to the utilisation of Biosimilars in eastern European countries: a survey of experts. Biomed Res Int 2018:1–9. https://doi.org/10.1155/2018/9597362

Iskit AB (2021) Key concepts in biosimilar medicines: what physicians must know. Northern Clinics of Istanbul 9(1):86–91. https://doi.org/10.14744/nci.2021.84669

Iwata H, Saji S, Ikeda M, Inokuchi M, Uematsu T, Toyama T, Horii R, Yamauchi C (2019) The Japanese breast cancer society clinical practice guidelines, 2018 edition: the tool for shared decision making between doctor and patient. Breast Cancer 27(1):1–3. https://doi.org/10.1007/s12282-019-01021-x

Jha A, Upton A, Dunlop WCN, Akehurst R (2015) The budget impact of biosimilar infliximab (Remsima®) for the treatment of autoimmune diseases in five European countries. Adv Ther 32(8):742–756. https://doi.org/10.1007/s12325-015-0233-1

Jois R, Mukherjee S, Rajeswari S, Rath P, Goyal V, Gupta D (2020) Similar biologics in India: A story of access or potential for compromise? Indian J Med Res 152(5):456–456. https://doi.org/10.4103/ijmr.ijmr_43_18

Joshi D, Khursheed R, Gupta S, Wadhwa D, Singh TG, Sharma S, Porwal S, Gauniyal S, Vishwas S, Goyal S, Gupta G, Eri RD, Williams KA, Dua K, Singh SK (2022) Biosimilars in oncology: latest trends and regulatory status. Pharmaceutics 14(12):2721. https://doi.org/10.3390/pharmaceutics14122721

Jurczak W, Długosz Danecka M, Buske C (2019) Rituximab biosimilars for lymphoma in Europe. Expert Opin Biol Ther 19(10):1045–1056. https://doi.org/10.1080/14712598.2019.1665017

Kabir ER, Moreino SS, Sharif Siam MK (2019) The breakthrough of Biosimilars: a twist in the narrative of biological therapy. Biomol Ther 9(9):410. https://doi.org/10.3390/biom9090410

Kang H-N, Knezevic I (2018) Regulatory evaluation of biosimilars throughout their product life-cycle. Bull World Health Organ 96(4):281–285. https://doi.org/10.2471/blt.17.206284

Kang H, Thorpe R, Knezevic I, Casas Levano M, Chilufya MB, Chirachanakul P, Chua HM, Dalili D, Foo F, Gao K, Habahbeh S, Hamel H, Kim GH, Perez Rodriguez V, Putri DE, Rodgers J, Savkina M, Semeniuk O, Srivastava S, Tavares Neto J (2021) Regulatory challenges with biosimilars: an update from 20 countries. Ann N Y Acad Sci 1491(1):42–59. https://doi.org/10.1111/nyas.14522

Kirchhoff CF, Wang XM, Conlon HD, Anderson S, Ryan AM, Bose A (2017) Biosimilars: key regulatory considerations and similarity assessment tools. Biotechnol Bioeng 114(12):2696–2705. https://doi.org/10.1002/bit.26438

Klein K, Gencoglu M, Heisterberg J, Acha V, Stolk P (2022) The global landscape of manufacturers of follow-on Biologics: an overview of five major biosimilar markets and 15 countries. BioDrugs 37(2):235–245. https://doi.org/10.1007/s40259-022-00568-0

Kohler M (2011) Regulatory pathways in the European Union. MAbs 3(3):241–242. https://doi.org/10.4161/mabs.3.3.15474

Konstantinidou S, Papaspiliou A, Kokkotou E (2019) Current and future roles of biosimilars in oncology practice (review). Oncol Lett 19(1):45–51. https://doi.org/10.3892/ol.2019.11105

Kumar R, Singh J (2014) Biosimilar drugs: current status. Int J Appl Basic Med Res 4(2):63. https://doi.org/10.4103/2229-516x.136774

Kurki P, Barry S, Bourges I, Tsantili P, Wolff-Holz E (2021) Safety, immunogenicity and interchangeability of biosimilar monoclonal antibodies and fusion proteins: a regulatory perspective. Drugs 81:1881–1896. https://doi.org/10.1007/s40265-021-01601-2

Kvien TK, Patel K, Strand V (2022) The cost savings of biosimilars can help increase patient access and lift the financial burden of health care systems. Semin Arthritis Rheum 52:151939. https://doi.org/10.1016/j.semarthrit.2021.11.009

Lemery SJ, Esteva FJ, Weise M (2016) Biosimilars: here and now. Am Soc Clin Oncol Educ Book 36:e151–e157. https://doi.org/10.1200/edbk_155954

Liu Y, Zhang N, Xie C, Jiang Y, Qin Y, Zhou L, Fan Y, Ren L, Yin C, Yang H, Xie W, Zhai Q, Li G, Chen H, Chen X (2022) Evolution of drug regulations and regulatory innovation for anticancer drugs in China. Acta Pharm Sin B 12(12):4365–4377. https://doi.org/10.1016/j.apsb.2022.08.004

Luo X, Du X, Li Z, Li J, Lv X, Li H, Guo Q, Wang C, Xue X, Le K, Jiang X, Huang L, Yang Y (2023) Clinical benefit, Price, and uptake for cancer Biosimilars vs reference drugs in China. JAMA Netw Open 6(10):e2337348–e2337348. https://doi.org/10.1001/jamanetworkopen.2023.37348

Makurvet FD (2020) Biologics versus small molecules: drug costs and patient access. Medicine in Drug Discovery 9(9):100075. https://doi.org/10.1016/j.medidd.2020.100075

Mariotto AB, Enewold L, Zhao J, Zeruto CA, Yabroff KR (2020) Medical care costs associated with cancer survivorship in the United States. Cancer Epidemiol Biomarkers Prev 29(7):1304–1312. https://doi.org/10.1158/1055-9965.epi-19-1534

Matli MC, Wilson AB, Rappsilber LM, Sheffield FP, Farlow ML, Johnson JL (2021) The first interchangeable biosimilar insulin: insulin glargine-yfgn. J Diabetes Sci Technol 17(2):490–494. https://doi.org/10.1177/19322968211067511

Medicines for Europe (2022). Reading listl biosimilar medicines. https://www.medicinesforeurope.com/wp-content/uploads/2022/04/M-Biosimilars-Reading-list-20220413_for_publication.pdf. Accessed 19 Jan 2024

Meher B, Balan S, Mohanty R, Jena M, Das S (2019) Biosimilars in India; current status and future perspectives. J Pharm Bioallied Sci 11(1):12. https://doi.org/10.4103/jpbs.jpbs_167_18

Messner K, Eickhoff C, Schulz M, Allemann S, Arnet I (2023) Knowledge and attitudes of German and Swiss community pharmacists towards biologicals and biosimilars—a prospective survey before and after the COVID-19 pandemic. BMC Health Serv Res 23(1):1432. https://doi.org/10.1186/s12913-023-10475-x

Morrow T, Felcone LH (2004) Defining the difference: what makes Biologics unique. Biotechnol Healthc 1(4):24–29. https://www.ncbi.nlm.nih.gov/pmc/articles/PMC3564302/

Mounho-Zamora B (2013) Regulatory standards for the approval of biosimilar products. Elsevier eBooks, pp 159–184. https://doi.org/10.1016/b978-0-12-394810-6.00006-x

Mysler E, Azevedo VF, Danese S, Alvarez D, Iikuni N, Ingram B, Mueller M, Peyrin-Biroulet L (2021) Biosimilar-to-biosimilar switching: what is the rationale and current experience? Drugs 81(16):1859–1879. https://doi.org/10.1007/s40265-021-01610-1

Nahleh Z, Lyman GH, Schilsky RL, Peterson DE, Tagawa ST, Chavez-MacGregor M, Rumble RB, Gupta S (2022) Use of biosimilar medications in oncology. JCO Oncology Practice 18(3):177–186. https://doi.org/10.1200/op.21.00771

Niazi SK (2022a) No two classes of biosimilars: urgent advice to the US congress and the FDA. J Clin Pharm Ther 47(9):1352–1361. https://doi.org/10.1111/jcpt.13743

Niazi SK (2022b) The coming of age of Biosimilars: a personal perspective. Biologics 2(2):107–127. https://doi.org/10.3390/biologics2020009

Niazi SK (2023) The FDA's new guideline 'generally accepted scientific knowledge' (GASK): an opportunity to expedite the approval of Biosimilars. Pharmaceuticals 16(11):1517–1517. https://doi.org/10.3390/ph16111517

Patel Y, Patel AA, Patel RJ (2022) A comparative study of biologic regulation in US, Canada, Australia, Europe and Singapore. Int. J. Drug Regul 10(2):28–34. https://doi.org/10.22270/ijdra.v10i2.518

Pfizer (2022) Biologics vs. Biosimilars: understanding the difference. Pfizer. www.pfizer.com. https://www.pfizer.com/news/articles/biologics_vs_biosimilars_understanding_the_

differences#:~:text=Biosimilars%20undergo%20rigorous%20testing%20and. Accessed 28 Nov 2023

Power D (2013) Licensing and prescribing biosimilars in Australia. GaBI Journal 2(3):152–154. https://doi.org/10.5639/gabij.2013.0203.030

Priyarega S, Natarajan R (2022) An overview of biosimilars for cancer, diabetes mellitus, rheumatoid arthritis and other immune-mediated diseases approved between 2016 and 2021. Results Chem 4:100356. https://doi.org/10.1016/j.rechem.2022.100356

Rak Tkaczuk KH, Jacobs IA (2014) Biosimilars in oncology: from development to clinical practice. Semin Oncol 41:S3–S12. https://doi.org/10.1053/j.seminoncol.2014.03.008

Rastogi S, Kalaiselvan V, Ali S, Ahmad A, Guru SA, Sarwat M (2021) Efficacy and safety of Filgrastim and its Biosimilars to prevent febrile neutropenia in cancer patients: a prospective study and meta-analysis. Biology 10(10):1069. https://doi.org/10.3390/biology10101069

Reinisch W, Smolen J (2015) Biosimilar safety factors in clinical practice. Semin Arthritis Rheum 44(6):S9–S15. https://doi.org/10.1016/j.semarthrit.2015.04.005

Roy PS, John S, Karankal S, Kannan S, Pawaskar P, Gawande J, Bagal B, Khattry N, Sengar M, Menon H, Gujral S, Nair R (2013) Comparison of the efficacy and safety of rituximab (Mabthera™) and its biosimilar (Reditux™) in diffuse large B-cell lymphoma patients treated with chemo-immunotherapy: a retrospective analysis. Indian Journal of Medical and Paediatric Oncology 34(04):292–298. https://doi.org/10.4103/0971-5851.125248

Rumore MM, Randy Vogenberg F (2016) Biosimilars: still not quite ready for prime time. P & T: a peer-reviewed journal for formulary management 41(6):366–375. https://www.ncbi.nlm.nih.gov/pmc/articles/PMC4894513/. Accessed 27 Nov. 2019

Schiestl M, Zabransky M, Sörgel F (2017) Ten years of biosimilars in Europe: development and evolution of the regulatory pathways. Drug Des Devel Ther 11:1509–1515. https://doi.org/10.2147/DDDT.S130318

Schoener B, Borger J (2022) Erythropoietin stimulating agents. PubMed. https://www.ncbi.nlm.nih.gov/books/NBK536997/#:~:text=Erythropoietin%20stimulating%20agents%20(ESAs)%20are

Schreiber S, Bandyopadhyay D, Gonçalves J, Mease PJ, Panaccione R, Emery P (2022) Critical appraisal and future outlook on anti-inflammatory biosimilar use in chronic immune-mediated inflammatory diseases. Semin Arthritis Rheum 55:152023. https://doi.org/10.1016/j.semarthrit.2022.152023

Sheridan M, Massich M, Ashourian N (2024) Biosimilars. J Infus Nurs 47(1):19–29. https://doi.org/10.1097/nan.0000000000000528

Singhania SS, Tripathi KK (2012) India: similar biologics guidelines and implementation. J Generic Med 9(4):181–186. https://doi.org/10.1177/1741134312466560

Smith MR (2003) Rituximab (monoclonal anti-CD20 antibody): mechanisms of action and resistance. Oncogene 22(47):7359–7368. https://doi.org/10.1038/sj.onc.1206939

Strickler JH, Hurwitz HI (2012) Bevacizumab-based therapies in the first-line treatment of metastatic colorectal cancer. Oncologist 17(4):513–524. https://doi.org/10.1634/theoncologist.2012-0003

Sung H, Ferlay J, Siegel RL, Laversanne M, Soerjomataram I, Jemal A, Bray F (2021) Global cancer statistics 2020: GLOBOCAN estimates of incidence and mortality worldwide for 36 cancers in 185 countries. CA Cancer J Clin 71(3):209–249

Sylvester K, Rocchio M, Beik N, Fanikos J (2013) Biosimilars: an emerging category of biologic drugs for emergency medicine practitioners. Curr Emerg Hosp Med Rep 1(4):226–235. https://doi.org/10.1007/s40138-013-0023-5

Tan CJ, Ke Y, Chan A (2020) Proceedings of the biosimilars workshop at the international symposium on oncology pharmacy practice 2019. J Oncol Pharm Pract 26(3_suppl):33–39. https://doi.org/10.1177/1078155219897959

Trautman H, Szabo E, James E, Tang B (2019) Patient-administered biologic and biosimilar Filgrastim may offer more affordable options for patients with nonmyeloid malignancies receiving chemotherapy in the United States: a budget impact analysis from the payer perspective. J Manag Care Spec Pharm 25(1):94–101. https://doi.org/10.18553/jmcp.2018.18094

Triantafyllidi E, Triantafillidis JK (2022) Systematic review on the use of Biosimilars of Trastuzumab in HER2+ breast cancer. Biomedicines 10(8):2045. https://doi.org/10.3390/biomedicines10082045

Triplitt C, Hinnen D, Valentine V (2017) How similar are Biosimilars? What do clinicians need to know about biosimilar and follow-on insulins? Clin Diabetes 35(4):209–216. https://doi.org/10.2337/cd16-0072

Ventola CL (2013) Biosimilars. Pharmacy and Therapeutics 38(5):270–287. https://www.ncbi.nlm.nih.gov/pmc/articles/PMC3737980/

Vulto AG, Jaquez OA (2017) The process defines the product: what really matters in biosimilar design and production? Rheumatology 56(suppl_4):iv14–iv29. https://doi.org/10.1093/rheumatology/kex278

White J, Goldman J (2018) Biosimilar and follow-on insulin: the ins, outs, and interchangeability. J Pharm Technol 35(1):25–35. https://doi.org/10.1177/8755122518802268

WHO (2018). WHO questions and answers: similar biotherapeutic products complementary document to the WHO guidelines on evaluation of similar biotherapeutic products (SBPs). https://cdn.who.int/media/docs/default-source/biologicals/qa_for_sbps_ecbs_2018.pdf?sfvrsn=23af9c60_3&download=true

Wolff-Holz E, Tiitso K, Vleminckx C, Weise M (2019) Evolution of the EU biosimilar framework: past and future. BioDrugs 33(6):621–634. https://doi.org/10.1007/s40259-019-00377-y

World Health Organization (2020) The top 10 causes of death. World Health Organization. https://www.who.int/news-room/fact-sheets/detail/the-top-10-causes-of-death

Wynne C, Schwabe C, Batra SS, Lopez-Lazaro L, Kankanwadi S (2018) A comparative pharmacokinetic study of DRL_BZ, a candidate biosimilar of bevacizumab, with Avastin® (EU and US) in healthy male subjects. Br J Clin Pharmacol 84(10):2352–2364. https://doi.org/10.1111/bcp.13691

Future Scope of Biosimilars for the Treatment of Various Cancers

16

Ashwin Subramanian, Gayathri Devi Muthukumarasamy, Saraswathi Venkataraman, Karthika Rangasamy, Ananya Nitin Kanade, Pavithra Vimala Arulrajan, Keerthana Saravanan, Suhashini Shanmuganathan, Gayathri Gopal, and Senthil Visaga Ambi

Abstract

Advancements in the identification of molecular drug targets and new therapeutic options for different cancer types have paved the way for target-based therapies for cancer. In this vein, biosimilars have revolutionized the cancer treatment options and have significant clinical outcomes. Because of the development of biologic medications, cancer treatment and supportive care have evolved dramatically. Biologics are increasingly being employed in cancer therapy and palliative care. However, these therapies continue to impose a significant financial burden on healthcare stakeholders. Cancer biologics are advanced treatments derived from the living organisms that are in practice in modern healthcare. But there is a shift towards biosimilars, which is driven by the aim to increase affordability while upholding the same high standards of quality, safety, and effectiveness for these life-saving therapies. The objective of this transition is to provide economically viable alternatives for patients without compromising the fundamental characteristics of these pioneering drugs, which are essential in the treatment of cancer. Biosimilars are found to play an increasingly important role in cancer treatment, assisting in improving access to novel treatment options by lowering costs. Since biosimilars are substantially similar but not identical to

A. Subramanian · G. D. Muthukumarasamy · S. Venkataraman · K. Rangasamy · A. N. Kanade · P. V. Arulrajan · K. Saravanan · S. Shanmuganathan
Biopharmaceutical Research Lab, Anusandhan Kendra-1, School of Chemical and Biotechnology, SASTRA Deemed-to-be-University, Thanjavur, Tamil Nadu, India

G. Gopal · S. V. Ambi (✉)
Biopharmaceutical Research Lab, Anusandhan Kendra-1, School of Chemical and Biotechnology, SASTRA Deemed-to-be-University, Thanjavur, Tamil Nadu, India

Department of Bioengineering, School of Chemical and Biotechnology, SASTRA Deemed-to-be-University, Thanjavur, Tamil Nadu, India
e-mail: senthilvisagaambi@scbt.sastra.edu

S. Bhatt et al. (eds.), *Biosimilars for Cancer Treatment*,
https://doi.org/10.1007/978-981-97-3001-8_16

reference products, rigorous regulatory procedures have been established for their approval. These regulatory frameworks specify the conditions for biosimilar approval, such as analytical tests comparing the biosimilar to the authorized biologic medicine. To determine the toxicity, efficacy, safety profiles, and pharmacokinetic/pharmacodynamic properties, nonclinical and clinical research are necessary for some of the biosimilars that have been approved by FDA including rituximab, trastuzumab, bevacizumab, filgrastim, pegfilgrastrim, and epoetins. When compared to chemotherapy alone, the addition of monoclonal antibodies such as bevacizumab and trastuzumab to the antineoplastic therapy has helped to significantly improve key outcomes such as progression-free survival (PFS) and overall survival (OS). Trastuzumab, for example, is designed to selectively inhibit the human epidermal growth factor 2 (HER2) receptor, whereas bevacizumab is designed to target vascular endothelial growth factor. Biosimilars not only increase the range of treatment options but also provide a safety net when the reference biologic is in short supply or unsuitable for specific patients. Additionally, the affordability of biosimilars contributes to the long-term sustainability of healthcare systems, especially in the face of the growing demand for cancer treatments. Finally, the development and production of biosimilars foster ongoing research and innovation in the fields of biotechnology and oncology, ultimately leading to improvements in treatment methods and patient outcomes. The development process for a biosimilar differs significantly from that of the original biologic. A better comprehension of the rigorous procedures involved in the development and approval of biosimilar drugs, along with increasing experience in their utilization, should alleviate these concerns. The future of biosimilars involves improved access to vital treatments, more diverse biosimilar choices, regulatory enhancements, and better patient education. These changes can substantially enhance patient results and alleviate the economic strain on healthcare, potentially revolutionizing the pharmaceutical field and boosting healthcare sustainability. This chapter highlights the benefits and challenges surrounding biosimilars in cancer treatment, emphasizing the need for continued research, education, and regulatory advancements to realize their potential.

Keywords

Biosimilars · Biologics · Drug targets · Cancer treatment · Regulatory advancements

16.1 Introduction

Cancer is considered as the chief cause of death around across the globe, making up one in every six deaths. There were 19.3 million new cases of cancer estimated in the year 2020 and ten million deaths reported. The cancer burden would be double the present number of cases in the coming years. The cost of cancer treatment has been increased over time. Nevertheless, the advancements in personalized medicine, the treatment cost is likely to be ignored. The introduction of biosimilars has increased market competition, which can benefit cancer patients by improving accessibility and lowering their financial burden (Joshi et al. 2022). Patients and prescribers need to be educated greatly about biosimilars because of their novelty and decreased emphasis on clinical testing. The American Society of Clinical Oncology (ASCO) informs and counsels the oncology community regarding the use of biosimilars, their efficacy and safety, substitutability, interchangeability, regulatory issues, and patient and provider education. There are certain challenges that have restricted biosimilars from being accepted in the clinical practice. Immunogenicity is the primary concern wherein the immune reactions towards biosimilars would have adverse effects and thereby hindering the efficacy of the product itself. This creates a need for safety and efficacy monitoring during clinical trials and even at the post-marketing phase. According to research, to be implemented in clinical practice, healthcare professionals must be made aware of the notion of biosimilars based on credible scientific data, such as clinical trial data (Winstead 2018). Only a handful of biosimilars are presently approved for use in supportive care and cancer treatment. Monoclonal antibodies (mAb), such as rituximab, trastuzumab, and bevacizumab, as well as supportive medicines including filgrastim, pegfilgrastim, epoetin α, and epoetin ζ, are all accessible as biosimilars (Peeters et al. 2021; Bachu et al. 2022). With this background, the chapter provides comprehensive information on the biosimilars used in various cancer types, mechanism of action, and the adverse effects associated with them. The main benefits and drawbacks of biosimilars are considered, along with their potential roles in medical oncology moving forward and the difficulties encountered with them. Further, the chapter would brief on the biosimilars that are in the process of clinical examination to be accepted for clinical practice, and these biosimilars would be accessed to the cancer patients as prescribed by the healthcare professionals, thereby increasing the accessibility and improving the health benefits.

16.2 Importance of Biosimilars in Oncology

Biosimilars have been formulated to address for various cancer biologics, which includes monoclonal antibodies and growth factors which are often used in treating hematological and solid cancers. The need for alternative therapies like biosimilars in cancers and its role has been briefly discussed below.

16.2.1 Burden of Cancer and Need for Accessible Therapies

Around the world, people are treated with questionable, complementary, or alternative methods against cancer. From early twentieth century, every span of 10 years had experienced a "miracle" therapy for cancer that achieved great attention and evenly fade away as the next one arose. The need for alternative cancer therapies comes from several sources. One of them is the mental frustration of the people suffering from cancer on the inability of the treatments against many cancers. Every year, there is growth in the count of cancer incidence, and no real therapies are available to treat them. One of the important reasons for public interest in alternative medicine nowadays is the dissatisfaction with the already existing methods such as chemotherapy and radiotherapy, which are still a big threat with respect to side effects in cancer patients and also nature of the modern medicine. Most of the cancer deaths happen due to metastasis, which is primarily because of achieving control of the initial tumor. But, other cancers like brain, head, and neck and cervix where the metastasis is a late or rare event, where the patient may die due to uncontrolled cancer in primary site. For most types of cancers, the initial chemotherapy administered to patients before surgery has failed to improve the control over tumor. The drugs like bevacizumab, rituximab, and trastuzumab which are utilized in the care of lung cancer, leukemia, and breast cancers are costly because of the difficulties in manufacturing and development, which adds to the already existing higher cost cancer treatment. Across the recent years, biosimilars have developed improving interest globally as a potent alternative to conventional drugs. Following that in 1980s, a new form of therapeutics was developed, with its first bio-medicine as recombinant human insulin. Nowadays, biotechnological techniques are employed to produce drugs witnessing significant growth after the advancement of genetic engineering methods for DNA manipulation. Biopharmaceuticals involve administering molecules into various living systems to study the protein expression relevant to therapeutic areas of interest (Konstantinidou et al. 2020).

16.2.2 Role of Biosimilars in Expanding Access to Cancer Treatment

A biosimilar will tend to establish increased resemblance to the product of reference in the aspect of efficacy, mechanism of action, safety, route of administration, and immunogenicity. Biosimilars can lower the current health care costs due to natural price competition with their product of reference. Currently, there are over 700 approved biosimilar medicines available all around the world. These medicines are used both directly and indirectly for neoplasm interventions. Filgrastim and epoetin alpha were one of the earliest biosimilar products that were used as supportive therapy in oncology. Filgrastim is rG-CSF factor that maintains the neutrophil production from bone marrow. It is administered to minimize the occurrence and longevity of febrile neutropenia in cancer patients. Nearly all recently introduced biosimilars are monoclonal antibodies, which are created especially for cancer therapy due to its increasing cost and other inflammatory diseases. The initial approval of a biosimilar for monoclonal antibody was in Europe for cancer therapy was the one for rituximab. It is a biologically modified chimeric monoclonal antibody employed for threating non-Hodgkin's lymphoma and lymphocytic leukemia. In the context of human epidermal growth factor receptor 2 (HER2)-positive metastatic breast cancer, some additional biosimilars are utilized either independently or in combination with chemotherapy or hormone therapy (e.g.., Ontruzant). CT-P13, a biosimilar for infliximab prepared in 2014, was the first biosimilar monoclonal antibody. It has been found that CT-P13 has the ability to cure rheumatological diseases through pivotal clinical trials since it acts as a tumor necrosis factor inhibitor. Overall, biosimilars have the ability to decrease the cost of care and improve access to cancer therapies. Hence, biosimilars are inexpensive drugs, which have the similar safety and adverse effects and no medical differences compared to the biologics. This will lead to reduction in the healthcare cost and make people accessible for new treatment options (Schellekens et al. 2016).

16.3 Biosimilar in Lung Cancer

Non-small cell lung cancer (NSCLC), a widespread and lethal lung malignancy, poses a significant strain on global healthcare resources. Avastin®, the approved bevacizumab, is authorized for diverse cancer treatments, incorporating its function in the primary treatment of advanced non-small cell lung cancer (NSCLC) alongside additional chemotherapy. International guidelines recommend bevacizumab's use in both first-line and maintenance settings for advanced NSCLC. Recent discoveries suggest new applications for bevacizumab, including combinations with emerging therapies and maintenance therapy post disease progression. MB02, a biosimilar adhering to rigorous international standards, replicates the physical, chemical, and biological properties of reference bevacizumab, substantiated by a comprehensive program, including a confirmatory clinical trial demonstrating comparable efficacy, safety, and the immunogenic response in the initial therapy for advanced non-squamous non-small cell lung cancer (NSCLC) using a combination of chemotherapy (carboplatin and paclitaxel) (Franke & Schenk 2023).

16.3.1 Mechanism of Action of Biosimilar

VEGF, known as vascular endothelial growth factor, assumes a critical role in facilitating tumor angiogenesis, contributing significantly to both tumor vascularization and metastasis. Increased levels of vascular endothelial growth factor (VEGF) are strongly linked to unfavorable results across various cancers, including NSCLC, CRC, glioblastoma, and ovarian cancer. Originally, sourced from mice and subsequently humanized, bevacizumab is a monoclonal antibody created to target VEGF-A, a specific variant of vascular endothelial growth factor (VEGF) that stimulates the growth and movement of endothelial cells. Its mechanism involves bevacizumab binding specifically to the VEGF-A protein, resulting in the inhibition of angiogenesis and impeding the formation of new blood vessels. Bevacizumab's mechanism of action includes attaching to a consistent epitope within VEGF, thus obstructing its interaction with the endothelial cell surface receptor (VEGFR). This binding disrupts subsequent signaling cascades, effectively hindering the formation of new blood vessels. Bevacizumab exhibits a wide range of anti-tumor effects, such as preventing the development of fresh blood vessels, causing regression of recently developed vessels, normalizing vasculature to improve the delivery of cytotoxic chemotherapy, and directly affecting tumor cells. Consistently inhibiting the VEGF ligand is essential to hinder tumor growth and potentially induce a gradual reduction in tumor size over time. Results from laboratory experiments emphasize the effectiveness of bevacizumab in hindering various VEGF-induced cellular processes. These activities encompass, among various aspects, restraining cell growth, lowering permeability, suppressing nitric oxide production, and obstructing cell migration. Clinical evidence indirectly indicates that the main effect of bevacizumab is cytostatic, primarily shown by its capacity to hinder the development of new blood vessels. This inference is further substantiated by pharmacodynamic biomarkers, such as circulating VEGF and PGF, which consistently demonstrate an elevation in reaction to bevacizumab treatment across diverse tumor types. In summation, the operational method of bevacizumab revolves around its precise targeting of VEGF and angiogenesis. This targeted approach consistently proves effective in impeding tumor progression across a spectrum of diverse malignancies (Melosky et al. 2018).

16.3.2 Various Biosimilars are Evaluated in Various Phases of Clinical Trial as Demonstrated in Table 16.1

16.3.3 Clinical Benefits

Recognizing factors that can predict for the efficacy of VEGF-pathway inhibitors like bevacizumab in lung cancer poses challenges. While VEGF levels correlated with treatment response, they lacked predictive value for survival. The E4599 research uncovered that there were low initial levels of ICAM linked to significantly improved overall and 1-year survival rates, suggesting potential benefits from bevacizumab in patients with low ICAM levels. High blood pressure onset during treatment, as observed in the ECOG 4599 study, was associated with enhanced outcomes. Polymorphisms in VEGF pathway components (e.g., VEGF-2578AA and VEGF-1154AA) were linked to improved overall survival in specific patient subsets. Ongoing research, including clinical trials, is essential to unravel and validate these potential biomarkers for predicting bevacizumab efficacy in cancer treatment (Planchard, 2011).

16.4 Biosimilar in Breast Cancer

The biosimilars involved in treating cancer such as breast, some of them are trastuzumab and bevacizumab. The main advantage in the case of biosimilars is that we are able to treat the diseases and the cost of them is cheaper than the original biologic drugs. In real-time examples, countries such as Europe and USA got approval

Table 16.1 Clinical studies comparing potential biosimilars with bevacizumab in patients with NSCLC

Product (manufacturer)	Trial registration ID	Design	n	Primary end point (Time frame)	Status	Outcome	References
ABP 215 (Mvasi) (Amgen)	NCT0196603	Randomized, DB parallel group	642	≤19 weeks	Completed	Similar efficacy, safety, and IG between ABP 215 and BEV	Gridelli et al. (2018)
BI 695502 (Boehringer Ingelheim)	NCT0227243	Randomized, DB parallel group	671	≤18 weeks	Ongoing, not recruiting	–	Planchard (2011)
BCD-021 (biocad)	NCT0176365	Randomized, DB parallel group	138	127 days	Active, not recruiting	(primary) similar efficacy, safety, and IG between BCD-021 and BEV	Sandler (2007)
FKB238 (Centus biotherapeutics)	NCT0281047	Randomized, DB, parallel group	730	≤12 months	Recruiting	–	Kazazi-Hyseni et al. (2010)
PF-06439535 (Pfizer)	NCT0236499	Randomized, DB parallel group	719	≤19 weeks	Ongoing, not recruiting	–	Greillier et al (2016)

BEV bevacizumab, *DB* double-blind, *ID* ClinicalTrials.gov identifier, *IG* immunogenicity, *NCT* national clinical trial for using the biosimilar trastuzumab (Triantafyllidi and Triantafillidis 2022). There are five biosimilars approved in Europe, whereas in the United States, two biosimilars are approved. Nearly 29,000 people in the world are using the biosimilars of trastuzumab, and it is expected that this count will increase gradually. The first approved biosimilar for trastuzumab by FDA is Ogivri (Alina Uifalean,Maria Ilies,Raul Nicoara,Lucia Maria Rus,Simona Codruta Hedhes 2018).

16.4.1 Development of Biosimilar

Recently, the rate of cases of invasive breast cancer among women is about 268,600. Roughly, about 42,260 women were succumbing to the disease, and it is noted that 40% of reduction in death rates from the breast cancer happens by the new therapies that are biosimilars and their action of mechanism access the presence or absence of markers molecules in breast cancer. Biosimilars are considered as copies of original biological drugs that are made from the living sources. The biosimilars do have similar structure and function to the original biological drugs (Bloomfield et al. 2022). The FDA has been approved that they were safe, and they also tested that they act as effective as the original biologic drugs. The trastuzumab biosimilars are the one to treat breast cancer by lowering the level of human epidermal growth factor receptor 2(HER). The HER protein located on the surface of the layers in the body and they are responsible for regulating and controlling the growth of breast cells. When the HER 2 gene is mutated, this condition makes them divide and grow more cells in the breast cells. The continuous growth and division lead to tumor formation, and this stage is said to be HER2 positive. The Herceptin binds to HER 2 protein, and this binding stops the cells from dividing and growing. Trastuzumab has become more costly, and many countries cannot afford it. So, the companies started to produce the biosimilars of trastuzumab (Bui and Taylor 2014). This has become a great opening for biosimilars development. The use of biosimilars of trastuzumab in the market is made to improve the health of breast cancer patients and increase their lifespan. The five approved biosimilar trastuzumab are: MYL-1401O (Ogivri), SB3(Ontruzant), ABP980 (Kanjinti), PF-05280014 (Trazimera). and CT-P6 (Herzuma) (Migliavacca Zucchetti et al. 2019).

16.4.2 Clinical Trial

The major aim of developing biosimilars is to establish the similarity to the reference product with some benefits to the patients. The methodical development process begins with comparative analytical characterization and also by evaluating the preclinical studies and assessment to determine the mode of action. They perform different phases to undergo various conclusion and results. Phase studies were performed in finding the common features in clinical pharmacokinetics and pharmacodynamics compared to the reference product. Till the clinical phase III, the testing of a drug does not ends (Doni Bloomfield, JD;Elvira D'Andrea, MD, MPH;Sarosh Nagar;Aaron Kesselheim, MD, JD 2022). Clinical trials phase III study is done to confirm the efficacy, risk factors in immunogenicity and safety of the product. The clinical trials of phase III only done when there is the difference between biosimilar and the reference product. Different dosage for trastuzumab exists, depending on usage such as an adjuvant, a metastatic, or a neoadjuvant. These regimens vary in terms of dose, combination of agents, and duration. Intravenous (IV) infusion is preferred method for trastuzumab administration (Bui and Taylor 2014). A dose such as loading and maintenance are very important in treating breast cancer.

16.4.3 Clinical Benefits

About 15–20% of women affected to the breast cancer due to over expression HER2 and leads to formation of tumor. To stop the overexpression of HER2, monoclonal antibody trastuzumab biosimilar is used. The mechanism of the trastuzumab biosimilar is same as the reference product. It has been designed that this biosimilar could bind the HER2 receptors that are located or present of the surface of the cancer cells. This binding blocks the overexpression of HER2 that are involved in causing the formation of tumor leading to breast cancer. Usage of trastuzumab biosimilar shows a significant decline in death rates and increases the life span of the patients. Globally, patients with HER2 breast cancer are increasingly in need of increased access to trastuzumab (Migliavacca Zucchetti et al. 2019). The safe and efficient availability made them to utilize for the needs. Thus, trastuzumab biosimilars may be advantageous for patients with HER2-positive metastatic gastric and gastroesophageal junction adenocarcinoma in addition to breast cancer. In introduction to trastuzumab biosimilars, the cost savings are expected but they are expected to be less than with generic medications and to be accompanied by a projected 20–30% discount. Gradually increase in usage of HER2-targeted biosimilars is due its cost, and it is finically benefited for the customers (Thill 2019).

Basically, the biosimilars products are less expensive than the original reference product. In generally speaking, based on the cost effectiveness, greater the number of patients were able to undergo this therapy. Reductions in the price of biosimilar trastuzumab can improve patient access to HER2-targeted treatment (Bui and Taylor 2014). However, the emerging biosimilars would definitely develop the nations, and it may yield higher economic returns from the countries. The factors such as quality, safety, and efficacy should be retained in biosimilars product as such the reference product. Therapeutic equivalency is established through clinical trials, guaranteeing the biosimilar's efficacy in comparison to trastuzumab, the product of origin. Biosimilar trastuzumab made an alternate route for treating the HER2-positive breast cancer patients (Rakesh Chopra 2017). The emergence of biosimilar product plays a crucial role when there is shortage in the original product. Biosimilars offer flexibility in treatment plans by making it easier to simplify the transition from reference product to the biosimilar. Clinical trials are necessary only if there is difference in the biosimilar trastuzumab and reference product. Comparison of trastuzumab biosimilars with reference product is shown in Table 16.2.

Table 16.2 Comparison of trastuzumab biosimilars with reference product

Product (Manufacture)	Comparison	n	Primary point (Time of Frame)	Design	Evaluation	Status	Outcome	Reference
Herzuma (CT-P6)	CT-p6 vs. Trastuzumab RP	475 549 549 103 254	11–12 months – 3 years 5 years	Equivalence Equivalence Equivalence –	Metastatic cancer Early breast cancer – Metastatic cancer Early breast cancer	Completed Completed – Completed	Equivalence margins are ±15% Neutropenia reaction is observed with 4% in CT-P6 and 5% in RF In CT-P6 therapy, patients with EBC achieved CHR when it is compared to RF. Results in similar efficacy	Triantafyllidi and Triantafillidis (2022)
ABP 980 (Herceptin)	ABP 980 Vs. Trastuzumab RP	827 725	– 3 weeks	Equivalence –	Early breast cancer Early breast cancer	Completed Completed	Predefined equivalence margins result in similar efficacy Rate of immunogenicity and side effects were similar	
SB3 (Ontruzant)	SB3 vs. Trastuzumab RP SB3 vs. Trastuzumab RP	875 800	438 days –	Equivalence –	Early breast cancer –	Completed –	The equivalent margins for SB3 group-45.8% and 35.8% for reference product Complete histological response rates 51.7% vs. 42.0% the results indicate that the SB3 is as similar to RP. Equivalence is observed between SB3 and RF indicates it is safe	
PF-05280014	PF-05280014 vs. Trastuzumab RP	707 4 2–349 (BS)and 353(RP) 352(BS) and 355(RP)	378 days – 33 weeks	Equivalence – –	Metastatic cancer and early breast cancer – Metastatic cancer	Completed Completed Completed	Equivalent results are observed with RP Both had similar pharmacokinetic parameters. Immunogenicity results in similar	

16.5 Biosimilar in Colorectal Cancer

In the dynamic landscape of colorectal cancer (CRC) treatment, biosimilars have emerged as transformative alternatives to established biologics. Bevacizumab biosimilars, such as Zirabev, Mvasi, and Acellbia, faithfully replicate the actions of the monoclonal antibody bevacizumab, offering cost-effective options for patients with metastatic CRC. Cetuximab and panitumumab biosimilars, identified generically, contribute to EGFR inhibition in CRC therapy. The introduction of biosimilars in colorectal cancer signifies a pivotal shift, not only providing therapeutic alternatives but also enhancing accessibility and affordability in the evolving landscape of cancer care.

16.5.1 Mechanism of Action

Bevacizumab, a key player in colorectal cancer treatment, operates by targeting the vascular endothelial growth factor (VEGF) pathway, crucial for angiogenesis in tumor growth. By specifically binding to VEGF, it prevents the interaction with endothelial cell receptors, preventing formation of new blood vessels and diminishing the tumor's blood supply. This disruption curtails the tumor's ability to thrive and metastasize. Clinical studies like AVF2107g suggest that incorporating bevacizumab into cytotoxic chemotherapy yields significant improvements in overall survival (OS), progression-free survival (PFS), and response rates, both in the initial and subsequent treatment phases. Its pivotal role in metastatic colorectal cancer treatment is underscored by increased median OS, especially upon reintroduction post-disease progression. Bevacizumab, targeting the VEGF pathway, stands as a cornerstone in the fight against colorectal cancer, contributing markedly to enhanced patient outcomes and survival rates (Rosen et al. 2017).

Panitumumab, a crucial monoclonal antibody in metastatic colorectal cancer treatment, targets the epidermal growth factor receptor (EGFR), hindering downstream signalling pathways that fuel cancer cell growth and survival. In patients with wild-type RAS, panitumumab, whether combined with chemotherapy or used alone, demonstrates significant efficacy across multiple treatment lines. However, caution is warranted in K-RAS mutated patients, where the addition of panitumumab may lack benefits and pose risks. Its combination with bevacizumab and chemotherapy requires careful consideration due to potential complications. Panitumumab's mechanism leads to low rates of severe infusional reactions, but vigilance is needed for mild to moderate toxicities like skin rash, diarrhoea, and hypomagnesemia. Comparative studies indicate similar efficacy between panitumumab and cetuximab. This targeted approach positions panitumumab as a valuable component in colorectal cancer treatment, particularly for patients with wild-type RAS (Rosen et al. 2017).

16.5.2 Clinical Studies of Potential Biosimilars to Benchmark Bevacizumab Have Been Shown in Table 16.3

[AUC 0–inf area under the curve from time zero to extrapolated infinite time, AUC tau area under the curve from time zero to last quantifiable concentration, BEV

Table 16.3 Clinical studies of potential biosimilars to bevacizumab

Biosimilar	Manufacturer	Design	Population (N)	Dose of bevacizumab	Primary end points (time frame)	Status	References/NCT No.
ABP 215	Amgen	Randomized, 3-arm, single-dose study to assess PK of ABP 215, BEV-EU, and BEV-US	Healthy adult males (NR)	Not reported (NR)	AUC0–inf; Cmax (NR)	Completed. Similar pharmacokinetics profiles	(Rosen, L.S.Et al.)
		Randomized, double-blind study to compare safety and efficacy of ABP 215 and BEV	Advanced non-squamous NSCLC (642)	15 mg/kg (every 3 weeks)	ORR (19 weeks)	Completed. Similar efficacy, safety, and immunogenicity profiles	(Rosen, L.S.Et al.)
PF-06439535	Pfizer	Randomized, double-blind, single-dose, 3-arm study to compare PK of PF-06439535, BEV-EU, and BEV-US	Healthy male volunteers (102)	5 mg/kg	Cmax; AUCtau; AUC0–inf (71 days)	Completed Similar pharmacokinetics and safety profiles	(Rosen, L.S.Et al.)
		Randomized, double-blind study to compare safety and efficacy of PF-06439535 and BEV-EU	Advanced non-squamous NSCLC (Est. 710)	15 mg/kg (every 3 weeks)	ORR (19 weeks)	Similar efficacy, safety, pharmacokinetics, immunogenicity	(Rosen, L.S.Et al.)
FKB238	Centus biotherapeutic	Randomized, double-blind study to compare safety and efficacy of FKB238 and BEV	Advanced/ recurrent non-squamous NSCLC (Est. 730)	15 mg/kg (every 3 weeks)	ORR (12 months)	Comparable in efficacy and safety	(Rosen, L.S.Et al.)

SB8	Samsung bioepis	Randomized, double-blind, single-dose, 3-arm study to compare PK, safety, and immunogenicity of SB8, BEV-EU, and BEV-US	Healthy male volunteers (119)	3 mg/kg	Cmax; AUCtau; AUC0–inf (85 days)	Completed	(Rosen, L.S.Et al.)
		Randomized, double-blind study to compare safety, efficacy, PK, and immunogenicity of SB8 and BEV-EU	Advanced non-squamous NSCLC (Est. 678)	15 mg/kg (every 3 weeks)	ORR (24 weeks)	Completed Similar pharmacokinetics profile and safety profiles	(Rosen, L.S.Et al.)

bevacizumab, BEV-EU bevacizumab sourced from the European Union, BEV-US bevacizumab sourced from the United States, C max maximum serum concentration, ORR objective/overall response rate].

16.5.3 Clinical Benefits

The clinical benefits of incorporating biosimilars into the treatment landscape of colorectal cancer are multifaceted and transformative. Biosimilars not only enhance treatment access and affordability, providing a broader array of effective therapies for patients, but also maintain rigorous regulatory standards to ensure equivalent efficacy and safety to reference biologics. Their versatility is demonstrated in the expansion of treatment options and facilitation of combination therapies, exemplified by studies such as BEACON CRC. Beyond the clinic, the introduction of biosimilars stimulates market competitiveness, fostering innovation and reducing overall costs. This cost-effectiveness contributes to the long-term sustainability of cancer care, optimizing resources and ensuring accessibility for a larger population. In summary, the clinical benefits of biosimilars in colorectal cancer encompass improved access, maintained efficacy and safety, expanded treatment options, market competitiveness, and long-term sustainability in cancer care (Zinzani et al. 2019).

16.6 Biosimilar in Liquid Cancer

Liquid cancer, also termed as blood cancers, is classified into leukemia and lymphoma. Lymphomas are cancers originating in the lymphatic system. It is categorized into 2 types: Hodgkin lymphoma and non-Hodgkin lymphoma. Non-Hodgkin lymphoma (NHL) is a form of cancer impacting White Blood Cells (WBC) specifically B cells and T cells. In NHL, B cells express CD20 and this form of cancer often tends to be aggressive. Biosimilars like GP2013, Ruxience, and Rituximab CT-P10 have emerged as a promising advancement in treating Non-Hodgkin Lymphoma (NHL), offering a comparable alternative to existing biologic drug Rituximab. These biosimilars are tailored specifically for targeting NHL, mimic the efficacy and safety of established treatments while potentially making them more accessible and cost-effective. Their arrival signifies a significant stride in providing effective and affordable care for individuals combating NHL.

16.6.1 Mechanism of Action

CD20 is a nonglycosylated phosphoprotein of 30-37 kDa and is expressed on the cell membrane of normal and malignant B lymphocytes. They are present on late pre-B lymphocytes and is then lost in terminally differentiated plasma cells. It comprises of 4 transmembrane domains, 1 intracellular domain and 2 extracellular domains (Uifălean et al. 2018). Additionally, CD20 is physically linked to B cell receptor (BCR), CD40, major histocompatibility complex II (MHC II), and Src kinase binding proteins (CBP). These CBP interacts with Src kinases like LYN, FYN, and LCK. Therefore, this promotes in cell proliferation. It has been reported that CD20 is phosphorylated at higher levels in proliferating malignant B cells compared to normal resting B cells. Therefore, in lymphoma especially non-Hodgkin lymphoma (NHL), the population of B cells expressing CD20, is high. Hence, the target for therapy is these kinds of cells. A very well-known medication uses monoclonal antibodies to treat NHL. Rituximab is a genetically engineered humanized chimeric anti CD20 monoclonal antibody IgG that is used for the treatment of NHL (Bachu et al. 2022; Migliavacca Zucchetti et al. 2019; Joshi et al. 2022). The binding of rituximab with CD20 induces complement mediated cytotoxicity where membrane activating complex (MAC) is formed and then is used to make perforations in the cell membrane of the B cells (Bachu et al. 2022; Bui and Taylor 2014). The binding also promotes cell-mediated toxicity where CD8+ cells kill the B lymphocytes, and antibody-dependent cell cytotoxicity (ADCC), where natural killer cells target and kill the flagged B lymphocytes (Bachu et al. 2022; Curigliano et al. 2016). Rituximab promotes intrinsic apoptosis, which is the mitochondrial pathway of caspase activation (Curigliano et al. 2016). Therefore, this leads to profound B cell depletion.

16.6.2 Clinical Trials of Potential Biosimilars of Liquid Cancer Have Been Shown in Table 16.4

16.6.3 Clinical Benefits

Rituximab can be used to treat patients with NHL with 40% efficacy (Dotan et al. 2010). A transition of B cell phenotype is observed. After B cell depletion therapy, the remaining population of B cells comprises of plasma cells and certain memory cells (cells without CD20) (Trouvin et al. 2015). It can also be used to treat patients suffering from chronic lymphocytic leukemia (CLL), but the efficacy is higher when used in combination with other chemotherapy drugs. Rituximab along with idelalisib is used to treat relapsed CLL, but the side effects are severe [(Furman et al. 2014)]. It is also used to treat patients suffering from rheumatoid arthritis (Mok 2013).

Table 16.4 Clinical studies of potential biosimilars of liquid cancer

Product	Trial registration ID	Design	n	Duration of study	Status
GP 2013	NCT01933516	Open	6	15 months	Phase 1
Ruxience	NCT01994382	Non-randomized	260	2664 days	Completed
Rituximab CT-P10	NCT02260804	Randomized	258	1395 days	Completed

Source: https://clinicaltrials.gov

16.7 Future Scope of Biosimilars in the Diagnosis and Treatment of Cancers

Biosimilars for the clinical management of various cancers is a booming field of research, but it has a long way to go before it becomes a mainstay in the world of medicine. There are a few challenges that it faces, which if overcome, can abet in paving way for biosimilars to become an important part of cancer treatment.

16.7.1 Current Challenges Faced by Biosimilars in Oncology

16.7.1.1 Interchangeability

"Interchangeability" is defined as the property of any two substances to be substituted in the place of each other, with marginal difference in its efficacy and safety. Although for known biosimilars, there should be no variations in efficacy upon interchanging, it is a point of concern, given the uncertainty of its impacts on immunogenicity. There still needs to be thorough evidence presented for it to be considered a viable process (Kos et al. 2018). For gaining approval for using biosimilars interchangeably, there has to be extensive evidence presented, even more stringer than those required for the approval of a biosimilar itself. Interchangeability is not accepted in several European nations, in order to avoid any complications regarding the traceability of the drug, in case of any untoward events. Although the recorded instances of negative reactions caused due to interchanging are found in trace, an absence of such evidence cannot be taken as proof for the safety of interchangeability—further scientific proof is required to back up this claim. In the recent times, biosimilars, especially in oncology, are in the form of monoclonal antibodies. Some monoclonal antibodies have a very minimal effect against cancer, and hence makes it difficult to correlate them with the first-generation biosimilars and their reference products. Moreover, they are used commonly along with chemotherapy, which renders the designing of studies difficult. Observing and proving that the differences in safety will be minor is a near-impossible task (Schellekens et al. 2016). Although many biosimilars have undergone testing for interchangeability, this data is not sufficient for establishing them as interchangeable for the long term (Soares and Cavalcanti 2021).

16.7.1.2 Extrapolation of Indications

Extrapolation of indications is when the biosimilars developed for one disease can also be indicated for the management of another illness. It is extremely sensitive and can be approved only under a very few conditions: there must be thorough clinical tests done on sensitive models to detect any possible variations between both indications, pharmacokinetics and pharmacodynamics in the pathology of interest must be similar to the one being extrapolated from and the characterization of the biosimilar must have been done extensively in such a way that the extrapolation of indication does not hold any risk (Kos et al. 2018). It is challenging to extrapolate results to oncology from the clinical trials data for other diseases because of the variations in the mechanisms of the diseases themselves and the method of treatment (Konstantinidou et al. 2020).

16.7.1.3 Lack of Awareness

Another major challenge faced in the use of biosimilars is the knowledge gap observed in both patients and health care professionals. This can be the cause of major hindrance in both manufacturing and usage of biosimilars. Research and manufacturing done diligently with a low acceptance rate by the target demographic will lead to a loss, without having achieved the purpose. Biosimilars, if accepted widely, can be a more affordable and viable option, so this lack of awareness can prove to be a major drawback (Konstantinidou et al. 2020).

16.7.2 How to Counter the Challenges

16.7.2.1 Interchangeability and Extrapolation of Indications

Maintenance of rigid protocols for the approval of biosimilar interchangeability gives reliable proof to both patients and doctors of the safety and efficacy of the treatment. Factors to be considered are: similar method of administration, same mechanism of action of drug, the dosage form of the drugs, and the manufacturing standards in comparison to the reference product. The above conditions must be studied extensively—different activities of the product must be focused upon in varying populations, and the changing effects and toxicities in each of those populations must be measured and studied. Structural similarity with the original product is also an important factor in determining interchangeability. Studying the glycosylation patterns of the biosimilar product with reference to the original biologic can be a reasonable comparison of structural similarity. Changes in the production processes can result in differences in the glycosylation patterns formed, thus altering the pharmacokinetics and pharmacodynamics of the final product. Mass spectroscopy can be employed to study these similarities and differences in the structure—hence being a reliable measurement of a potential candidate for an interchangeable biosimilar (Soares and Cavalcanti 2021).

International standards must be defined for deciding the comparison measures between the original product and the biosimilar, for maintaining uniformity in the data interpreted globally. Every biosimilar must be monitored closely after being released into the market for its long-term effects and a detailed report highlighting its prospects and risks must be presented to the World Health Organization (WHO), who will closely examine this before making their decision on the interchangeability of the product. A similar data collection and presentation are also required to prove the extrapolation of indications from the usage of the drug against one pathology to another. Extrapolation ensures that the development of a single drug can ensure applicability across multiple indications, thus mitigating the high cost of the drug and making it more accessible to a wider range of patients (Curigliano et al. 2016).

16.7.2.2 Lack of Awareness

Research groups working in the field of oncology can take active measures in spreading awareness to the people in need of biosimilars—patients, caregivers, and clinicians. This will have a beneficial impact on the attitude of both patients and healthcare providers alike, while discussing the consequences of a new treatment. This was highlighted in a study performed by (Ismailov et al. 2019), where the impact of increased awareness and education about biosimilars in oncology were examined. A brochure containing information was distributed, which contained a link to a quiz regarding the same information. People who answered positively when asked if they read the brochure responded with a good knowledge of biosimilars. Most participants who responded to the quiz were also willing to discuss the possibility of a biosimilar treatment with their healthcare providers. This kind of an educational initiative can promote patients and doctors to experience a "shared decision-making". The participants' preferred learning style is yet to be studied further and the long-term effects of such an education has not been explored as well. Further studies can involve a larger sample population and can be done to examine long-term effects.

Education can benefit professionals just as much as it can impact patients. A study conducted by (Williamson et al. 2019) showed the differences in the knowledge and the confidence of several field experts before and after attending a course about biosimilars. Although the percentage of people choosing the correct answers in a test increased drastically after the course, there was only a marginal difference in the confidence held by the professionals in using biosimilars in their fields pre- and post-course. This indicates that better education for healthcare professionals can prove to be highly useful in fostering healthy treatment for patients, and any further studies performed will require a larger demographic and a longer period of study.

16.7.3 Action Points for the Advancements

Funding authorities must seriously consider biosimilars as a field to invest in, especially as a treatment method for cancer. As this is a highly nuanced field, rigorous clinical trials must be performed before anything can be marketed as a cure for cancer. All the existing biosimilars must continuously undergo post-marketing trials where its long-term effects on the body are observed. This can help in identifying the changes in immunogenicity, which in turn will affect the interchangeability of the drug. Biosimilars developed for other diseases must be actively studied for extrapolation in oncology and existing extrapolated drugs must consistently undergo post-marketing clinical trials as well.

There must be deliberate efforts taken by professionals to improve knowledge and awareness regarding biosimilars—both in patients and in clinicians. This will encourage healthy dialogue between doctors and patients about the treatment. As the awareness about the value of biosimilars grows, there are higher chances of large-scale research and production, making it accessible to the public and hence, more affordable.

Cancer is a deadly disease that causes many fatalities annually, and the research around its cure is continuously changing and evolving. Biosimilars, if taken into serious consideration as a highly potential research area in the upcoming years, can prove to be a great turning point in the field of oncology.

16.8 Market Analysis and Economic Consideration

16.8.1 Market Competition

The perspective that biosimilars contribute to the enduring viability of the off-patent biologicals market is being trusted by many. The study by Cécile (Rémuzat et al. 2017) encompassed a comprehensive analysis of biosimilar price differentials in comparison to biologics. Incentive policies, representing the cumulative count of incentives available within each country, were examined in the study, covering aspects such as pharmaceutical price, quotas, prescription patterns, financial incentives, education/information, and patient co-payment. Market competition was integrated into the model, incorporating total count of biosimilars within each therapeutic class. The analysis focused on six therapeutic classes: G-CSF, EPO, insulin, Anti-TNF, gonadotropins, and HGH. The entry date into the market of the initial biosimilar in each therapeutic class was also a key parameter under investigation in the study. As a result, Poland demonstrated the most significant adoption of biosimilars compared to other nations, with utilization rates reaching 100% for HGH, 90% for EPO, and 81% for G-CSF. The market penetration was found to be higher when number of biosimilars is significantly greater in number. Additionally, an extended duration on the market provides an increased opportunity for a biosimilar to establish itself, leading to a higher market share compared to the originator. This trend is evident in the cases of the first three biosimilars launched—somatropin, epoetin, and filgrastim—which achieved the most substantial uptake, and with reduced cost-containment pressure. Policy decision-makers may find it beneficial to prioritize incentive policies as a means to promote the adoption of biosimilars. At the point of purchase, originators may sometimes adjust their prices to stay competitive in the market. Healthcare professionals are unlikely to be solely influenced by price in the short term when it comes to biosimilars, as the concept has not yet become fully ingrained in routine prescription practices. Additionally, the prescription of biosimilars requires active efforts, and manufacturers may opt to launch these products in countries with higher pricing and greater potential (Rémuzat et al. 2017). Biosimilars escape from the intensive clinical studies due to which the cost of biosimilar is less, and they are more preferred in the market. Many Indian Pharma companies focus on manufacturing and investing on Biosimilars. Companies like Amgen, Zydus, and Biocon join hands with other foreign companies for the development of biosimilars (Aayushi et al. 2021).

16.8.2 Market Trends and Projections

A significant amount of health care spending is allocated to biologics. Currently the expenses amount to $452 billion in worldwide spending. A move toward more biosimilar follow-on competition is being believed by many researchers given the recent expiration of biologics' patent protection, which would enable wider adoption and cheaper drug prices. Biologics risk competition from a subsequent class of medications known as biosimilars when their patent protection expires. Several nations have created regulations to encourage the introduction of biosimilars into the market. For instance, to provide a shortened route for biosimilars, the Biological Price Competition and Innovation Act was put into effect in the United States in 2010. Various studies were done to know the uptake of biosimilar in the market. In this study, the adoption of six ingredients, with at least one biosimilar available in all three countries, was observed, on average, Germany had the highest market share for biosimilars, followed by the United States and Switzerland. Nevertheless, in the United States, the market share for biosimilars increased at a more rapid rate compared to Germany and Switzerland. In the United States, newer biosimilars entering the market were predicted to have a quicker adoption, as indicated by their market share 1 year after launch. In the United States, the relative prices between the biosimilars and the reference products had a variation between 55% and 90% and a slight decrease gradually (Carl et al. 2022).

16.8.3 Cost Effectiveness and Affordability

Approval for the use of a growing number of biosimilars in the management and supportive care of cancer has been granted, and this trend is expected to continue globally. Treatment of cancer is a great concern to national healthcare budgets, posing a significant threat to overall healthcare fundings. The increasing costs of medications and hospital fees to treat cancer contribute to a greater financial burden. Recently, the prevalence of various cancers is on the rise, which further leads to the initiation of cancer medications at earlier stages due to advancements in early detection and prolonged treatment periods owing to improved patient survival. Between 2018 and 2019, oncology spending in health plans saw a substantial increase of 15–22%. The current pipeline includes numerous cancer medications undergoing clinical studies, with the oncology category projected to grow by 105% by 2024. Prices for certain chemotherapeutics and targeted therapies, such as bevacizumab, trastuzumab, and rituximab, have experienced notable cumulative increases—29%, 78%, and 85%, respectively, from 2005 to 2017.Biosimilars for these biologics present a significant opportunity to mitigate costs and enhance accessibility for a broader population of advanced cancer patients. A recent RAND (Research and Development) group report suggests that the adoption of biosimilars could potentially save the US health system up to USD 54 billion in the coming decade. The biologics are expensive due to the vast clinical studies they undergo and due to their physiochemical properties, patients suffer to pay high amounts for the drugs.

Despite numerous challenges, there is optimism among payers that biosimilars will drive down prices through competition, both with the original biologics and among other biosimilars. As familiarity with biosimilar products increases, regulatory restrictions may ease, leading to lower development and licensing costs. Widespread acceptance of biosimilars will ultimately lead to cost savings (Liu 2021).

16.9 Conclusion and Future Prospectives

Biopharmaceutical products are frequently employed in the treatment of cancer and tumors as palliative care and as prospective treatments. Since biological products are more expensive even if they are targeted agents, the global health system challenges are made worse. Furthermore, the patent expiry of these biological drugs has made it easier to sell biosimilars, which are more economical but have comparable efficacy and safety characteristics relative to the source. Consequently, most people can obtain cancer treatment, and biosimilars offer innovative therapies for a number of chronic and life-threatening illnesses. Compared to studies of their reference medications, oncology biosimilars have demonstrated comparable efficacy and safety, including in real-world situations, based on prespecified clinical efficacy comparability margins and physicochemical quality data. Within the next five years, more biosimilars for cancer will be accessible. Reducing the cost of efficient biological cancer treatments is expected to improve patient accessibility, hence minimizing disparities, and providing increased number of individuals with improved care. The usage of biosimilars continues to evolve quickly and will remain crucial to the treatment of cancer patients in the future. A thorough understanding of the safety and effectiveness of these medicines in the treatment of cancer provides the foundation for patient and provider acceptability, which will primarily determine the utilization of the many biosimilars that are anticipated to become available in the upcoming years. Consequently, awareness among the patient and provider would boost the trust concerning biosimilars and guarantee their effective incorporation into oncology practice.

References

Aayushi S. Ditani, Pragyan Paramita Mallick, Neelima Anup, Vishakha Tambe, Suryanarayana Polaka, Pinaki Sengupta, Kuldeep Rajpoot, Rakesh K. Tekade (2021) Biosimilars accessible in the market for the treatment of cancer, Journal of Controlled Release 336:112–129, ISSN 0168-3659. https://doi.org/10.1016/j.jconrel.2021.06.014

Bachu RD, Abou-Dahech M, Balaji S et al (2022) Oncology biosimilars: new developments and future directions. Cancer Rep 5:1–20. https://doi.org/10.1002/cnr2.1720

Bloomfield D, D'Andrea E, Nagar S, Kesselheim A (2022) Characteristics of clinical trials evaluating Biosimilars in the treatment of cancer: a systematic review and meta-analysis. JAMA Oncol 8(4):537–545. https://doi.org/10.1001/jamaoncol.2021.7230

Bui LA, Taylor C (2014) Developing clinical trials for biosimilars. Semin Oncol 41:S15–S25

Carl DL, Laube Y, Serra-Burriel M, Naci H, Ludwig WD, Vokinger KN (2022) Comparison of Uptake and Prices of Biosimilars in the US, Germany, and Switzerland. JAMA Netw Open 5(12):E2244670. https://doi.org/10.1001/jamanetworkopen.2022.44670
Curigliano G, O'Connor DP, Rosenberg JA, Jacobs I (2016) Biosimilars: extrapolation for oncology. Crit Rev Oncol Hematol 104:131–137. https://doi.org/10.1016/j.critrevonc.2016.06.002
Dotan E, Aggarwal C, Smith MR (2010) Impact of rituximab (Rituxan) on the treatment of B-cell non-Hodgkin's lymphoma. P T 35:148–157
Franke AJ, Schenk EL (2023) Re-ORIENT-ing antitumor immunity in EGFR-mutant non-small cell lung cancer: are antiangiogenics the key? In Chinese clinical oncology, vol 12, Issue 3. AME Publishing Company. https://doi.org/10.21037/cco-23-28
Furman RR, Sharman JP, Coutre SE et al (2014) Idelalisib and rituximab in relapsed chronic lymphocytic leukemia. N Engl J Med 370:997–1007. https://doi.org/10.1056/nejmoa1315226
Greillier L, Tomasini P, Barlesi F (2016) Bevacizumab in the treatment of nonsquamous non-small cell lung cancer: clinical trial evidence and experience. In: Therapeutic advances in respiratory disease, vol 10, Issue 5. SAGE Publications Ltd, pp 485–491. https://doi.org/10.1177/1753465816652083
Gridelli C, De Castro Carpeno J, Dingemans AMC, Griesinger F, Grossi F, Langer C, Ohe Y, Syrigos K, Thatcher N, Das-Gupta A, Truman M, Donica M, Smoljanovic V, Bennouna J (2018) Safety and efficacy of bevacizumab plus standard-of-care treatment beyond disease progression in patients with advanced non-small cell lung cancer: the avaall randomized clinical trial. JAMA Oncol, 4(12). https://doi.org/10.1001/jamaoncol.2018.3486
Ismailov RM, Khasanova ZD, Gascon P (2019) Knowledge and awareness of biosimilars among oncology patients in Colorado, USA. Future Oncol 15:2577–2584. https://doi.org/10.2217/fon-2019-0194
Joshi D, Khursheed R, Gupta S et al (2022) Biosimilars in oncology: latest trends and regulatory status. Pharmaceutics 14:2721. https://doi.org/10.3390/pharmaceutics14122721
Kazazi-Hyseni F, Beijnen JH, Schellens JHM (2010) Bevacizumab. The Oncologist 15(8):819–825. https://doi.org/10.1634/theoncologist.2009-0317
Konstantinidou S, Papaspiliou A, Kokkotou E (2020) Current and future roles of biosimilars in oncology practice (review). Oncol Lett 19:45–51. https://doi.org/10.3892/ol.2019.11105
Kos IA, Azevedo VF, Neto DE, Kowalski SC (2018) The biosimilars journey: current status and ongoing challenges. Drugs Context 7:1–13. https://doi.org/10.7573/dic.212543
Liu Y (2021) Utilizing oncology biosimilars to minimize the economic burden associated with cancer treatment: managed care considerations. The. Am J Manag Care 27(14 Suppl). https://doi.org/10.37765/ajmc.2021.88734
Melosky B, Reardon DA, Nixon AB, Subramanian J, Bair AH, Jacobs I (2018) Bevacizumab biosimilars: scientific justification for extrapolation of indications. In Future oncology, vol 14, Issue 24, Future Medicine Ltd, pp 2507–2520. https://doi.org/10.2217/fon-2018-0051
Migliavacca Zucchetti B, Nicolò E, Curigliano G (2019) Biosimilars for breast cancer. Taylor and Francis Ltd
Mok CC (2013) Rituximab for the treatment of rheumatoid arthritis: an update. Drug Des Devel Ther 8:87–100. https://doi.org/10.2147/DDDT.S41645
Peeters M, Planchard D, Pegram M et al (2021) Biosimilars in an era of rising oncology treatment options. Future Oncol 17:3881–3892. https://doi.org/10.2217/fon-2021-0546
Planchard D (2011) Bevacizumab in non-small-cell lung cancer: a review. Expert Rev Anticancer Ther 11(8):1163–1179. https://doi.org/10.1586/era.11.80
Rakesh Chopra GL (2017) Improving access to cancer treatment in Tajikistan | IAEA. Am Soc Clin Oncol 3(5):596–610
Rémuzat C, Dorey J, Cristeau O, Ionescu D, Radière G, Toumi M (2017) Key drivers for market penetration of biosimilars in Europe. J Mark Access Health Policy 5(1). https://doi.org/10.1080/20016689.2016.1272308
Rosen LS, Jacobs IA, Burkes RL (2017) Bevacizumab in colorectal cancer: current role in treatment and the potential of Biosimilars. Target Oncol 12:599–610. https://doi.org/10.1007/s11523-017-0518-1

Sandler A (2007) Bevacizumab in non-small cell lung cancer. Clinical Cancer Res 13(15). https://doi.org/10.1158/1078-0432.CCR-07-0647

Schellekens H, Smolen JS, Dicato M, Rifkin RM (2016) Safety and efficacy of biosimilars in oncology. Lancet Oncol 17:e502–e509. https://doi.org/10.1016/S1470-2045(16)30374-6

Soares JCS, Cavalcanti IDL, JLA V (2021) Can biosimilar products be interchangeable? Pharmaceutical perspective in the implementation of biosimilars in oncology. J Oncol Pharm Pract 27:1491–1502. https://doi.org/10.1177/10781552211016099

Thill M (2019) Biosimilar Trastuzumab in Clinical Trials. Breast care 14:17–22

Triantafyllidi E, Triantafillidis JK (2022) Systematic review on the use of Biosimilars of Trastuzumab in HER2+ breast cancer. Biomedicines 10:1–27. https://doi.org/10.3390/biomedicines10082045

Trouvin AP, Jacquot S, Grigioni S et al (2015) Usefulness of monitoring of B cell depletion in rituximab-treated rheumatoid arthritis patients in order to predict clinical relapse: a prospective observational study. Clin Exp Immunol 180:11–18. https://doi.org/10.1111/cei.12481

Uifălean A, Ilieş M, Nicoară R, Rus LM, Hegheş SC, Iuga CA (2018) Concepts and challenges of biosimilars in breast cancer: the emergence of trastuzumab biosimilars. Pharmaceutics 10(4). https://doi.org/10.3390/pharmaceutics10040168

Williamson C, Berger L, Sullivan TP et al (2019) Addressing oncologists' gaps in the use of biosimilar products. Am J Manag Care 25:SP188–SP191

Winstead E (2018) Biosimilars for cancer emerge as patents on widely used biological drugs expire subscribe

Zinzani PL, Dreyling M, Gradishar W et al (2019) Are Biosimilars the future of oncology and Haematology? Drugs 79:1609–1624. https://doi.org/10.1007/s40265-019-01193-y